TRANSPORT ANALYSIS

TRANSPORT ANALYSIS

Daniel Hershey

Department of Chemical Engineering
University of Cincinnati
Cincinnati, Ohio

PLENUM PRESS • NEW YORK – LONDON • 1973

660.28
H 572

Library of Congress Catalog Card Number 70-183564
ISBN 0-306-30555-0

© 1973 Plenum Press, New York
A Division of Plenum Publishing Corporation
227 West 17th Street, New York, N. Y. 10011

United Kingdom edition published by Plenum Press, London
A Division of Plenum Publishing Company, Ltd.
Davis House (4th Floor), 8 Scrubs Lane, Harlesden, London, NW10 6SE, England

Preface

It has been my experience in teaching graduate and undergraduate courses that if the students are conversant with the pertinent mathematical procedures, and can "think mathematically," there is almost no limit to their comprehension. Most courses that are considered difficult by students are either poorly taught or require a degree of mathematical sophistication that the students do not possess.

In *Transport Analysis*, I have culled some basic momentum transport (fluid flow) and mass transport phenomena and explicitly revealed the derivation of the governing equations. There is no mystery, no omitted steps or "it can be shown" phrases that are usually the bane of the student. There are chapters that review basic calculus, vector and matrix concepts, Laplace transform operations, and finite difference calculus. Ordinary differential and partial differential equations are derived and solved.

This book is intended for undergraduates and graduate students in engineering, chemistry, physics, and even biology and medicine. It is also intended for my non-engineering colleagues with whom I have collaborated during our cooperative research in the life sciences. If they knew what is contained in *Transport Analysis*, they probably wouldn't need me.

Acknowledgments

To Barbara and Michael, who helped keep me alert, happy, and fulfilled.

To Barbara, who deserves belated thanks for doing the drawings in *Everyday Science*.

To Anne Hagedorn, thanks for doing some of the typing.

To Gerry Denterlein, thanks for keeping tabs on the drawings.

To Edna Penn, special thanks for invaluable assistance in typing and in many other ways.

Contents

Part II. Transport Analysis in Continuous Processes

Chapter 3. Derivation of the Momentum Transport Equations

Chapter 4. Transport Analysis in Fluid Flow Phenomena

Chapter 5. Derivation of the Mass Transport Equations

Part III. Transport Analysis in Discrete Processes

Chapter 7. Finite Difference Calculus

Chapter 8. Transport Analysis in Cascaded Systems

Part I

INTRODUCTION

Chapter 1

Some Mathematical Concepts

In this chapter we shall introduce a few basic concepts involving units, Newton's second law of motion, and some "common sense" statements of basic transport relationships. After a brief calculus review, there will be presented statements of continuity, differentiability, and differentials, L'Hô-pital's rule, and the Leibnitz rule for differentiating integrals. We conclude with some vector and tensor operations, and a discussion of matrix algebra.

1.1. Elementary Transport Concepts

In the analysis of transport phenomena, whether simple or complex, it is usually helpful to establish the result desired, the driving force, and the resistance. Simply expressed, we have

$$\text{result} = K \frac{\text{driving force}}{\text{resistance}} \tag{1.1-1}$$

where K is a proportionality constant. Thus in considering ordinary mass transport by diffusion, the driving force is the concentration gradient and the resistance is related to the material through which the mass is being transported and the diffusion distance. Many other "laws," such as Ohm's law, are also represented by a form such as (1.1-1).

When working with the symbols which represent concentration, mass, pressure, or other physical entities, it should be kept in mind that they are composed of two intrinsic characteristics: a magnitude and the corresponding units. Thus in "solving" an equation, both the units and the numbers must balance:

$$N = D \frac{dC}{dx} \tag{1.1-2}$$

3

where

$$N = \text{mass flux, lbm/ft}^2\text{-hr}$$

$$D = \text{diffusivity, ft}^2/\text{hr}$$

$$dC/dx = \text{concentration gradient, (lbm/ft}^3)/\text{ft}$$

$$C = \text{concentration, lbm/ft}^3$$

$$x = \text{distance, ft}$$

With some specific numbers substituted into Eq. (1.1-2) we get

$$2\,\frac{\text{lbm}}{\text{ft}^2\text{-hr}} = \left(10^{-5}\,\frac{\text{ft}^2}{\text{hr}}\right)\left(2\times10^5\,\frac{\text{lbm/ft}^3}{\text{ft}}\right) \tag{1.1-3}$$

Note that in (1.1-3) both the units and the magnitudes balance.

A discussion of units and magnitudes may employ Newton's second law of motion as an illustration. In Newton's second law of motion, we use the definition that a one pound force (lbf) gives a one pound mass (lbm) an acceleration (g) of 32.17 ft/sec². In these units, Newton's second law is written as

$$F = \frac{mg}{g_c} \tag{1.1-4}$$

where g_c is a proportionality constant having both a magnitude and units. From (1.1-4) and the definition of a one pound force, we can evaluate g_c;

$$1\text{ lbf} = \frac{(1\text{ lbm})(32.17\text{ ft/sec}^2)}{g_c} \tag{1.1-5}$$

From (1.1-5) we can, in the units of lbf, lbm, ft, sec, arrive at the constant magnitude of g_c and its units. Rearranging (1.1-5), we get

$$g_c = \frac{(1\text{ lbm})(32.17\text{ ft/sec}^2)}{1\text{ lbf}} \tag{1.1-6}$$

or

$$g_c = 32.17\,\frac{\text{lbm ft/sec}^2}{\text{lbf}} \tag{1.1-7}$$

In another system of units, for instance, dynes, grams, centimeters, and seconds (the cgs system), g_c is still a constant, but of different magnitude and with different units.

Before leaving this subject, it is interesting to note the effect of multiplying by g/g_c, which from (1.1-4) is seen to have units of lbf/lbm. Thus, if one is balancing forces in an equation and there appears a density term, ϱ (lbm/ft^3), it is a relatively simple matter to convert density to a force term. All that is required is that we generate the quantity $\varrho(g/g_c)$, which produces the units of lbf/ft^3:

$$\varrho\left(\frac{\text{lbm}}{\text{ft}^3}\right)\frac{g}{g_c}\left(\frac{\text{lbf}}{\text{lbm}}\right) = \varrho\,\frac{g}{g_c}\left(\frac{\text{lbf}}{\text{ft}^3}\right) \tag{1.1-8}$$

At sea level, the acceleration due to gravity, g, is equal to 32.17 ft/sec^2, so that the ratio $g/g_c = 1.0$ lbf/lbm at sea level. As we travel away from the surface of the earth, the acceleration due to gravity decreases in magnitude and eventually diminishes to zero, indicating that the attractive force of the earth has become negligible.

Having introduced one basic "law," equation (1.1-1), we find it useful to conclude this section by mentioning another important relationship applicable to the derivation of the transport equations. It relates to the balancing of input, output, generation, depletion, and accumulation of mass, energy, or momentum. Equation (1.1-9) symbolically represents this relationship:

$$\sum_{\text{in}}\left(\begin{array}{c}\text{mass}\\ \text{or}\\ \text{energy}\\ \text{or}\\ \text{momentum}\end{array}\right) - \sum_{\text{out}}\left(\begin{array}{c}\text{mass}\\ \text{or}\\ \text{energy}\\ \text{or}\\ \text{momentum}\end{array}\right) + \left\{\begin{array}{c}+\text{ generation}\\ \text{or}\\ -\text{ depletion}\end{array}\right\}$$

$$+ \text{ external influences} = \text{rate of accumulation} \tag{1.1-9}$$

In (1.1-9), the symbol $\sum_{\text{in}}$ reads "the summation of all input streams." In dealing with mass transport, for example, we may have mass transfer by molecular diffusion as well as convection, while the generation (or depletion) term can account for chemical reactions. The rate of accumulation term shows the change in mass (with respect to time) in the volume element of interest. In heat transfer the "material" being transferred is energy, and in fluid flow phenomena it is momentum that is being transported.

1.2. Elementary Calculus Concepts

When working with differential equations, it is useful to have acquaintance with some basic concepts in calculus. For example a knowledge of the definition of continuity and differentiability allows us to formulate

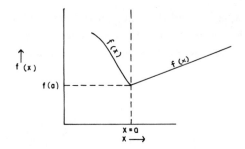

Fig. 1. A continuous function.

differential equations from the analysis of differential elements, as we shall see a little later.

We first examine the concept of continuity, illustrated in Figure 1. As the function $f(x)$ approaches $x = a$, it approaches the value $f(a)$ as we approach $x = a$ from the left $(x = a_-)$ or the right $(x = a_+)$. Mathematically this is expressed as

$$\lim_{x \to a_-} f(x) = \lim_{x \to a_+} f(x) \tag{1.2-1}$$

Equation (1.2-1) reads "the limit of the function of x as x approaches a from the left is equal to the limit as x approaches a from the right." If it is true that equation (1.2-1) holds, as shown in Figure 1, then we say that $f(x)$ is continuous at $x = a$. If the function $f(x)$ behaves as shown in Figure 2, we know from equation (1.2-1) that $f(x)$ is not continuous at $x = a$.

To say that a function is differentiable implies that the function has a derivative. From elementary calculus concepts we can write the definition of a derivative as

$$\lim_{\Delta x \to 0} \frac{f|_{x+\Delta x} - f|_x}{\Delta x} = \frac{df}{dx} \tag{1.2-2}$$

where the symbolism reads "the limit of the function f evaluated at a loca-

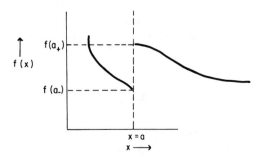

Fig. 2. A discontinuous function.

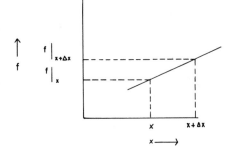

Fig. 3. The definition of a derivative.

tion $x + \Delta x$, minus the function f evaluated at x, divided by the increment Δx...." Figure 3 illustrates the behavior of the function f. As the size of increment Δx shrinks toward zero, we get the slope or derivative of the function f evaluated at location x. The question of whether a function f is actually differentiable (has a derivative) at point x is a bit more complex than implied by equation (1.2-2). For a function to have a derivative also requires a statement such as

$$\lim_{x \to a_-} \frac{df}{dx} = \lim_{x \to a_+} \frac{df}{dx} \qquad (1.2\text{-}3)$$

where we require not only that there be a derivative according to equation (1.2-2) but that its value be the same as we converge upon the point a from both the left and the right. Figure 4 shows an example where though a function $f(x)$ is continuous at $x = a$ [as defined by equation (1.2-1)], it is not differentiable. In Figure 4 the value of the slope df/dx for values of x less than $x = a$ might be equal to -1, but the value of df/dx for values of x greater than $x = a$ might have a value of $+1$. Thus, in this case we might find the same numerical value for the derivative as we approach $x = a$ from left and right, but the derivatives are not equal as required by equation (1.2-3). The function $f(x)$ is not differentiable (does not have a derivative) at $x = a$.

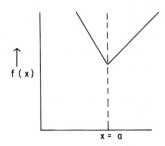

Fig. 4. A function which is continuous but not differentiable.

In evaluating limits of functions such as in equation (1.2-2), we often find an "indeterminacy" dilemma. For example, in trying to find the limit of a functional grouping such as $\lim_{x \to \infty} x/e^x$, we would get ∞/∞ if we naively substituted $x = \infty$. The symbol $x \to \infty$ reads "as x goes to infinity," whereas $x = \infty$ reads "where x is equal to infinity." A form like ∞/∞ does not appear amenable to immediate evaluation. Hence we say that $\lim_{x \to \infty} x/e^x$ is an indeterminate form. Similarly, $\lim_{x \to 0} (\sin x)/x$ yields $0/0$ if $x = 0$ is substituted into the relationship. This is also an indeterminate form since we cannot extract any direct information from such a result. Other forms such as $(\infty - \infty)/(\infty - \infty)$, $(\infty - \infty)/(0 - 0)$, and $(0 - 0)/(\infty - \infty)$ may similarly be indeterminate since $\infty - \infty$ may have values of 0, ∞, or values between these extremes. To resolve these questions of indeterminacy, we invoke L'Hôpital's rule:

$$\lim_{x \to a} \frac{f(x)}{g(x)} = \text{indeterminate form} = \lim_{x \to a} \frac{df/dx}{dg/dx} \qquad (1.2\text{-}4)$$

L'Hôpital's rule, as given by equation (1.2-4), states that should there arise an indeterminacy, we put this indeterminacy in an indeterminate form such as ∞/∞ or $0/0$. Next we differentiate the numerator and denominator separately and then substitute $x = a$. If we apply L'Hôpital's rule to the previous examples, we get

$$\lim_{x \to \infty} \frac{x}{e^x} = [\text{indeterminacy at } x = \infty] = \frac{\infty}{\infty} = \lim_{x \to \infty} \frac{(d/dx)(x)}{(d/dx)(e^x)}$$
$$= \lim_{x \to \infty} \frac{1}{e^x} = 0 \qquad (1.2\text{-}5)$$

$$\lim_{x \to 0} \frac{\sin x}{x} = [\text{indeterminacy at } x = 0] = \frac{0}{0} = \lim_{x \to 0} \frac{(d/dx)(\sin x)}{(d/dx)(x)}$$
$$= \lim_{x \to 0} \frac{\cos x}{1} = 1 \qquad (1.2\text{-}6)$$

L'Hôpital's rule is more general than indicated by (1.2-4). If after one differentiation procedure we still have an indeterminacy, we may repeat the differentiation procedure as many times as needed. However, before each differentiation the previous result must be put into an indeterminate form. Equation (1.2-7) extends (1.2-4) and illustrates the procedure:

$$\lim_{x \to a} \frac{f(x)}{g(x)} = \text{indeterminate form} = \lim_{x \to a} \frac{df/dx}{dg/dx} = \text{indeterminate form}$$
$$= \lim_{x \to a} \frac{d^2f/dx^2}{d^2g/dx^2} = \cdots \qquad (1.2\text{-}7)$$

Finally, in this review of some elementary calculus concepts and operations we introduce the Leibnitz rule for differentiating integrals. A concise statement of the Leibnitz rule is given by equations (1.2-8) and (1.2-9). Given an integral such as

$$I(r, s) = \int_{b(r,s)}^{a(r,s)} f(x, r)dx \qquad (1.2\text{-}8)$$

the Leibnitz rule for differentiating $I(r, s)$ is given symbolically by

$$\frac{\partial I(r, s)}{\partial r} = \int_{b(r,s)}^{a(r,s)} \frac{\partial f(x, r)}{\partial r}\, dx + f(a, r)\frac{\partial a(r, s)}{\partial r} - f(b, r)\frac{\partial b(r, s)}{\partial r} \qquad (1.2\text{-}9)$$

For example, if we wish to differentiate

$$I(r, s) = \int_{r+s}^{r^2 s} (x^2 + 2r)dx \qquad (1.2\text{-}10)$$

we invoke the Leibnitz rule as given by (1.2-9) and get

$$\frac{\partial I(r, s)}{\partial r} = \int_{r+s}^{r^2 s} \left[\frac{\partial}{\partial r}(x^2 + 2r)\right]dx + [(r^2 s)^2 + 2r]\frac{\partial(r^2 s)}{\partial r}$$

$$- [(r + s)^2 + 2r]\frac{\partial(r+s)}{\partial r}$$

$$= \int_{r+s}^{r^2 s} 2dx + [r^4 s^2 + 2r](2rs) - [(r^2 + 2rs + s^2) + 2r](1)$$

$$= 2(r^2 s - r - s) + 2r^5 s^3 + 4r^2 s - r^2 - 2rs - s^2 - 2r$$

$$= 6r^2 s - 4r - 2s + 2r^5 s^3 - r^2 - 2rs - s^2 \qquad (1.2\text{-}11)$$

This result can be verified by performing the integration indicated in equation (1.2-10) and then differentiating the results.

1.3. Some Elementary Vector and Tensor Operations

In many cases the governing equations describing transport phenomena can best be described in vector and tensor notation. This type of representation sometimes has the advantage of conciseness and clarity and many times pinpoints individual transport components which in sum constitute the complete description of the phenomena. When expressing equations in vector and tensor form it is necessary to understand the significance of some of the vector and tensor operations shown below.

Vectors and tensors can be represented in rectangular coordinates by notation such as that given by equation (1.3-1):

$$\bar{v} = v_1\bar{\delta}_1 + v_2\bar{\delta}_2 + v_3\bar{\delta}_3 = \sum_{i=1}^{3} v_i\bar{\delta}_i \qquad (1.3\text{-}1)$$

where $\bar{\delta}_i$ are the orthogonal unit vectors and v_i are the scalar components of the $\bar{v}$ vector. For example, in rectangular coordinates, let the subscript 1 correspond to x, the subscript 2 to y, and the subscript 3 to z. In analogous fashion, we may represent a tensor as follows:

$$\bar{\bar{\tau}} = \sum_i \sum_j \bar{\delta}_i\bar{\delta}_j\tau_{ij} \qquad (1.3\text{-}2)$$

Thus we find that the tensor $\bar{\bar{\tau}}$ has nine components:

$$\bar{\bar{\tau}} = \bar{\delta}_1\bar{\delta}_1\tau_{11} + \bar{\delta}_1\bar{\delta}_2\tau_{12} + \bar{\delta}_1\bar{\delta}_3\tau_{13} + \bar{\delta}_2\bar{\delta}_1\tau_{21} + \bar{\delta}_2\bar{\delta}_2\tau_{22} + \bar{\delta}_2\bar{\delta}_3\tau_{23}$$
$$+ \bar{\delta}_3\bar{\delta}_1\tau_{31} + \bar{\delta}_3\bar{\delta}_2\tau_{32} + \bar{\delta}_3\bar{\delta}_3\tau_{33}$$

where the term $\bar{\delta}_i\bar{\delta}_j$ is called a unit dyad signifying two vectors multiplied together.

With this type of notation we can perform some elementary vector and tensor operations. For example, the scalar or dot product of vectors can be obtained in a straightforward manner:

$$\bar{v} \cdot \bar{w} = \left(\sum_i \bar{\delta}_i v_i\right) \cdot \left(\sum_j \bar{\delta}_j w_j\right) = \sum_i \sum_j (\bar{\delta}_i \cdot \bar{\delta}_j) v_i w_j$$
$$= \sum_i \sum_j \delta_{ij} v_i w_j = \sum_i v_i w_i$$

where $\bar{\delta}_i \cdot \bar{\delta}_j = \delta_{ij}$ is the Kronecker delta with the properties

$$\begin{aligned} \delta_{ij} &= 1 \quad \text{when } i = j \\ \delta_{ij} &= 0 \quad \text{when } i \neq j \end{aligned} \qquad (1.3\text{-}3)$$

In obtaining the final result we have made use of the following relationships:

$$\begin{aligned} \bar{\delta}_1 \cdot \bar{\delta}_1 &= \bar{\delta}_2 \cdot \bar{\delta}_2 = \bar{\delta}_3 \cdot \bar{\delta}_3 = \bar{\delta}_i \cdot \bar{\delta}_i = \delta_{ii} = 1 \quad i = j \\ \bar{\delta}_1 \cdot \bar{\delta}_2 &= \bar{\delta}_2 \cdot \bar{\delta}_3 = \bar{\delta}_3 \cdot \bar{\delta}_1 = \bar{\delta}_i \cdot \bar{\delta}_j = \delta_{ij} = 0 \quad i \neq j \end{aligned} \qquad (1.3\text{-}4)$$

The vector or cross product can be obtained by similar methods. The steps in obtaining the vector product of two vectors are

$$\bar{v} \times \bar{w} = \left(\sum_j \bar{\delta}_j v_j\right) \times \left(\sum_k \bar{\delta}_k w_k\right)$$

$$= \sum_j \sum_k (\bar{\delta}_j \times \bar{\delta}_k) v_j w_k$$

$$= \sum_i \sum_j \sum_k \varepsilon_{ijk} \bar{\delta}_i v_j w_k \qquad (1.3\text{-}5)$$

where the vector product of two unit vectors is given by

$$(\bar{\delta}_j \times \bar{\delta}_k) = \sum_i \varepsilon_{ijk} \bar{\delta}_i \qquad (1.3\text{-}6)$$

and ε_{ijk} is called the alternating unit tensor, defined by

$$\varepsilon_{ijk} = +1 \quad \text{if } ijk = 123,\ 231,\ \text{or } 312 \text{ (cyclic permutations)}$$
$$\varepsilon_{ijk} = -1 \quad \text{if } ijk = 321,\ 132,\ \text{or } 213 \qquad (1.3\text{-}7)$$
$$\varepsilon_{ijk} = \ \ 0 \quad \text{if any two subscripts are alike}$$

From equations (1.3-5), (1.3-6), and (1.3-7) we can obtain the vector or cross product of $\bar{v} \times \bar{w}$:

$$\bar{v} \times \bar{w} = \varepsilon_{111} \bar{\delta}_1 v_1 w_1 + \varepsilon_{112} \bar{\delta}_1 v_1 w_2 + \varepsilon_{113} \bar{\delta}_1 v_1 w_3 + \varepsilon_{121} \bar{\delta}_1 v_2 w_1$$

$$+ \varepsilon_{122} \bar{\delta}_1 v_2 w_2 + \varepsilon_{123} \bar{\delta}_1 v_2 w_3 + \cdots \qquad (1.3\text{-}8)$$

The procedure used to obtain equation (1.3-8) from (1.3-5) is to first set $i = 1$, $j = 1$, and let $k = 1, 2, 3$. We then repeat the procedure with $i = 1$, $j = 2$, and $k = 1, 2, 3$. Continuing on with the summation, we run out the terms with $i = 1$, $j = 3$, $k = 1, 2, 3$; $i = 2$, $j = 1$, $k = 1, 2, 3$; $i = 2$, $j = 2$, $k = 1, 2, 3$, etc. After all this has been done the result is equation (1.3-8). It is possible with equation (1.3-7) to express the vector product $\bar{v} \times \bar{w}$ in equation (1.3-5) in a compact form:

$$\bar{v} \times \bar{w} = \begin{vmatrix} \bar{\delta}_1 & \bar{\delta}_2 & \bar{\delta}_3 \\ v_1 & v_2 & v_3 \\ w_1 & w_2 & w_3 \end{vmatrix} \qquad (1.3\text{-}9)$$

where the vector product is expressed in determinant form. If the determinant is expanded in the usual way, we get the familiar result for the vector product given in most textbooks.

The vector operator ∇, called the del operator, is defined as

$$\nabla = \sum_i \bar{\delta}_i \frac{\partial}{\partial x_i} \tag{1.3-10}$$

Since an operator must by definition operate upon something, let ∇ operate upon a scalar quantity, s. Thus from equation (1.3-10) we get, in rectangular coordinates,

$$\nabla s = \sum_i \bar{\delta}_i \frac{\partial s}{\partial x_i} = \bar{\delta}_1 \frac{\partial s}{\partial x_1} + \bar{\delta}_2 \frac{\partial s}{\partial x_2} + \bar{\delta}_3 \frac{\partial s}{\partial x_3} \tag{1.3-11}$$

We can also define a divergence vector operation in rectangular coordinates, using similar notation. Equation (1.3-12) shows how the divergence relationship is developed.

$$\begin{aligned}
\nabla \cdot \bar{v} &= \left(\sum_i \bar{\delta}_i \frac{\partial}{\partial x_i} \right) \cdot \left(\sum_j \bar{\delta}_j v_j \right) \\
&= \sum_i \sum_j (\bar{\delta}_i \cdot \bar{\delta}_j) \frac{\partial}{\partial x_i} v_j \\
&= \sum_i \sum_j \delta_{ij} \frac{\partial}{\partial x_i} v_j \\
&= \sum_i \frac{\partial v_i}{\partial x_i} \tag{1.3-12}
\end{aligned}$$

Note that $\delta_{ij} = 0$ when $i \neq j$.

The vector cross product $\nabla \times \bar{v}$, called curl, can also be obtained in rectangular coordinates by this technique:

$$\begin{aligned}
\nabla \times \bar{v} &= \left(\sum_j \bar{\delta}_j \frac{\partial}{\partial x_j} \right) \times \left(\sum_k \bar{\delta}_k v_k \right) \\
&= \sum_j \sum_k (\bar{\delta}_j \times \bar{\delta}_k) \frac{\partial}{\partial x_j} v_k \\
&= \begin{vmatrix} \bar{\delta}_1 & \bar{\delta}_2 & \bar{\delta}_3 \\ \dfrac{\partial}{\partial x_1} & \dfrac{\partial}{\partial x_2} & \dfrac{\partial}{\partial x_3} \\ v_1 & v_2 & v_3 \end{vmatrix} \\
&= \bar{\delta}_1 \left(\frac{\partial v_3}{\partial x_2} - \frac{\partial v_2}{\partial x_3} \right) + \bar{\delta}_2 \left(\frac{\partial v_1}{\partial x_3} - \frac{\partial v_3}{\partial x_1} \right) \\
&\quad + \bar{\delta}_3 \left(\frac{\partial v_2}{\partial x_1} - \frac{\partial v_1}{\partial x_2} \right) \tag{1.3-13}
\end{aligned}$$

The determinant form of the cross product in equation (1.3-13) is obtained by analogy with equation (1.3-9).

One final vector operation, called the Laplacian, is defined as $\nabla \cdot \nabla = \nabla^2$. Specifically, if the Laplacian operator is applied in rectangular coordinates to a scalar, s, we get

$$\nabla \cdot \nabla s = \nabla^2 s = \left(\sum_i \bar{\delta}_i \frac{\partial}{\partial x_i} \right) \cdot \left(\sum_j \bar{\delta}_j \frac{\partial s}{\partial x_j} \right)$$

$$= \sum_i \sum_j \delta_{ij} \frac{\partial}{\partial x_i} \frac{\partial s}{\partial x_j}$$

$$= \sum_i \frac{\partial^2 s}{\partial x_i^2} \qquad (1.3\text{-}14)$$

We have used equations (1.3-10) and (1.3-11) in arriving at equation (1.3-14).

Tensors can be represented in rectangular coordinates by a form such as

$$\bar{\bar{\tau}} = \begin{pmatrix} \tau_{11} & \tau_{12} & \tau_{13} \\ \tau_{21} & \tau_{22} & \tau_{23} \\ \tau_{31} & \tau_{32} & \tau_{33} \end{pmatrix}$$

or, in another type of notation,

$$\bar{\bar{\tau}} = \sum_i \sum_j \bar{\delta}_i \bar{\delta}_j \tau_{ij} \qquad (1.3\text{-}15)$$

where $\bar{\delta}_i \bar{\delta}_j$ are the unit dyads. These dyads can be considered as two vectors multiplied together with the properties

$$\bar{\delta}_1 \bar{\delta}_1 = \begin{pmatrix} 1 & 0 & 0 \\ 0 & 0 & 0 \\ 0 & 0 & 0 \end{pmatrix} \qquad (1.3\text{-}16)$$

$$\bar{\delta}_1 \bar{\delta}_2 = \begin{pmatrix} 0 & 1 & 0 \\ 0 & 0 & 0 \\ 0 & 0 & 0 \end{pmatrix} \qquad (1.3\text{-}17)$$

$$\bar{\delta}_1 \bar{\delta}_3 = \begin{pmatrix} 0 & 0 & 1 \\ 0 & 0 & 0 \\ 0 & 0 & 0 \end{pmatrix} \qquad (1.3\text{-}18)$$

With equations (1.3-4) and (1.3-15) it is possible to show some elementary tensor operations in rectangular coordinates. First we must utilize equation (1.3-4) to develop a series of basic relationships:

$$\bar{\delta}_i \bar{\delta}_j : \bar{\delta}_k \bar{\delta}_l = \delta_{il}\delta_{jk} \tag{1.3-19}$$

$$\bar{\delta}_i \bar{\delta}_j \cdot \bar{\delta}_k = \bar{\delta}_i \delta_{jk} \tag{1.3-20}$$

$$\bar{\delta}_i \cdot \bar{\delta}_j \bar{\delta}_k = \delta_{ij}\bar{\delta}_k \tag{1.3-21}$$

$$\bar{\delta}_i \bar{\delta}_j \cdot \bar{\delta}_k \bar{\delta}_l = \delta_{jk}\bar{\delta}_i\bar{\delta}_l \tag{1.3-22}$$

Now, if we are given two tensors in the notation of equation (1.3-15),

$$\bar{\bar{\sigma}} = \sum_i \sum_j \bar{\delta}_i \bar{\delta}_j \sigma_{ij}$$

$$\bar{\bar{\tau}} = \sum_k \sum_l \bar{\delta}_k \bar{\delta}_l \tau_{kl} \tag{1.3-23}$$

we can perform a basic tensor operation:

$$\bar{\bar{\sigma}} : \bar{\bar{\tau}} = \left(\sum_i \sum_j \bar{\delta}_i \bar{\delta}_j \sigma_{ij}\right) : \left(\sum_k \sum_l \bar{\delta}_k \bar{\delta}_l \tau_{kl}\right)$$

$$= \sum_i \sum_j \sum_k \sum_l (\bar{\delta}_i \bar{\delta}_j : \bar{\delta}_k \bar{\delta}_l)\sigma_{ij}\tau_{kl}$$

$$= \sum_i \sum_j \sum_k \sum_l \delta_{il}\delta_{jk}\sigma_{ij}\tau_{kl}$$

$$= \sum_i \sum_j \sigma_{ij}\tau_{ji} \tag{1.3-24}$$

where we have made use of equations (1.3-19) and (1.3-3) to get the final result.

Similarly we can show the single dot operation for rectangular coordinates:

$$\bar{\bar{\sigma}} \cdot \bar{\bar{\tau}} = \left(\sum_i \sum_j \bar{\delta}_i \bar{\delta}_j \sigma_{ij}\right) \cdot \left(\sum_k \sum_l \bar{\delta}_k \bar{\delta}_l \tau_{kl}\right)$$

$$= \sum_i \sum_j \sum_k \sum_l (\bar{\delta}_i \bar{\delta}_j \cdot \bar{\delta}_k \bar{\delta}_l)\sigma_{ij}\tau_{kl}$$

$$= \sum_i \sum_j \sum_k \sum_l \delta_{jk}\bar{\delta}_i \bar{\delta}_l \sigma_{ij}\tau_{kl}$$

$$= \sum_i \sum_l \bar{\delta}_i \bar{\delta}_l \left(\sum_j \sigma_{ij}\tau_{jl}\right) \tag{1.3-25}$$

To obtain the final result in equation (1.3-25) the property of the Kronecker delta was used and the subscript k was set equal to j.

Without showing the details, we can also, for rectangular coordinates, produce the following results:

$$\bar{v}\bar{w} = \sum_i \sum_j \bar{\delta}_i \bar{\delta}_j v_i w_j \tag{1.3-26}$$

$$\bar{\bar{\tau}} \cdot \bar{v} = \sum_i \bar{\delta}_i \left(\sum_j \tau_{ij} v_j \right) \tag{1.3-27}$$

$$\nabla \cdot \bar{\bar{\tau}} = \sum_k \bar{\delta}_k \left(\sum_i \frac{\partial}{\partial x_i} \tau_{ik} \right) \tag{1.3-28}$$

1.4. Linear Operations with Functions, Vectors, and Matrices

In knowing whether a system is linear or not, we have an idea of the response of a system to certain inputs. Linearity implies that if we have two functions such as

$$y_1(t) = f_1(t) \tag{1.4-1}$$

and

$$y_2(t) = f_2(t) \tag{1.4-2}$$

then

$$y_1(t) + y_2(t) = [f_1(t) + f_2(t)] \qquad \text{(superposition)} \tag{1.4-3}$$

and

$$ky_1(t) = kf_1(t) \tag{1.4-4}$$

In dealing with linear systems, many times the governing equations can be represented in matrix notation. This is advantageous for digital computer operations used in obtaining the solution to these equations. A matrix can be represented in the form

$$\bar{\bar{A}} = \begin{pmatrix} a_{11} & a_{12} & \cdots & a_{1n} \\ a_{21} & a_{22} & \cdots & a_{2n} \\ \vdots & \vdots & & \vdots \\ a_{m1} & a_{m2} & \cdots & a_{mn} \end{pmatrix} \tag{1.4-5}$$

where there are m rows and n columns in $\bar{\bar{A}}$. This $\bar{\bar{A}}$ matrix is, by convention, referred to as an m(row)$\times n$(column) matrix. A square matrix is of course $n \times n$. An identity matrix is one which when multiplied by another matrix gives the same matrix back again. The identity matrix, $\bar{\bar{I}}$, is usually written, for 3×3 matrices, as

$$\bar{I} = \begin{pmatrix} 1 & 0 & 0 \\ 0 & 1 & 0 \\ 0 & 0 & 1 \end{pmatrix} \qquad (1.4\text{-}6)$$

The transpose of a matrix $\bar{A}$, written as $\bar{A}^T$, implies an interchange of rows and columns. For example, if

$$\bar{A} = \begin{pmatrix} 2 & 0 & -1 \\ 1 & 1 & 4 \end{pmatrix} = \begin{pmatrix} a_{11} & a_{12} & a_{13} \\ a_{21} & a_{22} & a_{23} \end{pmatrix} \qquad (1.4\text{-}7)$$

then the transpose of $\bar{A}$ is

$$\bar{A}^T = \begin{pmatrix} 2 & 1 \\ 0 & 1 \\ -1 & 4 \end{pmatrix} \qquad (1.4\text{-}8)$$

A symmetric matrix has the property $\bar{A} = \bar{A}^T$; for example,

$$\bar{A} = \begin{pmatrix} 0 & 1 & 2 \\ 1 & 2 & 3 \\ 2 & 3 & 4 \end{pmatrix} = \begin{pmatrix} a_{11} & a_{12} & a_{13} \\ a_{21} & a_{22} & a_{23} \\ a_{31} & a_{32} & a_{33} \end{pmatrix} \qquad (1.4\text{-}9)$$

Since $a_{12} = a_{21} = 1$, $a_{13} = a_{31} = 2$, and $a_{23} = a_{32} = 3$, this is a symmetric matrix.

Another useful definition is that of the adjoint matrix $\bar{\bar{A}}$ whose elements are obtained from the transposed matrix of the cofactors of matrix $\bar{A}$. An example will clarify this definition. Suppose $\bar{A}$ is given by

$$\bar{A} = \begin{pmatrix} x & 1 & y \\ 1 & 2 & 1 \\ 0 & 3 & 2 \end{pmatrix} = \begin{pmatrix} a_{11} & a_{12} & a_{13} \\ a_{21} & a_{22} & a_{23} \\ a_{31} & a_{32} & a_{33} \end{pmatrix} \qquad (1.4\text{-}10)$$

Then some of the cofactors of matrix $\bar{A}$ are

$$a_{11} = \begin{vmatrix} 2 & 1 \\ 3 & 2 \end{vmatrix} = 4 - 3 = 1$$

$$a_{12} = \begin{vmatrix} 1 & 1 \\ 0 & 2 \end{vmatrix} = 2 - 0 = 2$$

$$\vdots \qquad \qquad \vdots$$

$$a_{32} = \begin{vmatrix} x & y \\ 1 & 1 \end{vmatrix} = x - y$$

where the symbol $|\ |$ indicates the determinant form.

Next we construct a matrix of the cofactors of matrix $\bar{\bar{A}}$:

$$\begin{pmatrix} 1 & -2 & - \\ - & - & - \\ - & -(x-y) & - \end{pmatrix}$$

The sign of a_{ij} is determined from the following general criterion:

$$\text{if } i+j = \text{even, use } (+) \text{ coefficient}$$
$$\text{if } i+j = \text{odd, use } (-) \text{ coefficient}$$

The transpose of this matrix is

$$\begin{pmatrix} 1 & - & - \\ -2 & - & -(x-y) \\ - & - & - \end{pmatrix}$$

and it is this transpose that is referred to as the adjoint of $\bar{\bar{A}}$, written as $\bar{\bar{A}}$. For the matrix $\bar{\bar{A}}$ given in (1.4-10)

$$\bar{\bar{A}} = \begin{pmatrix} 1 & -(2-3y) & (1-2y) \\ -2 & 2x & -(x-y) \\ 3 & -3x & (2x-1) \end{pmatrix} \tag{1.4-11}$$

Finally, the inverse matrix, $\bar{\bar{A}}^{-1}$, is defined by the equation

$$\bar{\bar{A}}^{-1}\bar{\bar{A}} = \bar{\bar{I}} \tag{1.4-12}$$

A convenient way of calculating $\bar{\bar{A}}^{-1}$ is

$$\bar{\bar{A}}^{-1} = \frac{\bar{\bar{A}}}{\det \bar{\bar{A}}} \qquad \text{if } \det \bar{\bar{A}} \neq 0 \tag{1.4-13}$$

where

$$\bar{\bar{A}} = \text{adjoint of } \bar{\bar{A}}$$
$$\det \bar{\bar{A}} = \text{determinant of } \bar{\bar{A}}$$

For example, if

$$\bar{\bar{A}} = \begin{pmatrix} 8 & 4 & 2 \\ 2 & 8 & 4 \\ 1 & 2 & 8 \end{pmatrix} \tag{1.4-14}$$

then

$$\det \bar{\bar{A}} = \begin{vmatrix} 8 & 4 & 2 \\ 2 & 8 & 4 \\ 1 & 2 & 8 \end{vmatrix} = 392 \tag{1.4-15}$$

$$\bar{\bar{A}} = \begin{pmatrix} 56 & -28 & 0 \\ -12 & 62 & -28 \\ -4 & 4 & 56 \end{pmatrix} \tag{1.4-16}$$

and

$$\bar{\bar{A}}^{-1} = \begin{pmatrix} \dfrac{56}{392} & \dfrac{-28}{392} & \dfrac{0}{392} \\ \dfrac{-12}{392} & \dfrac{62}{392} & \dfrac{-28}{392} \\ \dfrac{-4}{392} & \dfrac{4}{392} & \dfrac{46}{392} \end{pmatrix} \tag{1.4-17}$$

A matrix consisting of a single column is called a column vector, and a matrix consisting of a single row is called a row vector. These vectors are illustrated in Figure 5.

Two matrices may be multiplied only if the are conformable, i.e., if the number of columns in one equals the number of rows in the other, e.g.,

$$(1 \times 5)(5 \times 2) = (1 \times 2) \tag{1.4-18}$$

or in general,

$$(n \times m)(m \times r) = (n \times r) \tag{1.4-19}$$

In other words, a matrix consisting of n row and m columns can be multiplied only by a matrix that has m rows and any number of columns. The resulting matrix has the row size of the first matrix and the column size of the second matrix.

The multiplication of two matrices can also be denoted symbolically by

$$\bar{\bar{A}}\bar{\bar{B}} = \bar{\bar{C}} \tag{1.4-20}$$

$$\begin{pmatrix} a_1 \\ a_2 \\ \vdots \\ a_m \end{pmatrix} \begin{matrix} \text{column} \\ \text{vector} \end{matrix} \qquad (b_1, b_2, \ldots, b_m)$$

row vector

Fig. 5. Column and row vectors (single column or row matrices).

where the ik component of matrix $\bar{\bar{C}}$ is

$$C_{ik} = \sum_{j=1}^{n} a_{ij}b_{jk} \tag{1.4-21}$$

and

$$\bar{\bar{A}} = (a_{ij}) \quad \text{and} \quad \bar{\bar{B}} = (b_{jk}) \tag{1.4-22}$$

The symbol (a_{ij}) signifies a matrix with elements a_{ij}. For example, if

$$\bar{\bar{A}} = \begin{pmatrix} 1 & 0 & 2 \\ 2 & 1 & 1 \\ 0 & 1 & 2 \end{pmatrix} \quad \bar{\bar{B}} = \begin{pmatrix} 0 & 1 & 3 \\ 2 & 1 & 0 \\ 3 & 2 & 1 \end{pmatrix} \tag{1.4-23}$$

then

$$\bar{\bar{C}} = \bar{\bar{A}}\bar{\bar{B}} = \begin{pmatrix} 6 & 5 & 5 \\ 4 & 5 & 7 \\ 8 & 5 & 2 \end{pmatrix} \tag{1.4-24}$$

where the typical calculation of one element by equation (1.4-21) is

$$\begin{aligned} C_{23} &= \sum_{j=1}^{3} a_{2j}b_{j3} \\ &= a_{21}b_{13} + a_{22}b_{23} + a_{23}b_{33} \\ &= (2)(3) + (1)(0) + (1)(1) = 6 + 1 = 7 \end{aligned} \tag{1.4-25}$$

Matrices can also be operated upon by differentiation and integration:

$$\frac{d\bar{\bar{A}}(x)}{dx} = \frac{d(a_{ij}(x))}{dx} \tag{1.4-26}$$

and

$$\int_{x_1}^{x_2} \bar{\bar{A}}(x)dx = \int_{x_1}^{x_2} (a_{ij}(x))dx \tag{1.4-27}$$

where $\bar{\bar{A}} = (a_{ij})$.

The rank of a matrix is the order of the highest nonzero determinant contained in the matrix.

1.5. Matrix Solutions of Sets of Linear Equations

The matrix properties of Section 1.4 find application in the solution of some equations describing cascaded or multicomponent systems (which are discussed in Chapter 8). If there is a set of governing equations such as

$$a_{11}x_1 + a_{12}x_2 + \cdots + a_{1n}x_n = b_1$$
$$\vdots \qquad \vdots \qquad\qquad \vdots \qquad \vdots \qquad\qquad (1.5\text{-}1)$$
$$a_{n1}x_1 + a_{n2}x_2 + \cdots + a_{nn}x_n = b_n$$

where the a_{ij} terms are constant, then equations (1.5-1) can be represented in matrix form by

$$\bar{\bar{A}}\bar{X} = \bar{B} \qquad\qquad (1.5\text{-}2)$$

where

$$\bar{\bar{A}} = \begin{pmatrix} a_{11} & a_{12} & \cdots & a_{1n} \\ \vdots & \vdots & & \vdots \\ a_{n1} & a_{n2} & \cdots & a_{nn} \end{pmatrix} \qquad (n \times n \text{ matrix})$$

$$\bar{X} = \begin{pmatrix} x_1 \\ \vdots \\ x_n \end{pmatrix} \qquad (n \times 1 \text{ matrix})$$

and

$$\bar{B} = \begin{pmatrix} b_1 \\ \vdots \\ b_n \end{pmatrix} \qquad (n \times 1 \text{ matrix})$$

Note that we have conformable matrix operations in equation (1.5-2):

$$(n \times n)(n \times 1) = (n \times 1)$$

For a set of linear equations to have a solution, the rank of the matrix $\bar{\bar{A}}$ and the rank of the augmented matrix $\bar{\bar{A}} + \bar{B}$ must be the same. If this is so, then the solution to equation (1.5-2) is

$$\bar{X} = \bar{\bar{A}}^{-1}\bar{B} \qquad\qquad (1.5\text{-}3)$$

where $\det \bar{\bar{A}} \neq 0$. For example, suppose we have a set of algebraic equations

$$2x_1 + 3x_2 + 4x_3 + 5x_4 = 1$$
$$3x_1 + 7x_2 + 5x_3 + 4x_4 = 1$$
$$x_1 + 4x_2 + 9x_3 + 2x_4 = 1 \qquad (1.5\text{-}4)$$
$$5x_1 + 2x_2 + 7x_3 + \;x_4 = 1$$

From equations (1.5-4) and (1.5-2) we get

$$\bar{\bar{A}} = \begin{pmatrix} 2 & 3 & 4 & 5 \\ 3 & 7 & 5 & 4 \\ 1 & 4 & 9 & 2 \\ 5 & 2 & 7 & 1 \end{pmatrix} \qquad \bar{B} = \begin{pmatrix} 1 \\ 1 \\ 1 \\ 1 \end{pmatrix}$$

From $\bar{\bar{A}}$ and equation (1.4-13) we get

$$\bar{\bar{A}}^{-1} = -\frac{1}{690} \begin{pmatrix} 5 & -35 & 130 & -145 \\ 122 & -164 & -2 & 50 \\ -9 & 63 & -96 & -15 \\ -206 & 62 & 26 & 40 \end{pmatrix}$$

Finally, from equations (1.4-21) and (1.5-3) we get

$$\bar{X} = \begin{pmatrix} x_1 \\ \vdots \\ \vdots \\ x_4 \end{pmatrix} = -\frac{1}{690} \begin{pmatrix} -45 \\ +6 \\ -57 \\ -78 \end{pmatrix} \qquad (1.5\text{-}5)$$

Many times a set of equations can be represented in compact matrix form as

$$\bar{\bar{A}}\bar{X} = \lambda\bar{X} \qquad (1.5\text{-}6)$$

or

$$(\bar{\bar{A}} - \lambda\bar{\bar{I}})\bar{X} = 0 \qquad (1.5\text{-}7)$$

where λ is a scalar multiplier and $\bar{\bar{I}}$ is the identity matrix. For equation (1.5-7) to be true, it is necessary that either the determinant of $(\bar{\bar{A}} - \lambda\bar{\bar{I}})$ is zero or $\bar{X} = 0$ (trivial solution). The matrix $(\bar{\bar{A}} - \lambda\bar{\bar{I}})$ is called the characteristic matrix and the determinant of $(\bar{\bar{A}} - \lambda\bar{\bar{I}})$ is called the characteristic equation,

$$\det(\bar{\bar{A}} - \lambda\bar{\bar{I}}) = \begin{vmatrix} (a_{11} - \lambda) & a_{12} & \cdots & a_{1n} \\ a_{21} & (a_{21} - \lambda) & \cdots & a_{2n} \\ \vdots & \vdots & & \vdots \\ a_{n1} & a_{n2} & \cdots & (a_{nn} - \lambda) \end{vmatrix} \qquad (1.5\text{-}8)$$

When $\det(\bar{\bar{A}} - \lambda\bar{\bar{I}})$ is expanded, we get the characteristic polynomial $P(\lambda)$:

$$P(\lambda) = \lambda^n + P_1\lambda^{n-1} + \cdots + P_{n-1}\lambda + P_n = 0 \qquad (1.5\text{-}9)$$

The values of λ that satisfy equation (1.5-9) are called eigenvalues. Associated with each value λ_i is a column vector $\bar{X}_i$ which satisfies equation (1.5-7). These $\bar{X}_i$ vectors are called eigenvectors. For example, let

$$\bar{A} = \begin{pmatrix} 1 & 2 \\ 2 & 1 \end{pmatrix} \quad \text{and} \quad \bar{I} = \begin{pmatrix} 1 & 0 \\ 0 & 1 \end{pmatrix}$$

then

$$\det(\bar{A} - \lambda\bar{I}) = \begin{vmatrix} 1 - \lambda & 2 \\ 2 & 1 - \lambda \end{vmatrix} = \lambda^2 - 2\lambda - 3$$

so that from equation (1.5-9) we get

$$P(\lambda) = \lambda^2 - 2\lambda - 3 \tag{1.5-10}$$

Equation (1.5-10) has roots $\lambda_1 = -1$ and $\lambda_2 = 3$. To find the eigenvectors corresponding to $\lambda_1 = -1$, we get from equation (1.5-7)

$$(\bar{A} + \bar{I})\bar{X}_1 = 0 \quad \lambda_1 = -1 \tag{1.5-11}$$

or

$$\begin{pmatrix} 1 + 1 & 2 \\ 2 & 1 + 1 \end{pmatrix} \begin{pmatrix} X_1 \\ X_2 \end{pmatrix} = 0 \tag{1.5-12}$$

which, considering equation (1.4-21), yields

$$\begin{pmatrix} 2X_1 + 2X_2 \\ 2X_1 + 2X_2 \end{pmatrix} = 0 \quad \lambda_1 = -1 \tag{1.5-13}$$

Note that in equation (1.5-12) a (2×2) matrix was multiplied with one of size (2×1), yielding a (2×1) matrix by the conformable requirement $(2 \times 2)(2 \times 1) = (2 \times 1)$. It should also be pointed out that the right-hand side of equation (1.5-13) is actually $\begin{pmatrix} 0 \\ 0 \end{pmatrix}$. Thus from equation (1.5-13) we get for $\lambda_1 = -1$ the relationship

$$X_1 + X_2 = 0 \quad \lambda_1 = -1 \tag{1.5-14}$$

Using $\lambda_2 = 3$, and repeating the above steps, we get

$$\begin{pmatrix} 1 - 3 & 2 \\ 2 & 1 - 3 \end{pmatrix} \begin{pmatrix} X_1 \\ X_2 \end{pmatrix} = 0 \quad \lambda_2 = 3 \tag{1.5-15}$$

or

$$X_1 - X_2 = 0 \quad \lambda_2 = 3 \tag{1.5-16}$$

From equation (1.5-14) we get

$$\bar{X} = \begin{pmatrix} X_1 \\ -X_1 \end{pmatrix} \qquad \lambda_1 = -1 \qquad (1.5\text{-}17)$$

which in relative terms is

$$\bar{X}_1 = \begin{pmatrix} 1 \\ -1 \end{pmatrix} \qquad \lambda_1 = -1 \qquad (1.5\text{-}18)$$

or any scalar multiple of equation (1.5-18). Similarly we get

$$\bar{X}_2 = \begin{pmatrix} 1 \\ 1 \end{pmatrix} \qquad \lambda_2 = 3 \qquad (1.5\text{-}19)$$

1.6. Matrix Solutions of Linear Simultaneous Differential Equations

Sets of linear simultaneous differential equations with constant coefficients can be handled by matrix methods by first reducing the equations to matrix notation. For example, suppose we have a set of differential equations

$$\begin{aligned}
\dot{y}_1 &= a_{11} y_1 + a_{12} y_2 + \cdots + a_{1n} y_n \\
\dot{y}_2 &= a_{21} y_1 + a_{22} y_2 + \cdots + a_{2n} y_n \\
\vdots & \qquad \vdots \qquad \vdots \qquad\qquad \vdots \\
\dot{y}_n &= a_{n1} y_1 + a_{n2} y_2 + \cdots + a_{nn} y_n
\end{aligned} \qquad (1.6\text{-}1)$$

where $\dot{y}_1 = dy_1/dt$, $\dot{y}_2 = dy_2/dt$, etc. Equation (1.6-1) arises frequently in kinetics expressions and can be reduced to a matrix equation:

$$\bar{\dot{Y}} = \frac{d\bar{Y}}{dt} = \bar{A}\,\bar{Y} \qquad (1.6\text{-}2)$$

where

$$\bar{Y} = \begin{pmatrix} y_{11} & \cdots & y_{n1} \\ \vdots & & \vdots \\ y_{n1} & \cdots & y_{nn} \end{pmatrix} \qquad \bar{\dot{Y}} = \begin{pmatrix} \dfrac{dy_{11}}{dt} & \cdots & \dfrac{dy_{n1}}{dt} \\ \vdots & & \vdots \\ \dfrac{dy_{n1}}{dt} & \cdots & \dfrac{dy_{nn}}{dt} \end{pmatrix} \qquad \bar{A} = \begin{pmatrix} a_{11} & \cdots & a_{1n} \\ \vdots & & \vdots \\ a_{n1} & \cdots & a_{nn} \end{pmatrix}$$

The solution to equation (1.6-2) is[1]

$$\bar{Y} = (e^{\bar{A}t})\,\bar{Y}_0 \qquad (1.6\text{-}3)$$

[1] D. M. Himmelblau and K. B. Bischoff, *Process Analysis and Simulation*, John Wiley, New York (1968), p. 331.

where

$$e^{\bar{A}t} = \bar{I} + At + \cdots + \frac{\bar{A}^n t^n}{n!} \tag{1.6-4}$$

and

$$\bar{Y}(0) = \bar{Y}_0 = \bar{Y} \qquad \text{at } t = 0 \tag{1.6-5}$$

Now suppose there is an inhomogeneous set of equations

$$
\begin{aligned}
\dot{y}_1 &= a_{11} y_1 + \cdots + a_{1n} y_n + x_1(t) \\
\vdots \quad &\quad \vdots \qquad\qquad \vdots \quad\quad \vdots \\
\dot{y}_n &= a_{n1} y_1 + \cdots + a_{nn} y_n + x_n(t)
\end{aligned}
\tag{1.6-6}
$$

which can be written in matrix form as

$$\frac{d\bar{Y}}{dt} + \bar{A}\bar{Y} = \bar{X}(t) \tag{1.6-7}$$

By analogy with scalar operations, it can be shown[2] that we can make use of the integrating factor $e^{\bar{A}(t-t_0)}$ to get from equation (1.6-7) to

$$\frac{d}{dt}(e^{\bar{A}(t-t_0)}\bar{Y}) = (e^{\bar{A}(t-t_0)})\bar{X}(t) \tag{1.6-8}$$

Each side of equation (1.6-8) may now be integrated from t_0 to t using the initial condition $\bar{Y} = \bar{Y}_0$ at $t = t_0$. The result is

$$e^{\bar{A}(t-t_0)}\bar{Y} - \bar{I}\bar{Y}_0 = \int_{t_0}^{t} (e^{\bar{A}(t'-t_0)})\bar{X}(t')dt' \tag{1.6-9}$$

or

$$\bar{Y} = e^{-\bar{A}(t-t_0)}\bar{Y}_0 + e^{-\bar{A}(t-t_0)}\int_{t_0}^{t} (e^{\bar{A}(t'-t_0)})\bar{X}(t')dt' \tag{1.6-10}$$

If $\bar{Y} = 0$ at $t - t_0 = 0$, then equation (1.6-10) becomes

$$\bar{Y} = e^{-\bar{A}t}\int_0^t e^{\bar{A}t'}\bar{X}(t')dt' = \int_0^t e^{\bar{A}(t'-t)}\bar{X}(t')dt' \tag{1.6-11}$$

Returning to the example of equation (1.6-1), if $\bar{H}$ is the square matrix formed by the column eigenvectors $\bar{X}$ of equation (1.6-2),

$$\bar{H} = (\bar{X}_1, \bar{X}_2, \ldots, \bar{X}_n) \tag{1.6-12}$$

---

[2] *Ibid.*, p. 332.

it can be used in the solution of (1.6-2):

$$\bar{\bar{Y}} = e^{\bar{\bar{A}}t}\,\bar{Y}_0 = \left(\bar{I} + \bar{\bar{A}}t + \frac{(\bar{\bar{A}}t)^2}{2!} + \cdots\right)\bar{Y}_0 \tag{1.6-13}$$

where equations (1.6.3) and (1.6-4) have been used. We will also use the identity matrix property

$$\bar{\bar{H}}\bar{I} = \bar{\bar{H}} \tag{1.6-14}$$

or

$$\bar{\bar{H}}\bar{I}\bar{\bar{H}}^{-1} = \bar{\bar{H}}\bar{\bar{H}}^{-1} = \bar{I} \tag{1.6-15}$$

where we have post-multiplied equation (1.6-14) by $\bar{\bar{H}}^{-1}$ and made use of the property that a matrix multiplied by its inverse yields the identity matrix. From equation (1.6-2) let

$$\bar{\dot{Y}} = \bar{\bar{A}}\,\bar{Y} = \lambda\,\bar{Y} \tag{1.6-16}$$

so that

$$(\bar{\bar{A}} - \lambda\bar{I})\,\bar{Y} = 0 \tag{1.6-17}$$

or

$$\bar{\bar{A}} = \lambda\bar{I} \tag{1.6-18}$$

By matrix algebra, from equation (1.6-18) we can get

$$\bar{\bar{A}} = \bar{\bar{H}}(\lambda\bar{I})\bar{\bar{H}}^{-1} \tag{1.6-19}$$

and

$$\bar{\bar{A}}^2 = (\bar{\bar{H}}(\lambda\bar{I})\bar{\bar{H}}^{-1})(\bar{\bar{H}}(\lambda\bar{I})\bar{\bar{H}}^{-1}) = \bar{\bar{H}}(\lambda\bar{I})^2\bar{\bar{H}}^{-1} \tag{1.6-20}$$

Substituting equations (1.6-19) and (1.6-20) into (1.6-13) yields after some manipulation

$$\begin{aligned}\bar{\bar{Y}} &= \left(\bar{I} + H(\lambda\bar{I})\bar{\bar{H}}^{-1}t + \frac{\bar{\bar{H}}(\lambda\bar{I})^2\bar{\bar{H}}^{-1}t^2}{2!} + \cdots\right)\bar{Y}_0 \\ &= \left(\bar{\bar{H}}\left(\bar{I} + \lambda\bar{I}t + \frac{(\lambda\bar{I})^2t^2}{2!} + \cdots\right)\bar{\bar{H}}^{-1}\right)\bar{Y}_0 \\ &= (\bar{\bar{H}}e^{\lambda\bar{I}t}\bar{\bar{H}}^{-1})\,\bar{Y}_0 \end{aligned} \tag{1.6-21}$$

where equation (1.6-4) has been used.

Comparing equations (1.6-21) and (1.6-13), we get

$$e^{\bar{\bar{A}}t} = \bar{\bar{H}}e^{\lambda\bar{I}t}\bar{\bar{H}}^{-1} \tag{1.6-22}$$

so that

$$e^{\bar{A}t} = \bar{H} \begin{pmatrix} e^{\lambda_1 t} & 0 & \cdots & 0 \\ 0 & e^{\lambda_2 t} & \cdots & 0 \\ \cdot & \cdot & & \cdot \\ \cdot & \cdot & & \cdot \\ \cdot & \cdot & & \cdot \\ 0 & 0 & \cdots & e^{\lambda_n t} \end{pmatrix} \bar{H}^{-1} \tag{1.6-23}$$

where

$$\lambda \bar{I} = \begin{pmatrix} \lambda_1 & 0 & \cdots & 0 \\ 0 & \lambda_2 & \cdots & 0 \\ \cdot & \cdot & & \cdot \\ \cdot & \cdot & & \cdot \\ \cdot & \cdot & & \cdot \\ 0 & 0 & \cdots & \lambda_n \end{pmatrix} \tag{1.6-24}$$

If now $\bar{H}^{-1} \bar{Y}_0 = \bar{b}$, then the solution of equation (1.6-2), which is (1.6-3) with (1.6-23), is now given by

$$\bar{Y} = \bar{H} \begin{pmatrix} e^{\lambda_1 t} & 0 & \cdots & 0 \\ 0 & e^{\lambda_2 t} & \cdots & 0 \\ \cdot & \cdot & & \cdot \\ \cdot & \cdot & & \cdot \\ \cdot & \cdot & & \cdot \\ 0 & 0 & \cdots & e^{\lambda_n t} \end{pmatrix} \begin{pmatrix} b_1 \\ b_2 \\ \cdot \\ \cdot \\ \cdot \\ b_n \end{pmatrix} \tag{1.6-25}$$

or

$$\bar{Y} = (\bar{X}_1, \bar{X}_2, \ldots, \bar{X}_n) \begin{pmatrix} e^{\lambda_1 t} & 0 & \cdots & 0 \\ 0 & e^{\lambda_2 t} & \cdots & 0 \\ \cdot & \cdot & & \cdot \\ \cdot & \cdot & & \cdot \\ \cdot & \cdot & & \cdot \\ 0 & 0 & \cdots & e^{\lambda_n t} \end{pmatrix} \begin{pmatrix} b_1 \\ b_2 \\ \cdot \\ \cdot \\ \cdot \\ b_n \end{pmatrix} \tag{1.6-26}$$

where equation (1.6-12) has also been used.

For example, a differential equation such as

$$\frac{d^2 y}{dt^2} - 3 \frac{dy}{dt} + 2y = e^{-t} \tag{1.6-27}$$

is to be solved, with initial conditions

$$y = 0 \qquad \text{at } t = 0 \tag{1.6-28}$$

$$\frac{dy}{dt} = 1 \qquad \text{at } t = 0 \tag{1.6-29}$$

Let

$$y_1 = y$$

$$y_2 = \frac{dy}{dt}$$

$$y_3 = e^{-t}$$

so that the following are true by definition:

$$\dot{y}_1 = y_2 \tag{1.6-30}$$

$$\dot{y}_2 = -2y_1 + 3y_2 + y_3 \tag{1.6-31}$$

$$\dot{y}_3 = -y_3 \tag{1.6-32}$$

and from the initial conditions we can get

$$y_1 = 0 \quad \text{at } t = 0 \tag{1.6-33}$$

$$y_2 = 1 \quad \text{at } t = 0 \tag{1.6-34}$$

$$y_3 = 1 \quad \text{at } t = 0 \tag{1.6-35}$$

Equations (1.6-30) through (1.6-35) can be written in matrix form:

$$\bar{\bar{Y}} = \begin{pmatrix} 0 & 1 & 0 \\ -2 & 3 & 1 \\ 0 & 0 & -1 \end{pmatrix} \bar{Y} = \bar{A}\bar{Y} \tag{1.6-36}$$

where

$$\bar{Y} = \begin{pmatrix} y_1 \\ y_2 \\ y_3 \end{pmatrix} \qquad \bar{\bar{Y}} = \begin{pmatrix} \dfrac{dy_1}{dt} \\ \dfrac{dy_2}{dt} \\ \dfrac{dy_3}{dt} \end{pmatrix} \qquad \bar{A} = \begin{pmatrix} 0 & 1 & 0 \\ -2 & 3 & 1 \\ 0 & 0 & -1 \end{pmatrix}$$

We next find the eigenvalues λ_i of (1.6-36) using (1.6-17):

$$\det(\bar{A} - \lambda \bar{I}) = 0 \tag{1.6-37}$$

or

$$\begin{vmatrix} (0-\lambda) & 1 & 0 \\ -2 & (3-\lambda) & 1 \\ 0 & 0 & (-1-\lambda) \end{vmatrix} = 0 \tag{1.6-38}$$

where equation (1.5-8) has been used. From equation (1.6-38) we get a polynomial in λ which yields $\lambda_1 = 1$, $\lambda_2 = 2$, $\lambda_3 = -1$. From $(\bar{A} - \lambda_n \bar{I})\bar{X}_n = 0$ we get, for $\lambda_1 = 1$,

$$\begin{pmatrix} (0-1) & 1 & 0 \\ -2 & (3-1) & 1 \\ 0 & 0 & (-1-1) \end{pmatrix}(\bar{X}_1) = 0 \tag{1.6-39}$$

or the eigenvector

$$\bar{X}_1 = \begin{pmatrix} 1 \\ 1 \\ 0 \end{pmatrix} \qquad \lambda_1 = 1 \tag{1.6-40}$$

Similarly, for λ_2 and λ_3 we get

$$\begin{pmatrix} (0-2) & 1 & 0 \\ -2 & (3-2) & 1 \\ 0 & 0 & (-1-2) \end{pmatrix} (\bar{X}_2) = 0 \tag{1.6-41}$$

or

$$\bar{X}_2 = \begin{pmatrix} 1 \\ 2 \\ 0 \end{pmatrix} \qquad \lambda_2 = 2 \tag{1.6-42}$$

and

$$\bar{X}_3 = \begin{pmatrix} 1 \\ -1 \\ 6 \end{pmatrix} \qquad \lambda_3 = -1 \tag{1.6-43}$$

This yields for $\bar{\bar{H}} = (\bar{X}_1, \bar{X}_2, \bar{X}_3)$ from equation (1.6-12)

$$\bar{\bar{H}} = \begin{pmatrix} 1 & 1 & 1 \\ 1 & 2 & -1 \\ 0 & 0 & 6 \end{pmatrix} \tag{1.6-44}$$

and

$$\bar{\bar{H}}^{-1} = \frac{1}{6} \begin{pmatrix} 12 & -6 & -3 \\ -6 & 6 & 0 \\ 0 & 0 & 1 \end{pmatrix}$$

Thus from (1.6-18), (1.6-19), and (1.6-24) we get $\lambda \bar{I} = \bar{\bar{H}} \bar{\bar{A}} \bar{\bar{H}}^{-1}$, or

$$\begin{pmatrix} \lambda_1 & 0 & 0 \\ 0 & \lambda_2 & 0 \\ 0 & 0 & \lambda_3 \end{pmatrix} = \begin{pmatrix} 1 & 0 & 0 \\ 0 & 2 & 0 \\ 0 & 0 & -1 \end{pmatrix} \tag{1.6-45}$$

and $\lambda_1 = 1$, $\lambda_2 = 2$, $\lambda_3 = -1$.

Finally the solution of equation (1.6-36) from (1.6-21) is

$$\bar{Y}(t) = (\bar{\bar{H}} e^{\lambda \bar{I} t} \bar{\bar{H}}^{-1}) \bar{Y}_0$$

$$= \frac{1}{6} \begin{pmatrix} 1 & 1 & 1 \\ 1 & 2 & -1 \\ 0 & 0 & 6 \end{pmatrix} \begin{pmatrix} e^t & 0 & 0 \\ 0 & e^{2t} & 0 \\ 0 & 0 & e^{-t} \end{pmatrix} \begin{pmatrix} 12 & -6 & -3 \\ -6 & 6 & 2 \\ 0 & 0 & 1 \end{pmatrix} \begin{pmatrix} 0 \\ 1 \\ 1 \end{pmatrix} \tag{1.6-46}$$

or

$$\begin{pmatrix} y_1 \\ y_2 \\ y_3 \end{pmatrix} = \begin{pmatrix} -\frac{3}{2}e^t + \frac{8}{3}e^{2t} + \frac{1}{6}e^{-t} \\ -\frac{3}{2}e^t + \frac{8}{3}e^{2t} - \frac{1}{6}e^{-t} \\ 0 \quad + 0 \quad + e^{-t} \end{pmatrix} \tag{1.6-47}$$

In equation (1.6-46) $\bar{\bar{Y}}_0$ is obtained from the initial conditions on y_1, y_2, and y_3.

Assignments in Chapter 1

1.1. Verify equation (1.2-11) by working directly with equation (1.2-10).

1.2. Verify equation (1.3-8).

1.3. Verify the result in equation (1.3-13).

1.4. Show the result in equations (1.3-26), (1.3-27), and (1.3-28).

1.5. Complete equation (1.4-11).

1.6. Confirm equations (1.4-15) and (1.4-17).

1.7. Confirm equation (1.4-24).

1.8. Show that equation (1.5-2) is the same as equation (1.5-1).

1.9. Verify equation (1.5-5).

1.10. In equation (1.6-4), compare $\bar{A}$ with $\bar{A}^2$.

1.11. Verify equations (1.6-19) and (1.6-20).

1.12. Show that equation (1.6-23) is true.

1.13. Show that equation (1.6-36) is true.

1.14. Verify that $\lambda_1 = 1$, $\lambda_2 = 2$, and $\lambda_3 = -1$ in equation (1.6-38).

1.15. Verify equations (1.6-40), (1.6-42), and (1.6-43).

1.16. Starting with $\bar{H}$ in equation (1.6-44), get $\bar{H}^{-1}$.

1.17. Verify equation (1.6-47).

1.18. Given a set of equations,

$$\dot{y}_1 = y_2 - y_3 \qquad\qquad y_1(0) = 1$$
$$\dot{y}_2 = 2y_2 + y_3 \qquad\qquad y_2(0) = 1$$
$$\dot{y}_3 = 4y_1 - 2y_2 + 5y_3 \qquad y_3(0) = -2$$

obtain the solution for

$$\bar{\bar{Y}}(t) = \begin{pmatrix} y_1 \\ y_2 \\ y_3 \end{pmatrix}$$

by the matrix methods of Section 1.6.

For Further Reading

Advanced Calculus, by Angus E. Taylor, Ginn and Company, New York, 1955.

Transport Phenomena, by R. B. Bird, W. E. Stewart, and E. N. Lightfoot, John Wiley, New York, 1960.

Mathematical Methods in Chemical Engineering, by V. G. Jenson and G. V. Jeffreys, Academic Press, New York, 1963.

Process Analysis and Simulation, by D. M. Himmelblau and K. B. Bischoff, John Wiley, New York, 1968.

Chapter 2

Laplace Transforms

In this chapter we shall show how the Laplace transform operational methods apply to the analysis of transport phenomena. We shall trace the definition of the Laplace transform to its application with various functional forms. The Laplace transform technique will then be shown in the handling of ordinary and partial differential equations. Finite difference equations, integral, integrodifferential, and differential-difference equations are also treated. A discussion of methods of inverting Laplace transforms leads to a review of complex variables, since one of the methods for finding inverse Laplace transforms involves integration in the complex plane. Many examples are shown illustrating the procedures for finding the inverse Laplace transform. This chapter concludes with a discussion of the difficulties encountered when the Laplace transform technique for solving differential equations is applied to nonlinear equations.

2.1. Definitions and Basic Operations

The definition of the Laplace transform is given by

$$\mathscr{L}_{t \to s} f(t) = \bar{f}(s) = \int_0^\infty f(t) e^{-st}\, dt \qquad (2.1\text{-}1)$$

which reads: "take the Laplace transform of $f(t)$ with respect to t ($\mathscr{L}_{t \to s}$), with the variable t going over to the parameter s." The result of this operation is symbolically represented as $\bar{f}(s)$, with the bar over the function f to indicate that the original function $f(t)$ has been transformed to $\bar{f}(s)$. If the function f is dependent upon two independent variables, such as $f(t, x)$, we could if we wished apply the Laplace transform twice, indicated symbolically by

$$\mathscr{L}_{t \to s} f(t, x) = \bar{f}(s, x) = \int_0^\infty f(t, x)e^{-st}\, dt \qquad (2.1\text{-}2)$$

and

$$\mathscr{L}_{x \to p} \bar{f}(s, x) = \bar{\bar{f}}(s, p) = \int_0^\infty \bar{f}(s, x)e^{-px}\, dx \qquad (2.1\text{-}3)$$

where $\bar{\bar{f}}(s, p)$ indicates that the Laplace transform has been taken twice. This bookkeeping system and its notation are important, as we shall see later in solving partial differential equations by Laplace transform techniques.

In performing Laplace transform operations defined by equation 2.1-1, some translational properties may be used as short cuts. For example, it can be shown[1] by the integration by parts technique that

$$\mathscr{L}_{t \to s} e^{at} f(t) = \bar{f}(s - a) \qquad (2.1\text{-}4)$$

is an identity, where

$$\mathscr{L}_{t \to s} f(t) = \bar{f}(s) \qquad (2.1\text{-}5)$$

As an illustration of the relationship given by equation (2.1-4), suppose we wish to obtain the Laplace transform of $e^{-t} \cos 2t$. The result is

$$\mathscr{L}_{t \to s} e^{-t} \cos 2t = \frac{s + 1}{(s + 1)^2 + 4} \qquad (2.1\text{-}6)$$

where

$$\mathscr{L}_{t \to s} \cos 2t = \frac{s}{s^2 + 4} \qquad (2.1\text{-}7)$$

A second shifting formula, somewhat the inverse of equation (2.1-4), states that if there are some functional relationships such as

$$\mathscr{L}_{t \to s} f(t) = \bar{f}(s) \qquad (2.1\text{-}8)$$

and

$$g(t) = f(t - a) \qquad t > 0 \qquad (2.1\text{-}9)$$

then

$$\mathscr{L}_{t \to s} g(t) = e^{-as} \bar{f}(s) \qquad (2.1\text{-}10)$$

---

[1] M. R. Spiegel, *Laplace Transforms*, Schaum, New York (1965), p. 3.

For example, suppose we know that

$$\mathscr{L}_{t\to s}t^3 = \frac{3!}{s^4} \qquad (2.1\text{-}11)$$

is true; then from equation (2.1-10) we also know that the following is true:

$$\mathscr{L}_{t\to s}(t-2)^3 = e^{-2s}\frac{3!}{s^4} \qquad (2.1\text{-}12)$$

where

$$g(t) = f(t-2) = (t-2)^3 \qquad (2.1\text{-}13)$$

and

$$f(t) = t^3 \qquad (2.1\text{-}14)$$

In most cases the Laplace transform operation can be obtained in two ways: (1) look up the result in the very extensive tables available[2] or (2) perform the integral operations indicated by equation (2.1-1). Obviously route (1) is more desirable. Once the Laplace transform operation has been performed, and we have a relationship involving $\bar{f}(s)$, it remains only to invert the process and transform $\bar{f}(s)$ back into $f(t)$. This inverse Laplace transform operation is represented symbolically by

$$\mathscr{L}^{-1}_{s\to t}\bar{f}(s) = f(t) \qquad (2.1\text{-}15)$$

so that the complete Laplace transform operation consists of the cyclic equations

$$\mathscr{L}_{t\to s}f(t) = \bar{f}(s) \qquad (2.1\text{-}16)$$

$$\mathscr{L}^{-1}_{s\to t}\bar{f}(s) = f(t) \qquad (2.1\text{-}17)$$

The mathematical description of equation (2.1-17) is more complex than that shown by equation (2.1-1) and is given later in this chapter. For the moment it is sufficient to say that in most cases the inverse Laplace transform may also be looked up in the very extensive tables available.[3]

One of the most common Laplace transform operations involves the breaking down of derivatives into algebraic quantities, and this is represented by

$$\mathscr{L}_{t\to s}\frac{df(t)}{dt} = s\bar{f}(s) - f(0_-) \qquad (2.1\text{-}18)$$

[2] G. E. Roberts and H. Kaufman, *Table of Laplace Transforms*, W. B. Saunders Co., Philadelphia (1966).
[3] *Ibid.*

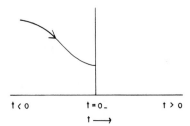

Fig. 6. Illustration of $t = 0_-$.

$t < 0$ $t = 0_-$ $t > 0$

$t \longrightarrow$

and

$$\mathscr{L}_{t \to s} \frac{d^2 f(t)}{dt^2} = s^2 \bar{f}(s) - s f(0_-) - \left. \frac{df(t)}{dt} \right|_{t=0} \qquad (2.1\text{-}19)$$

where $f(0_-)$ indicates $f(t)\,|_{t=0_-}$ and $t = 0_-$ refers to the approach of t to zero from the direction shown in Figure 6.

One can generalize the results given above for derivatives:

$$\mathscr{L}_{t \to s} \frac{d^n f}{dt^n} = s^n \bar{f}(s) - s^{n-1} f(0) - s^{n-2} f'(0) - \cdots - s f^{n-2}(0) - f^{n-1}(0)$$
$$(2.1\text{-}20)$$

All of these results and many others involving Laplace transforms are obtained by integration by parts.[4] For example, the first derivative operation is given by

$$\mathscr{L}_{t \to s} \frac{df}{dt} = \int_0^\infty \frac{df}{dt} e^{-st} \, dt = s \bar{f}(s) - f(0_-) \qquad (2.1\text{-}21)$$

using

$$u = e^{-st} \qquad\qquad dV = \frac{df}{dt} \, dt$$

$$du = -s e^{-st} \, dt \qquad V = f$$

Another important functional form which is handled easily by the Laplace transform operation is given by

$$\mathscr{L}_{t \to s} \int_0^t f(u) \, du = \frac{\bar{f}(s)}{s} \qquad (2.1\text{-}22)$$

For example, if

$$\mathscr{L}_{t \to s} \sin 2t = \frac{s}{s^2 + 4} \qquad (2.1\text{-}23)$$

[4] M. R. Spiegel, *op. cit.*, p. 4.

is true, then from equation (2.1-22), we can get

$$\mathscr{L}_{t\to s}\int_0^t \sin 2u\, du = \frac{s/(s^2+4)}{s} \qquad (2.1\text{-}24)$$

The Laplace transform of an associated function such as $t^n f(t)$ is also handled by integration by part[5] and is given by

$$\mathscr{L}_{t\to s}t^n f(t) = (-1)^n \bar{f}^{(n)}(s) \qquad (2.1\text{-}25)$$

where

$$\bar{f}^{(n)}(s) = \frac{d^n \bar{f}(s)}{ds^n} \qquad (2.1\text{-}26)$$

An application of equations (2.1-25) and (2.1-26) follows. If

$$\mathscr{L}_{t\to s}e^{2t} = \frac{1}{s-2} \qquad (2.1\text{-}27)$$

then

$$\mathscr{L}_{t\to s}t^2 e^{2t} = (-1)^2 \frac{d^2}{ds^2}\left(\frac{1}{s-2}\right) = \frac{2}{(s-2)^3} \qquad (2.1\text{-}28)$$

If the independent variable, t, appears in the denominator of an associated function, such as $f(t)/t$, then the Laplace transform operation is given by

$$\mathscr{L}_{t\to s}\frac{f(t)}{t} = \int_s^\infty \bar{f}(u)\, du \qquad (2.1\text{-}29)$$

For example, if

$$\mathscr{L}_{t\to s}\sin t = \frac{1}{s^2+1} \qquad (2.1\text{-}30)$$

is true, then from equation (2.1-29) we can get

$$\mathscr{L}_{t\to s}\frac{\sin t}{t} = \int_s^\infty \frac{1}{u^2+1}\, du = \tan^{-1}\left(\frac{1}{s}\right) \qquad (2.1\text{-}31)$$

We can also operate on periodic functions by Laplace transform techniques, where a periodic function is defined by $f(t+T)=f(t)$ where T is the time required for $f(t)$ to return to its original value. Without showing the details,[6] we present the Laplace transform of a periodic function:

$$\mathscr{L}_{t\to s}f(t) = \frac{\int_0^T e^{-st}f(t)\, dt}{1-e^{-st}} \qquad (2.1\text{-}32)$$

[5] *Ibid.*, p. 5.
[6] *Ibid.*

In dealing with Laplace transforms applied to differential equations we can at times invoke some properties which are used either as checks on the verity of our equations or as conditions for evaluating arbitrary constants. These properties are summarized by the following equations:

(1) If

$$\mathscr{L}_{t \to s} f(t) = \bar{f}(s) \tag{2.1-33}$$

then

$$\lim_{s \to \infty} \bar{f}(s) = 0 \tag{2.1-34}$$

(2) *Initial Value Theorem*

$$\lim_{t \to 0_+} f(t) = \lim_{s \to \infty} s \bar{f}(s)$$

(3) *Final Value Theorem*

$$\lim_{t \to \infty} f(t) = \lim_{s \to 0} s \bar{f}(s) \tag{2.1-36}$$

2.2. The Inverse Laplace Transform

All of the preceding discussion centered on finding the Laplace transform of some function designated generally as $f(t)$. Equations (2.1-16) and (2.1-17) indicated the cyclic nature of the transform method, i.e., after finding $\bar{f}(s)$, we must reverse the process and extract $f(t)$. The easiest way of finding the inverse Laplace transform is to look it up in the tables. The inverse Laplace transform definition is given by

$$\mathscr{L}^{-1}_{s \to t} \bar{f}(s) = f(t) = \frac{1}{2\pi i} \int_{\text{Br}} \bar{f}(s) e^{st} \, ds \tag{2.2-1}$$

Equation (2.2-1) is a complex variables equation, where the integration is performed over a Bromwhich path (Br). In a later section we will review some complex variables principles and operations which will clarify equation (2.2-1).

Here, without going into the theory involved, we list the common methods of finding the inverse Laplace transform of $\bar{f}(s)$, seeking to produce $f(t)$:

1. Look up $\bar{f}(s)$ in the tables.[7]
2. By a method of partial fractions, take the complicated-looking

----

[7] G. E. Roberts and H. Kaufman, *op. cit.*

expression for $\bar{f}(s)$ and break it up into simpler forms that can be looked up in the tables.

3. Use the Heaviside expansion. If we can represent $\bar{f}(s)$ as

$$\bar{f}(s) = \frac{P(s)}{Q(s)} \tag{2.2-2}$$

where $P(s)$ and $Q(s)$ are polynomials and $P(s)$ is of degree less than $Q(s)$, then the inverse is given by

$$\mathscr{L}^{-1}_{s \to t} \frac{P(s)}{Q(s)} = \sum_{k=1}^{n} \frac{P(\alpha_k)}{Q'(\alpha_k)} e^{\alpha_k t} \tag{2.2-3}$$

where α_k are the n distinct roots of $Q(s)$, and $Q'(\alpha_k)$ is defined by

$$Q'(\alpha_k) = \frac{dQ(s)}{ds}\bigg|_{s=\alpha_k} \tag{2.2-4}$$

4. Use the convolution integral. We apply this technique when $\bar{f}(s)$ is given in a form such as

$$\bar{f}(s) = \bar{g}(s)\bar{h}(s) \tag{2.2-5}$$

If $\bar{f}(s)$ is expressed in this manner, as the product of two functions $\bar{g}(s)$ and $\bar{h}(s)$, then by use of the convolution integral, given by

$$\mathscr{L}^{-1}_{s \to t}\bar{f}(s) = f(t) = \mathscr{L}^{-1}_{s \to t}\bar{g}(s)\bar{h}(s) = \int_0^t g(u)h(t-u)\,du \tag{2.2-6}$$

where

$$\mathscr{L}^{-1}_{s \to t}\bar{g}(s) = g(t) \tag{2.2-7}$$

and

$$\mathscr{L}^{-1}_{s \to t}\bar{h}(s) = h(t) \tag{2.2-8}$$

we can produce $f(t)$; in equation (2.2-6) the independent variable t in $h(t)$ is replaced by $(t-u)$. Equation (2.2-6) can also be written as

$$\mathscr{L}^{-1}_{s \to t}\bar{f}(s) = f(t) = \mathscr{L}^{-1}_{s \to t}\bar{g}(s)\bar{h}(s) = \int_0^t g(t-u)h(u)\,du \tag{2.2-9}$$

so that the choice of which function, g or h, is written with the $(t-u)$ variable depends upon which result is simpler to manipulate.

5. Perform an integration in the complex plane, using equation (2.2-1). This will be shown in detail later in this chapter.

2.3. Application of Laplace Transforms to Ordinary Differential Equations

We are now in a position to apply the techniques of Sections 2.1 and 2.2 to some ordinary differential equations. For example, suppose there is a differential equation as given by equation (2.3-1) with initial conditions as shown in equations (2.3-2) and (2.3-3):

$$\frac{d^2y}{dt^2} + y = t \tag{2.3-1}$$

$$y = 1 \qquad \text{at } t = 0 \tag{2.3-2}$$

$$\frac{dy}{dt} = -2 \qquad \text{at } t = 0 \tag{2.3-3}$$

Taking the Laplace transform, $\mathscr{L}_{t \to s}$, of both sides of equation (2.3-1) according to equation (2.1-19), we get

$$s^2 \bar{y}(s) - sy(0) - \frac{dy}{dt}\bigg|_{t=0} + \bar{y}(s) = \frac{1}{s^2} \tag{2.3-4}$$

or

$$\bar{y}(s) = \frac{1}{s^2} + \frac{s}{s^2 + 1} - \frac{3}{s^2 + 1} \tag{2.3-5}$$

In arriving at equation (2.3-5), we have used equations (2.3-2) and (2.3-3) and a partial fractions procedure to simplify the relationship found for $\bar{y}(s)$. Equation (2.3-5) can be inverted simply by looking up the individual terms in the tables. The final result is

$$\mathscr{L}^{-1}_{s \to t} \bar{y}(s) = y(t) = t + \cos\cdot t - 3 \sin t \tag{2.3-6}$$

If we are attempting to solve a differential equation in $y(t)$ which has an unspecified forcing function in it, such as

$$\frac{d^2y}{dt^2} + a^2y = f(t) \tag{2.3-7}$$

the Laplace transform technique is quite useful in finding the solution. Suppose equation (2.3-7) has a set of initial conditions

$$y = 1 \qquad \text{at } t = 0 \tag{2.3-8}$$

$$\frac{dy}{dt} = -2 \qquad \text{at } t = 0 \tag{2.3-9}$$

associated with it. The solution proceeds in a manner similar to the preceding example. The result for $\bar{y}(s)$ is

$$s^2\bar{y}(s) - s + 2 + a^2\bar{y}(s) = \bar{f}(s) \qquad (2.3\text{-}10)$$

where

$$\mathscr{L}_{t\to s} f(t) = \bar{f}(s) \qquad (2.3\text{-}11)$$

From equation (2.3-10) we can express $\bar{y}(s)$ explicitly:

$$\bar{y}(s) = \frac{s-2}{s^2 + a^2} + \frac{\bar{f}(s)}{s^2 + a^2} \qquad (2.3\text{-}12)$$

The inversion of equation (2.3-12) is easily accomplished with the use of the tables and the convolution integral, equation (2.2-6). The inversion may be shown symbolically as

$$\mathscr{L}^{-1}_{s\to t}\bar{y}(s) = y(t) = \mathscr{L}^{-1}_{s\to t}\left\{\frac{s-2}{s^2 + a^2}\right\} + \mathscr{L}^{-1}_{s\to t}\left\{\bar{f}(s)\left(\frac{1}{s^2 + 2^2}\right)\right\} \qquad (2.3\text{-}13)$$

or

$$y(t) = \cos at - \frac{2\sin at}{a} + \frac{1}{a}\int_0^t f(u) \sin a(t-u)\, du \qquad (2.3\text{-}14)$$

If the ordinary differential equation has the independent variable multiplied in explicitly, as for instance in

$$t\frac{d^2y}{dt^2} + \frac{dy}{dt} + 4ty = 0 \qquad (2.3\text{-}15)$$

equation (2.1-25) can be used to obtain the solution in a straightforward manner. Thus equation (2.3-15) becomes

$$-\frac{d}{ds}\left\{s^2\bar{y}(s) - sy\big|_{t=0} - \frac{dy}{dt}\Big|_{t=0}\right\}$$

$$+ \{s\bar{y}(s) - y\big|_{t=0}\} + 4(-1)\frac{d}{ds}\{\bar{y}(s)\} = 0 \qquad (2.3\text{-}16)$$

or

$$(s^2 + 4)\frac{d\bar{y}(s)}{ds} - s\bar{y}(s) = 0 \qquad (2.3\text{-}17)$$

where the initial conditions are $y = 3$, $t = 0$, and $dy/dt = 0$, $t = 0$. By methods given in texts on differential equations, we can solve equation

(2.3-17) for $\bar{y}(s)$, obtaining

$$\bar{y}(s) = \frac{C}{\sqrt{s^2 + 4}} \tag{2.3-18}$$

From the tables we extract the inverse Laplace transform

$$y(t) = CJ_0(2t) \tag{2.3-19}$$

where J_0 is the zero-order Bessel function.

Not all initial and boundary conditions are immediately applicable since the Laplace transform method requires a knowledge of the dependent variable at the initial conditions. For example, in the previous problems we always had y and dy/dt known at $t = 0$. Though the solution is more easily accomplished if the conditions are available at $t = 0$, we can nevertheless proceed even though the conditions are not presented in this form. For example, suppose a problem is posed as follows:

$$t\frac{d^2y}{dt^2} + 2\frac{dy}{dt} + ty = 0 \tag{2.3-20}$$

$$y = 1 \qquad \text{at } t = 0 \tag{2.3-21}$$

$$y = 0 \qquad \text{at } t = \pi \tag{2.3-22}$$

$$\frac{dy}{dt} = C \qquad \text{at } t = 0 \tag{2.3-23}$$

Taking the Laplace transform of equation (2.3-20) we get

$$\frac{d\bar{y}(s)}{ds} = -\frac{1}{s^2 + 1} \tag{2.3-24}$$

where equations (2.3-21) and (2.3-23) have been used. Note that equation (2.3-22) has not yet been applied. Equation (2.3-24) can be solved by separation of variables techniques to yield

$$\bar{y}(s) = -\tan^{-1} s + A \tag{2.3-25}$$

All that remains now is to evaluate the constant A in equation (2.3-25) from the third boundary condition, equation (2.3-22). Equation (2.3-22) is in terms of $y(t)$ and t, not $\bar{y}(s)$ and s as required in equation (2.3-25). We overcome this dilemma by taking the Laplace transform of equation

(2.3-22), which yields

$$\bar{y}(s) = 0 \qquad \text{at} \quad \frac{1}{s^2} = \frac{\pi}{s} \tag{2.3-26}$$

The Laplace transform has been performed on both sides of each equation. From equation (2.3-22) we get the transformed boundary condition, equation (2.3-26), which will be substituted into equation (2.3-25) in order to evaluate the constant A. When equation (2.3-26) is substituted into equation (2.3-25), the result is

$$0 = -\tan^{-1} \frac{1}{\pi} + A \tag{2.3-27}$$

which yields $A = \tan^{-1}(1/\pi)$. Thus equation (2.3-25), with some trigonometric manipulations, can be inverted to yield the final solution

$$y(t) = \frac{\sin t}{t} \tag{2.3-38}$$

One last example we present illustrating the utility of the Laplace transform in ordinary differential equations involves simultaneous equations such as

$$\frac{dx}{dt} = 2x - 3y \tag{2.3-29}$$

$$\frac{dy}{dt} = y - 2x \tag{2.3-30}$$

with

$$x = 8 \qquad \text{at } t = 0 \tag{2.3-31}$$

and

$$y = 3 \qquad \text{at } t = 0 \tag{2.3-32}$$

Taking the Laplace transforms of equations (2.3-29) and (2.3-30), $\mathscr{L}_{t \to s}$, with the initial conditions (2.3-31) and (2.3-32), we get

$$s\bar{x}(s) - 8 = 2\bar{x}(s) - 3\bar{y}(s) \tag{2.3-33}$$

and

$$s\bar{y}(s) - 3 = \bar{y}(s) - 2\bar{x}(s) \tag{2.3-34}$$

which are two algebraic equations which can be solved for $\bar{x}(s)$ and $\bar{y}(s)$, yielding

$$\bar{x}(s) = \frac{5}{s+1} + \frac{3}{s-4} \tag{2.3-35}$$

and

$$\bar{y}(s) = \frac{5}{s+1} - \frac{2}{s-4} \tag{2.3-36}$$

Inverting equations (2.3-35) and (2.3-36) using the tables gives the following result:

$$\mathscr{L}^{-1}_{s\rightarrow t}\bar{x}(s) = x(t) = 5e^{-t} + 3e^{4t} \tag{2.3-37}$$

$$\mathscr{L}^{-1}_{s\rightarrow t}\bar{y}(s) = y(t) = 5e^{-t} - 2e^{4t} \tag{2.3-38}$$

2.4. Application of Laplace Transforms to Partial Differential Equations

In turning to the solution of partial differential equations there is a need for more attention to nomenclature and terminology. For partial differential equations, with two or more independent variables, it becomes necessary to express the various Laplace transform operations in the following manner:

$$\mathscr{L}_{t\rightarrow s}U(x, t) = \bar{U}(x, s) \tag{2.4-1}$$

and if we take the transform again, we get

$$\mathscr{L}_{x\rightarrow p}\bar{U}(x, s) = \bar{\bar{U}}(p, s) \tag{2.4-2}$$

Thus, one bar over the dependent variable indicates that the Laplace transform has been taken once. A second bar indicates a second transform. This is a bookkeeping procedure which is essential in dealing with partial differential equations. Before showing how the solutions to some partial differential equations are effected by the Laplace transform technique, we must introduce the following equation:

$$\begin{aligned}
\mathscr{L}_{t\rightarrow s}\frac{\partial U(x, t)}{\partial x} &= \int_0^\infty \frac{\partial U(x, t)}{\partial x} e^{-st}\, dt \\
&= \frac{\partial}{\partial x}\int_0^\infty U(x, t)e^{-st}\, dt \\
&= \frac{\partial \bar{U}(x, s)}{\partial x}
\end{aligned} \tag{2.4-3}$$

The result in equation (2.4-3) is predicated upon the independence of s and x and the utilization of the Leibnitz rule for differentiating an integral. On the other hand, the operations indicated by

$$\mathscr{L}_{t\rightarrow s}\frac{\partial U(x, t)}{\partial t} = \int_0^\infty \frac{\partial U(x, t)}{\partial t} e^{-st}\, dt = s\bar{U}(x, s) - U(x, 0) \tag{2.4-4}$$

where $U(x, 0) = U(x, t)|_{t=0}$, are straightforward and are the same as those operations performed with ordinary differential equations. In writing the dependent variable both independent variables are carried along at all times as $U(x, t)$.

Now we are in a position to solve an equation such as

$$\frac{\partial U(x, t)}{\partial t} = \frac{\partial^2 U(x, t)}{\partial x^2} \tag{2.4-5}$$

with the initial and boundary conditions

$$U(x, t) = 3 \sin 2\pi x \quad \text{at } t = 0 \quad (\text{or } U(x, 0) = 3 \sin 2\pi x) \tag{2.4-6}$$

$$U(x, t) = 0 \quad \text{at } x = 0 \quad (\text{or } U(0, t) = 0) \tag{2.4-7}$$

$$U(x, t) = 0 \quad \text{at } x = 1 \quad (\text{or } U(1, t) = 0) \tag{2.4-8}$$

If we take the Laplace transform, $\mathcal{L}_{t \to s}$, of equation (2.4-5), the result is

$$s\bar{U}(x, s) - U(x, 0) = \frac{d^2 \bar{U}(x, s)}{dx^2}$$

which yields

$$\frac{d^2 \bar{U}(x, s)}{dx^2} - s\bar{U}(x, s) = -3 \sin 2\pi x \tag{2.4-9}$$

when (2.4-6) is used. By standard methods of solution this ordinary second-order differential equation can be solved to yield

$$\bar{U}(x, s) = C_1 e^{\sqrt{s}\, x} + C_2 e^{-\sqrt{s}\, x} + \frac{3}{s + 4\pi^2} \sin 2\pi x \tag{2.4-10}$$

In order to solve for the arbitrary constants C_1 and C_2 we need relations between $\bar{U}(x, s)$ and x. We get these from equations (2.4-7) and (2.4-8), but first it is necessary to transform $U(x, t)$ in equations (2.4-7) and (2.4-8) into relations with $\bar{U}(x, s)$. The operations to obtain these relationships are indicated by the following equations:

$$\mathcal{L}_{t \to s}\{\text{eq. (2.4-7)}\} = \{\bar{U}(x, s) = 0 \text{ at } x = 0\} \tag{2.4-11}$$

and

$$\mathcal{L}_{t \to s}\{\text{eq. (2.4-8)}\} = \{\bar{U}(x, s) = 0 \text{ at } x = 1\} \tag{2.4-12}$$

Applying equations (2.4-11) and (2.4-12) to (2.4-10), we get $C_1 = C_2 = 0$, so that

$$\bar{U}(x, s) = \frac{3}{s + 4\pi^2} \sin 2\pi x \tag{2.4-13}$$

results. The inverse is obtained readily by looking up the right-hand term in the tables. The final solution is

$$U(x, t) = 3e^{-4\pi^2 t} \sin 2\pi x \qquad (2.4\text{-}14)$$

We could also have solved equation (2.4-9) by taking the Laplace transform again, $\mathscr{L}_{x \to p}\{$eq. (2.4-9)$\}$, which would yield an algebraic equation in $\bar{U}(p, s)$. This would then be inverted twice. The choice of the order of inversions, $\mathscr{L}_{p \to x}^{-1}$ and then $\mathscr{L}_{s \to t}^{-1}$, or $\mathscr{L}_{s \to t}^{-1}$ and then $\mathscr{L}_{p \to x}^{-1}$, is dependent upon the complexity of $\bar{U}(p, s)$ and the presence of various functions of s and p. The following problem illustrates the consecutive transform approach. From

$$\frac{\partial U(x, t)}{\partial t} = \frac{\partial^2 U(x, t)}{\partial x^2} \qquad (2.4\text{-}15)$$

we take the transform $\mathscr{L}_{x \to p}$, which gives

$$\frac{\partial \bar{U}(p, t)}{\partial t} = p^2 \bar{U}(p, t) - \frac{dU(x, t)}{dx}\bigg|_{x=0} \qquad (2.4\text{-}16)$$

where the initial condition $U(0, t) = 0$ was used. Applying the Laplace transform, $\mathscr{L}_{t \to s}$, again to equation (2.4-16), we get

$$s\bar{U}(p, s) - \bar{U}(p, t)|_{t=0} = p^2 \bar{\bar{U}}(p, s) - \bar{F}(s) \qquad (2.4\text{-}17)$$

where

$$\bar{F}(s) = \mathscr{L}_{t \to s} \frac{dU(x, t)}{dx}\bigg|_{x=0} = \mathscr{L}_{t \to s} F(t) \qquad (2.4\text{-}18)$$

At this point we need information in order to evaluate $\bar{U}(p, t)|_{t=0}$ and $\bar{F}(s)$. If we are given an initial condition such as

$$U(x, t)|_{t=0} = 3 \sin 2\pi x \qquad (2.4\text{-}19)$$

then it is a relatively simple matter to operate upon equation (2.4-19) as follows:

$$\mathscr{L}_{x \to p}\{\text{eq. (2.4-19)}\} = \left\{\bar{U}(p, t)|_{t=0} = \frac{6\pi}{p^2 + 4\pi^2}\right\} \qquad (2.4\text{-}20)$$

Thus we have evaluated one of the unknown functions in equation (2.4-17). Combining equation (2.4-17) and (2.4-20), and with some rearrangement, we get

$$\bar{\bar{U}}(p, s) = -\left(\frac{6\pi}{p^2 - s}\right)\left(\frac{1}{p^2 + 4\pi^2}\right) + \left(\frac{1}{p^2 - s}\right)(\bar{F}(s)) \qquad (2.4\text{-}21)$$

Now we begin to "peel off" the inverse transforms. It is apparent from an inspection of equation (2.4-21) that it would be more propitious to begin with $\mathscr{L}^{-1}_{p\to x}$. Applying this inversion to equation (2.4-21), we get[8]

$$\bar{U}(x, s) = -6\pi\left\{\frac{2\pi \sinh \sqrt{s}\,x - \sqrt{s}\,\sin(2\pi x)}{2\pi \sqrt{s}\,(4\pi^2 + s)}\right\} + \frac{\sinh \sqrt{s}\,x}{\sqrt{s}}\,\bar{F}(s)$$

(2.4-22)

Note that in performing these operations p and s are independent. Hence s behaves as a constant when the inverse $\mathscr{L}^{-1}_{p\to x}$ is performed. We need now to determine $\bar{F}(s)$ before the final inverse operation is performed on equation (2.4-22). A final boundary condition which is available could be given by a form such as

$$U(1, t) = 0$$

which yields upon transformation, $\mathscr{L}_{t\to s}$,

$$\bar{U}(1, s) = 0$$

(2.4-23)

If now equation (2.4-23) is applied to equation (2.4-22), the result is

$$0 = -6\pi\left\{\frac{2\pi \sinh \sqrt{s} - \sqrt{s}\,\sin 2\pi}{2\pi \sqrt{s}\,(4\pi^2 + s)}\right\} + \frac{\sinh \sqrt{s}}{\sqrt{s}}\,\bar{F}(s)$$

$$= \sinh \sqrt{s}\left(\frac{-6\pi}{4\pi^2 + s} + \bar{F}(s)\right)$$

(2.4-24)

or

$$\bar{F}(s) = \frac{6\pi}{4\pi^2 + s}$$

(2.4-25)

which allows us to evaluate $\bar{F}(s)$. Thus equation (2.4-22) becomes

$$\bar{U}(x, s) = \frac{3}{s + 4\pi^2}\,\sin 2\pi x$$

(2.4-26)

which finally, upon applying the last inversion, $\mathscr{L}^{-1}_{s\to t}$, yields

$$U(x, t) = 3e^{-4\pi^2 t}\,\sin 2\pi x$$

(2.4-27)

[8] G. E. Roberts and M. Kaufman, op. cit., p. 197.

2.5. Laplace Transforms Applied to Other Equation Forms

Laplace transform techniques can be used easily for various specific equation forms. For example, the convolution integral form

$$y(t) = f(t) + \int_0^t K(t - u)y(u) \, du \tag{2.5-1}$$

can be solved readily by taking the Laplace transform:

$$\mathscr{L}_{t \to s}\{\text{eq. (2.5-1)}\} = \{\bar{y}(s) = \bar{f}(s) + \bar{K}(s)\bar{y}(s)\} \tag{2.5-2}$$

From equation (2.5-2) we can solve explicitly for $\bar{y}(s)$:

$$\bar{y}(s) = \frac{\bar{f}(s)}{1 - \bar{K}(s)} \tag{2.5-3}$$

and the inverse solution is readily found. As an illustration of the above procedure, suppose we have an equation such as

$$y(t) = t^2 + \int_0^t y(u) \sin(t - u) \, du \tag{2.5-4}$$

and with the Laplace transform, $\mathscr{L}_{t \to s}$, obtain

$$\bar{y}(s) = \frac{2}{s^3} + \bar{y}(s)\left(\frac{1}{s^2 + 1}\right) \tag{2.5-5}$$

When equation (2.5-5) is rearranged, $\bar{y}(s)$ is expressed explicitly,

$$\bar{y}(s) = \frac{2}{s^3} + \frac{2}{s^5} \tag{2.5-6}$$

and can be inverted to yield the solution

$$y(t) = 2\left(\frac{t^2}{2!}\right) + 2\left(\frac{t^4}{4!}\right) \tag{2.5-7}$$

If there is a derivative in addition to an integral in a given equation, it is referred to as an integrodifferential equation. For example, suppose the equation

$$\frac{dy}{dt} + 5\int_0^t \cos 2(t - u)y(u) \, du = 10 \tag{2.5-8}$$

is to be solved subject to the initial condition

$$y(t) = 2 \quad \text{at } t = 0 \tag{2.5-9}$$

Operating on equation (2.5-8), $\mathscr{L}_{t \to s}$, we get

$$s\bar{y}(s) - y(0) + 5\left(\frac{s}{s^2 + 4}\right)(\bar{y}(s)) = \frac{10}{s} \tag{2.5-10}$$

Substituting equation (2.5-9) into (2.5-10), and with some rearrangements, we get[9]

$$\bar{y}(s) = \frac{2s^3 + 10s^2 + 8s + 40}{s^2(s^2 + 9)} \tag{2.5-11}$$

which after some algebraic and partial fractions manipulations and the inverse operation, yields

$$y(t) = \frac{1}{27}(24 + 120t + 30\cos 3t + 50\sin 3t) \tag{2.5-12}$$

Another type of equation that is encountered frequently, particularly in cascaded types of systems, is represented by finite difference equations such as

$$3y(t) - 4y(t - 1) + y(t - 2) = t \tag{2.5-13}$$

with

$$y(t) = 0 \quad \text{for } t < 0 \tag{2.5-14}$$

(In this chapter we will show how these finite difference equations are solved by Laplace transform techniques. In Chapters 7 and 8 finite difference calculus will be introduced, and finite difference equations will be solved by that technique. Thus the reader will be afforded a choice of solution techniques).

In equation (2.5-13), the notation $y(t - 1)$ indicates the value of the y function at a time $t - 1$. In using the Laplace transforms for this case, we need to perform some transform operations defined by

$$\mathscr{L}_{t \to s} y(t - 1) = \int_0^\infty y(t - 1)e^{-st} dt \tag{2.5-15}$$

If $t - 1 = u$, equation (2.5-15) becomes, with some manipulation, a simple

---

[9] M. R. Spiegel, *op. cit.*, p. 120.

expression:

$$\mathscr{L}_{t\to s}\, y(t-1) = \int_{-1}^{\infty} y(u)e^{-s(u+1)}\, du$$

$$= e^{-s}\int_{-1}^{0} y(u)e^{-su}\, du + e^{-s}\int_{0}^{\infty} y(u)e^{-su}\, du$$

$$= e^{-s}\bar{y}(s) \qquad\qquad (2.5\text{-}16)$$

where $y(u) = 0$ for $u < 0$, and by definition

$$\int_{0}^{\infty} y(u)e^{-su}\, du = \bar{y}(s) \qquad\qquad (2.5\text{-}17)$$

We can perform these same manipulations for $y(t-2)$:

$$\mathscr{L}_{t\to s}\, y(t-2) = \int_{0}^{\infty} y(t-2)e^{-st}\, dt \qquad\qquad (2.5\text{-}18)$$

which (letting $u = t - 2$) yields

$$\mathscr{L}_{t\to s}\, y(t-2) = e^{-2s}\bar{y}(s) \qquad\qquad (2.5\text{-}19)$$

Applying equations (2.5-16) and (2.5-19) to (2.5-13), we get

$$3\bar{y}(s) - 4e^{-s}\bar{y}(s) + e^{-2s}\bar{y}(s) = \frac{1}{s^2} \qquad\qquad (2.5\text{-}20)$$

which, after some manipulation, yields

$$\bar{y}(s) = \frac{1}{s^2(1 - e^{-s})(3 - e^{-s})} \qquad\qquad (2.5\text{-}21)$$

Equation (2.5-21) can be rearranged as follows:

$$\bar{y}(s) = \frac{1}{2s^2}\left\{\frac{1}{1 - e^{-s}} - \frac{1}{3(1 - e^{-s}/3)}\right\}$$

$$= \frac{1}{2s^2}\left\{(1 + e^{-s} + e^{-2s} + \cdots) - \frac{1}{3}\left(1 + \frac{e^{-s}}{3} + \frac{e^{-2s}}{3^2} + \cdots\right)\right\}$$

$$= \frac{1}{3s^2} + \frac{1}{2}\sum_{n=1}^{\infty}\left(1 - \frac{1}{3^{n+1}}\right)\frac{e^{-ns}}{s^2} \qquad\qquad (2.5\text{-}22)$$

By looking up equation (2.5-22) in the inversion tables, we finally get the solution to equation (2.5-13):

$$y(t) = \frac{t}{3} + \frac{1}{2}\sum_{n=1}^{[t]}\left(1 - \frac{1}{3^{n+1}}\right)(t - n) \qquad (2.5\text{-}23)$$

where $[t]$ is defined as the greatest integer less than or equal to t.

A variation on this finite difference theme is the differential-difference equation, such as

$$\frac{dy}{dt} + y(t - 1) - t^2 = 0 \qquad (2.5\text{-}24)$$

with

$$y(t) = 0 \qquad \text{for } t \leq 0 \qquad (2.5\text{-}25)$$

If we take the Laplace transform of equation (2.5-24) using the procedures shown in equations (2.5-13) through (2.5-22), the result, without showing the details, is[10]

$$s\bar{y}(s) - y(0) + e^{-s}\bar{y}(s) = \frac{2}{s^3} \qquad (2.5\text{-}26)$$

or

$$\bar{y}(s) = \frac{2}{s^3(s + e^{-s})} \qquad (2.5\text{-}27)$$

By a series expansion we can convert equation (2.5-27) to

$$\bar{y}(s) = \frac{2}{s^4}\left(1 - \frac{e^{-s}}{s} + \frac{e^{-2s}}{s^2} - \frac{e^{-3s}}{s^3} + \cdots\right)$$

$$= \frac{2}{s^4} - \frac{2e^{-s}}{s^5} + \frac{2e^{-2s}}{s^6} - \frac{2e^{-3s}}{s^7} + \cdots$$

$$= 2\sum_{n=0}^{\infty}(-1)^n\frac{e^{-ns}}{s^{n+4}} \qquad (2.5\text{-}28)$$

The inverse transform of equation (2.5-28) obtained from the tables yields[11]

$$\mathcal{L}^{-1}_{s \to t}\,\bar{y}(s) = y(t) = 2\sum_{n=0}^{[t]}\frac{(t - n)^{n+3}}{(n + 3)!} \qquad (2.5\text{-}29)$$

where $[t]$ again denotes the greatest integer less than or equal to t.

[10] M. R. Spiegel, *op. cit.*, p. 122.
[11] *Ibid.*

Continuing the discussion of finite differences, we can in some situations end up with governing equations such as

$$a_{n+2} - 5a_{n+1} + 6a_n = 0 \tag{2.5-30}$$

with

$$a_0 = 0 \qquad a_1 = 1 \tag{2.5-31}$$

This type of equation arises particularly in stagewise operations and is discussed in detail in Chapters 7 and 8. Equation (2.5-30) can be converted to a form which is readily amenable to Laplace transform techniques. If we define the function

$$y(t) = a_n \qquad n \le t < n + 1 \qquad n = 0, 1, 2, 3, \ldots \tag{2.5-32}$$

then equation (2.5-30) becomes more familiar:

$$y(t + 2) - 5y(t + 1) + 6y(t) = 0 \tag{2.5-33}$$

Proceeding as in the previous examples, equation (2.5-33) with (2.5-31) becomes

$$e^{2s}\bar{y}(s) - \frac{e^s(1 - e^{-s})}{s} - 5e^s\bar{y}(s) + 6\bar{y}(s) = 0 \tag{2.5-34}$$

or

$$
\begin{aligned}
\bar{y}(s) &= \frac{e^s(1 - e^{-s})}{s(e^{2s} - 5e^s + 6)} = \frac{e^s(1 - e^{-s})}{s}\left\{\frac{1}{(e^s - 3)(e^s - 2)}\right\} \\
&= \frac{e^s(1 - e^{-s})}{s}\left\{\frac{1}{(e^s - 3)} - \frac{1}{(e^s - 2)}\right\} \\
&= \frac{(1 - e^{-s})}{s}\left\{\frac{1}{(1 - 3e^{-s})} - \frac{1}{(1 - 2e^{-s})}\right\}
\end{aligned} \tag{2.5-35}
$$

Equation (2.5-35) is inverted to yield, finally,

$$a_n = 3^n - 2^n \qquad n = 0, 1, 2, \ldots \tag{2.5-36}$$

In Chapter 7 we will show that the result laboriously obtained here by Laplace transforms can be easily and quickly arrived at by finite difference calculus.

2.6. Complex Variables Applied to the Inverse Laplace Transform

Thus far we have been inverting the transformed function $\bar{f}(s)$ by methods (1) through (4) in Section 2.2. If these relatively simple methods of finding the inverse Laplace transform are not possible then we must resort to integration in the complex plane, method (5) in Section 2.2, which is defined by equation (2.2-1). From Section 2.2 the complete statement of the Laplace transform operation is defined by

$$\mathscr{L}_{t\rightarrow s}f(t) = \bar{f}(s) = \int_0^\infty f(t)e^{-st}\,dt \tag{2.6-1}$$

$$\mathscr{L}_{s\rightarrow t}^{-1}\bar{f}(s) = f(t) = \frac{1}{2\pi i}\int_{Br}\bar{f}(s)e^{st}\,ds \tag{2.6-2}$$

The term Br refers to the Bromwhich path, which is an integration in the complex plane. Before proceeding with the integration indicated by equation (2.6-2) it is useful to review some complex variables algebra and calculus.

A complex number may be represented as

$$Z_1 = x_1 + iy_1 = r_1(\cos\theta_1 + i\sin\theta_1) \tag{2.6-3}$$

where

$$r_1 = \sqrt{x_1{}^2 + y_1{}^2} \tag{2.6-4}$$

Figure 7 shows how this notation is aplied.

We may require the product of two complex numbers, or the division of one complex number by another. These are shown in equations (2.6-5) and (2.6-6):

$$Z_1 Z_2 = r_1 r_2\{\cos(\theta_1 + \theta_2) + i\sin(\theta_1 + \theta_2)\} \tag{2.6-5}$$

and

$$\frac{Z_1}{Z_2} = \frac{r_1}{r_2}\{\cos(\theta_1 - \theta_2) + i\sin(\theta_1 - \theta_2)\} \tag{2.6-6}$$

We may also definite

$$e^{i\theta} = \cos\theta + i\sin\theta \tag{2.6-7}$$

so that equation (2.6-7), applied to equation (2.6-3), yields a more con-

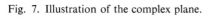

Fig. 7. Illustration of the complex plane.

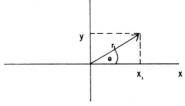

venient form for a complex number:

$$Z_1 = r_1 e^{i\theta_1} \tag{2.6-8}$$

and hence

$$Z_1 Z_2 = r_1 r_2 e^{i(\theta_1 + \theta_2)} \tag{2.6-9}$$

and

$$\frac{Z_1}{Z_2} = \frac{r_1}{r_2} e^{i(\theta_1 - \theta_2)} \tag{2.6-10}$$

We can also extract roots of complex numbers, which becomes readily apparent from equation (2.6-8). Equation (2.6-11) shows the result:

$$Z^{1/n} = r^{1/n} e^{i\theta/n} \tag{2.6-11}$$

or more generally

$$Z^{1/n} = r^{1/n} e^{i(\theta + 2k\pi)/n} \qquad k = 0, 1, 2, \ldots \tag{2.6-12}$$

From equation (2.6-7) and (2.6-12) we can also write $Z^{1/n}$ as

$$Z^{1/n} = r^{1/n} \left\{ \cos\left(\frac{\theta + 2k\pi}{n} \right) + i \sin\left(\frac{\theta + 2k\pi}{n} \right) \right\} \qquad k = 0, 1, 2, \ldots \tag{2.6-13}$$

We can also work with functions of complex variables, $f(Z)$, where $Z = x + iy$. Equation (2.6-14) shows symbolically the relationship between the complex function and the real values of x and y:

$$w = f(Z) = u(x, y) + iv(x, y) \tag{2.6-14}$$

For example, if $w = Z^2$, then equation (2.6-15) shows how to determine the u and v functions:

$$w = Z^2 = (x + iy)^2 = (x^2 - y^2) + i(2xy) \tag{2.6-15}$$

with $u = (x^2 - y^2)$ and $v = 2xy$. The function u is called the real (Re) part of w, and v is the imaginary (Im) part. We can also write

$$e^Z = e^{x+iy} = e^x e^{iy} = e^x(\cos y + i \sin y) \tag{2.6-16}$$

from equation (2.6-7). Note that

$$e^{2k\pi i} = \cos 2k\pi + i \sin 2k\pi = 1 + 0 = 1 \tag{2.6-17}$$

where $k = 0, 1, 2, \ldots.$ It is also true that

$$\ln Z = \ln(re^{i\theta}) = \ln r + i\theta \tag{2.6-18}$$

or, more generally,

$$\ln Z = \ln(re^{i(\theta + 2k\pi)}) = \ln r + i(\theta + 2k\pi) \tag{2.6-19}$$

Complex variables lend themselves also to some basic calculus techniques such as differentiation. For instance,

$$\frac{d}{dZ}(Z^n) = nZ^{n-1} \tag{2.6-20}$$

and

$$\frac{d}{dZ}(\sin Z) = \cos Z \tag{2.6-21}$$

In order for these basic calculus techniques to be applicable, the $w = f(Z)$ function must satisfy what are called the Cauchy–Riemann conditions. Given equation (2.6-14), the Cauchy–Riemann conditions are defined by

$$\frac{\partial u}{\partial x} = \frac{\partial v}{\partial y} \qquad \frac{\partial u}{\partial y} = -\frac{\partial v}{\partial x} \tag{2.6-22}$$

If $w = f(Z)$ satisfies equation (2.6-22), then we say w is an analytic function and can be differentiated, integrated, and subjected to many of the operations we associate with real variables.

A very important theorem in connection with the inverse Laplace transform is the Cauchy theorem. It states that if $w = f(Z)$ is analytic (satisfies the Cauchy–Riemann conditions) within a region enclosed by line C, then the following equation holds:

$$\oint_C f(Z)\, dZ = 0 \tag{2.6-23}$$

where $\oint_C$ indicates that the integration is performed around a closed path C. For example, if $f(Z) = 2Z$, we can show that $f(Z)$ is analytic everywhere [satisfies equation (2.6-22)], so that from equation (2.6-23) we have

$$\oint_C 2Z\, dZ = 0 \tag{2.6-24}$$

If we wish to integrate not over a closed path, but from Z_1 to Z_2, this can be done directly. For example, if $Z_1 = 2i$ and $Z_2 = 1 + i$, then the integration from Z_1 to Z_2 is performed as follows:

$$\int_{Z_1=2i}^{Z_2=1+i} 2Z \, dZ = Z^2 \Big|_{2i}^{1+i} = (1 + i)^2 - (2i)^2 = 2i + 4 \qquad (2.6\text{-}25)$$

There is an additional consequence of equation (2.6-23), namely, that the integration of an analytical function from Z_1 to Z_2 is independent of path.

Building upon the Cauchy theorem [equation (2.6-23)] we next introduce a fundamental theorem of integration in the complex plane, the Cauchy integral theorem, defined by

$$f(a) = \frac{1}{2\pi i} \oint_C \frac{f(Z)}{Z - a} \, dZ \qquad (2.6\text{-}26)$$

where $f(Z)$ is analytic within the closed curve C and the point a is an interior point within C. The symbol, $\bigcirc$ indicates that the integration is performed in a counterclockwise direction. A generalization of the Cauchy integral theorem is given by

$$f^{(n)}(a) = \frac{n!}{2\pi i} \oint_C \frac{f(Z)}{(Z - a)^{n+1}} \, dZ \qquad (2.6\text{-}27)$$

where $f^{(n)}(a)$ indicates the nth derivative of the f function, evaluated at $Z = a$. In evaluating the inverse Laplace transform we shall need to make use of expressions such as

$$\frac{1}{2\pi i} \oint_C \frac{f(Z)}{Z - a} \, dZ$$

or

$$\frac{1}{2\pi i} \oint_C \frac{f(Z)}{(Z - a)^{n+1}} \, dZ$$

which can be evaluated from equations (2.6-26) and (2.6-27).

Before getting into the inverse Laplace transform operations, we need also to introduce the concept of a singularity or singular point. If $f(Z)$ is not analytic at a point, then that point is said to be a singular point. Usually the function $f(Z)$ goes to infinity at the singular point. For example, $f(Z) = 1/(Z - 3)^2$ contains a singular point at $Z = 3$. We also include "pole" in the terminology, an expression which is somewhat synonymous with singularity. A pole is a singularity which causes $f(Z)$ to go to infinity. For example, given

$$f(Z) = \frac{\varphi(Z)}{(Z - a)^n} \qquad (2.6\text{-}28)$$

if $\varphi(Z)$ is analytic and n is a positive integer, then there is a singularity at $Z = a$ (a pole of order n). If $n = 1$ then $Z = a$ is a "simple" pole. Suppose we have a function

$$f(Z) = \frac{Z}{(Z - 3)^2(Z + 1)} \qquad (2.6\text{-}29)$$

For this case there are singularities at $Z = 3$ (pole of order 2) and $Z = -1$ (pole of order 1). Another example,

$$f(Z) = \frac{3Z - 1}{(Z^2 + 4)} = \frac{3Z - 1}{(Z + 2i)(Z - 2i)} \qquad (2.6\text{-}30)$$

yields simple poles at $Z = \pm 2i$. A function such as

$$f(Z) = \frac{\sin Z}{Z} \qquad (2.6\text{-}31)$$

is said to have a removable singularity at $Z = 0$ since $\lim_{z \to 0} f(Z) = 1$, as determined by the application of L'Hôpital's rule. A more complicated function such as

$$f(Z) = \sqrt{Z} \qquad (2.6\text{-}32)$$

is said to have a "branch" point at $Z = 0$ and is multivalued. Branch points will be discussed later in this section.

We are now ready to evaluate some integrals by performing the integration in the complex plane. These examples will lead directly to the evaluation of the inverse Laplace transform. A common integral encountered in the inverse transform operations is the integral $\oint_C dZ/(Z-a)^n$. With the Cauchy integral formula, equation (2.6-26) or (2.6-27), it is an easy matter to evaluate this integral. The answer is quickly written as

$$\oint_C \frac{dZ}{(Z - a)^n} = \begin{Bmatrix} 0 & \text{if } n \neq 1 \\ 2\pi i & \text{if } n = 1 \end{Bmatrix} \qquad (2.6\text{-}33)$$

To get the results shown in equation (2.6-33) we note that comparing equation (2.6-33) with the Cauchy integral formula yields $f(Z) = 1$ in equation (2.6-26) or (2.6-27). If $n = 1$ we go to equation (2.6-26) and get the result shown in equation (2.6-33). If $n \neq 1$ we go to equation (2.6-27) and find that any derivative of $f(Z) = 1$ yields zero.

Suppose as another example we wish to evaluate the integral $\oint_C dZ/(Z - 3)$ where the path C is a circle of radius 1 with the origin at $Z = 0$. For this closed path C the value of Z can never be $Z = 3$, so that within this region there are no singularities even though the integral suggests that $Z = 3$ is a singularity. Since there are no singularities within path C, the integral under question is equal to zero, as obtained from equation (2.6-23). A few examples will help to establish the procedure in evaluating complex integrals.

Example 2.6-1

Evaluate $\oint_C (\cos Z)/(Z - \pi)\, dZ$, where C is a circle of radius 3 with the origin at $Z = 1$. Figure 8 shows the integration path.

Since $Z = \pi$ lies within C, by equation (2.6-26) we get

$$\oint_C \frac{\cos Z}{Z - \pi}\, dZ = 2\pi i \cos \pi = -2\pi i$$

Example 2.6-2

Evaluate $\oint_C e^Z/[Z(Z + 1)]\, dZ$ over the same path as in Example 2.6-1. In this example there are singularities at $Z = 0$ and $Z = -1$ and by the Cauchy integral formula we get

for $Z = 0$:

$$\oint_C \frac{e^Z}{Z(Z + 1)}\, dZ = \oint_C \frac{e^Z/(Z + 1)}{Z}\, dz = \frac{2\pi i e^0}{(0 + 1)} = 2\pi i$$

for $Z = -1$:

$$\oint_C \frac{e^Z}{Z(Z + 1)}\, dZ = \oint_C \frac{(e^Z/Z)}{(Z + 1)}\, dz = 2\pi i\left(\frac{e^{-1}}{-1}\right) = -2\pi i e^{-1}$$

Note that $f(Z)$ in the Cauchy integral formula [equation (2.6-26)] is the entire integrand except for the singularity, which is omitted. Since there are

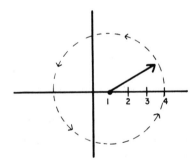

Fig. 8. Integration path for example (2.6.1).

two singularities in this example, we make use of the "residue" theorem, which states that the final result is the sum of the results obtained from each singularity. Thus in Example 2.6-2, we get

$$\oint_C \frac{e^Z}{Z(Z+1)} \, dZ = 2\pi i - 2\pi i e^{-1} = 2\pi i(1 - e^{-1})$$

Example 2.6-3

Evaluate $\oint_C (5Z^2 - 3Z + 2)/(Z - 1)^3 \, dZ$ on a path C which encloses $Z = 1$. In this example the poles are not simple poles. Comparing this integral with the Cauchy integral formula [equation (2.6-27)], we note $n = 2$, $f(Z) = 5Z^2 - 3Z + 2$, and $d^2 f(Z)/dZ^2 = 10$. Hence the final result is immediately determined from equation (2.6-27):

$$\oint_C \frac{5Z^2 - 3Z + 2}{(Z - 1)^3} \, dZ = \frac{2\pi i(10)}{2!} = 10\pi i$$

Example 2.6-4

Determine the poles of the integrand

$$\frac{2Z^3 - Z + 1}{(Z - 4)^2(Z - i)(Z - 1 + 2i)}$$

In this example,

$$Z = 4 \qquad \text{(pole of order 2)}$$
$$Z = i \qquad \text{(simple pole)}$$
$$Z = 1 - 2i \qquad \text{(simple pole)}$$

Example 2.6-5

Determine the poles of $(\sin mZ)/(Z^2 + 2Z + 2)$. By the quadratic formula we can determine the simple poles,

$$Z = \frac{-2 \pm \sqrt{4 - 8}}{2} = -1 + i, \quad -1 - i$$

Example 2.6-6

Determine the poles of $(1 - \cos Z)/Z$. In searching for the poles in this case, it would appear that $Z = 0$ is a singularity. However since $\lim_{z \to 0}(1 - \cos Z)/Z = 0$ by L'Hôpital's rule, we conclude that the point $Z = 0$ is a removable singularity and hence is not a pole.

Example 2.6-7

Evaluate the integral $\oint_C Z^2/[(Z-2)(Z^2+1)]\,dZ$. Here we have an integrand that should be written as

$$\oint_C \frac{Z^2}{(Z-2)(Z-i)(Z+i)}\,dZ$$

where $Z=\pm i$ is obtained from $Z^2+1=0$. Thus from equation (2.6-26) and the residue theorem we can evaluate the integral:

for $Z=2$:

$$\oint_C \frac{[Z^2/(Z-i)(Z+i)]}{(Z-2)}\,dZ = 2\pi i\left(\frac{4}{5}\right)$$

for $Z=i$:

$$\oint_C \frac{[Z^2/(Z-2)(Z+i)]}{(Z-i)}\,dZ = 2\pi i\,\frac{i^2}{(i-2)(i+i)} = 2\pi i\left(\frac{1-2i}{10}\right)$$

for $Z=-i$:

$$\oint_C \frac{[Z^2/(Z-2)(Z-i)]}{(Z+i)}\,dZ = 2\pi i\,\frac{i^2}{(-i-2)(-2i)} = 2\pi i\left(\frac{1+2i}{10}\right)$$

The final result is the sum of the three residues:

$$\oint_C \frac{Z^2}{(Z-2)(Z^2+1)}\,dZ = 2\pi i\left\{\frac{4}{5} + \left(\frac{1-2i}{10}\right) + \left(\frac{1+2i}{10}\right)\right\}$$

Example 2.6-8

Evaluate $\oint_C 1/[Z(Z+2)^3]\,dZ$. The singularities are $Z=0$ (pole of order 1) and $Z=-2$ (pole of order 3)

for $Z=0$:

$$\oint_C \frac{(1/(Z+2)^3)}{Z}\,dZ = 2\pi i\left(\frac{1}{8}\right)$$

for $Z=-2$:

$$\oint_C \frac{(1/Z)}{(Z+2)^3}\,dZ = 2\pi i\left[\frac{1}{2!}\,\frac{d^2}{dZ^2}\left(\frac{1}{Z}\right)\Big|_{Z=-2}\right] = 2\pi i\left(-\frac{1}{8}\right)$$

The final result is the sum of the residues of the singularities:

$$\oint_C \frac{1}{Z(Z+2)^3}\,dZ = 2\pi i\left(\frac{1}{8} - \frac{1}{8}\right) = 0$$

Example 2.6-9

Evaluate $\oint_C Ze^{Zt}/(Z-3)^2\, dZ$. The singularities are at $Z=3$ (pole of order 2). The value of the integral is therefore

$$\oint_C \frac{Ze^{Zt}}{(Z-3)^2}\, dZ = 2\pi i\left[\frac{1}{1!}\ \frac{d}{dZ}\ (Ze^{Zt})\Big|_{Z=3}\right]$$
$$= 2\pi i(e^{3t}+3te^{3t})$$

Example 2.6-10

Evaluate $\oint_C 1/(\tan Z)\, dZ$ for a path C enclosing $Z=5\pi$. In this example we consider a slightly different type of singularity. The integral can be rewritten as $\oint_C (\cos Z)/(\sin Z)\, dZ$. In this form the use of the Cauchy integral formula is not readily apparent. We can however put the integral in a form analogous to equation (2.6-26) by writing the integral in the following way for the simple pole at $Z=5\pi$:

$$\oint_C \frac{\cos Z}{\sin Z}\, dZ = \oint_C \frac{\left(\lim\limits_{Z\to 5\pi} \dfrac{(Z-5\pi)\cos Z}{\sin Z}\right)}{(Z-5\pi)}\, dZ$$

This arrangement has the same result as in the previous examples where the numerator $f(Z)$ did not contain the singularity: in Example 2.6-1 we could have written the integral as

$$\oint_C \frac{\cos Z}{Z-\pi}\, dZ = \oint_C \frac{\left(\lim\limits_{Z\to \pi} \dfrac{(Z-\pi)\cos Z}{Z-\pi}\right)}{(Z-\pi)}\, dZ$$

In Example 2.6-1 the altered form shown above is trivially different since $(Z-\pi)$ will cancel from the term in parentheses. However, in Example 2.6-10 the singularity is implicitly tied into $\sin Z$. The integrand is evaluated by using L'Hôpital's rule as shown below:

$$\oint_C \frac{\left(\lim\limits_{Z\to 5\pi} \dfrac{(Z-5\pi)\cos Z}{\sin Z}\right)}{Z-5\pi}\, dZ = 2\pi i \lim_{Z\to 5\pi} \frac{(Z-5\pi)(\cos Z)}{\sin Z}$$

$$= 2\pi i\left\{\lim_{Z\to 5\pi} \frac{Z-5\pi}{\sin Z}\ \lim_{Z\to 5\pi} \cos Z\right\}$$

$$= 2\pi i\left\{\lim_{Z\to 5\pi} \frac{(d/dZ)(Z-5\pi)}{d/dZ \sin Z}\ \lim_{Z\to 5\pi} \cos Z\right\}$$

$$= 2\pi i\left\{\frac{1}{\cos Z}\Big|_{Z\to 5\pi} \cdot (-1)\right\}$$

$$= 2\pi i\{1\}$$

We shall see later in this section how this technique can be applied where the implicit singularities are infinite in number.

Example 2.6-11

Evaluate $\oint_C dZ/(Z^4 + 1)$. The singularities are found from the equation $Z^4 + 1 = 0$. From $Z^4 = -1$ we get $Z = (-1)^{1/4}$. Noting that $-1 = e^{\pi i}$, we find the four roots (which are the singularities) to be

$$Z = (e^{\pi i})^{1/4} = e^{i\left(\frac{\pi + 2k\pi}{4}\right)} \qquad k = 0, 1, 2, \ldots$$

Thus the four singularities for $Z^4 + 1 = 0$ are

$$Z_1 = e^{i(\pi/4)} \qquad\qquad \text{Pole of order 1}$$
$$Z_2 = e^{i(\pi/4 + \pi/2)} = e^{i(3\pi/4)} \qquad \text{Pole of order 1}$$
$$Z_3 = e^{i(\pi/4 + \pi)} = e^{i(5\pi/4)} \qquad \text{Pole of order 1}$$
$$Z_4 = e^{i(\pi/4 + 3/2\pi)} = e^{i(7\pi/4)} \qquad \text{Pole of order 1}$$

The original complex integral can now be evaluated:

$$\oint_C \frac{dZ}{Z^4 + 1} = 2\pi i \left\{ \lim_{Z \to Z_1} (Z - e^{\pi i/4}) \cdot \frac{1}{Z^4 + 1} + \lim_{Z \to Z_2} (Z - e^{3\pi i/4}) \cdot \frac{1}{Z^4 + 1} \right.$$

$$\left. + \lim_{Z \to Z_3} (Z - e^{5\pi i/4}) \cdot \frac{1}{Z^4 + 1} + \lim_{Z \to Z_4} (Z - e^{7\pi i/4}) \cdot \frac{1}{Z^4 + 1} \right\}$$

$$= 2\pi i \left\{ \lim_{Z \to Z_1} \frac{1}{4Z_3} + \lim_{Z \to Z_2} \frac{1}{4Z_3} + \lim_{Z \to Z_3} \frac{1}{4Z_3} + \lim_{Z \to Z_4} \frac{1}{4Z^3} \right\}$$

where L'Hôpital's rule has been used. By substituting in the appropriate Z_1, Z_2, Z_3, and Z_4 values, the problem is solved.

2.7. Inverse Laplace Transforms by Integration in the Complex Plane

The material of Section 2.6 prepared the way for a discussion of the method for finding the inverse Laplace transform which utilizes integration in the complex plane. Equations (2.6-1) and (2.6-2) were presented as the definitions of the forward and inverse Laplace transform operations. For the inverse Laplace transform, integration in the complex plane was required, around a path referred to as the Bromwich path (Br). Specifically,

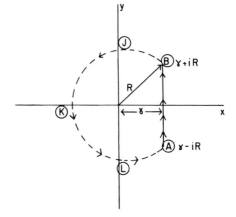

Fig. 9. The Bromwhich path $A \to B$.

the inverse transform is now written as

$$\mathscr{L}^{-1}_{s \to t}\tilde{f}(s) = f(t) = \frac{1}{2\pi i} \int_{Br} \tilde{f}(s)e^{st}\, ds$$

$$= \lim_{R \to \infty} \frac{1}{2\pi i} \int_{\gamma-iR}^{\gamma+iR} \tilde{f}(s)e^{st}\, ds$$

$$= \frac{1}{2\pi i} \int_{\gamma-i\infty}^{\gamma+i\infty} \tilde{f}(s)e^{st}\, ds \qquad (2.7\text{-}1)$$

Figure 9 shows the Bromwhich path of integration required in equation (2.7-1) and also shows the contour which can close the region. The path from point B back to A is by a circle of radius R. Thus we can write in the complex plane an integral over a closed path C composed of $ABJKLA$:

$$\oint_C f(Z)\, dZ = \lim_{R \to \infty} \int_{\gamma-iR}^{\gamma+iR} f(Z)\, dZ + \lim_{R \to \infty} \int_{BJKLA} f(Z)\, dZ \qquad (2.7\text{-}2)$$

It can be shown[12] that for most conditions encountered, the integral $\int_{BJKLA} f(Z)\, dZ$ is equal to zero as $R \to \infty$. It therefore works out that the inverse Laplace transform defined by equation (2.7-1) is evaluated from equation (2.7-2) if we can evaluate an integral such as $\oint_C f(Z)\, dZ$. This is precisely what was done in Section 2.6. If we simply choose the value of γ in the Bromwhich path to include all the singularities of $f(Z)$ and solve by the methods of Section 2.6, the result is related to the inverse Laplace transform. Combining equations (2.7-1) and (2.7-2), we obtain the formula

---

[12] R. V. Churchill, *Complex Variables*, McGraw-Hill, New York (1948), p. 126.

for the inverse Laplace transform by integration in the complex plane:

$$f(t) = \mathscr{L}_{s \to t}^{-1} \bar{f}(s) = \frac{1}{2\pi i} \int_{\gamma - i\infty}^{\gamma + i\infty} \bar{f}(s) e^{st} \, ds = \frac{1}{2\pi i} \oint_C \bar{f}(Z) e^{Zt} \, dZ \qquad (2.7\text{-}3)$$

Some examples which should help clarify this procedure are presented below.

Example 2.7-1

Find the inverse Laplace transform of $\bar{f}(s) = 1/[(s + 1)(s - 2)^2]$ by integration in the complex plane.

From equation (2.7-3) the inverse Laplace transform is

$$f(t) = \mathscr{L}_{s \to t}^{-1} \bar{f}(s) = \frac{1}{2\pi i} \int_{\gamma - i\infty}^{\gamma + i\infty} \frac{e^{st}}{(s + 1)(s - 2)^2} \, ds$$

$$= \frac{1}{2\pi i} \oint_C \frac{e^{Zt}}{(Z + 1)(Z - 2)^2} \, dZ$$

From the Cauchy integral theorem, equations (2.6-26) and (2.6-27), and the residue theorem, we can evaluate the integral as shown below. There are singularities at $Z = -1$ (pole of order 1) and $Z = 2$ (pole of order 2):

for $Z = -1$:

$$\oint_C \frac{[e^{Zt}/(Z - 2)^2]}{(Z + 1)} \, dZ = 2\pi i [e^{Zt}/(Z - 2)^2]\big|_{Z = -1} = \frac{2\pi i e^{-t}}{(-3)^2} = 2\pi i (\tfrac{1}{9} e^{-t})$$

for $Z = 2$:

$$\oint_C \frac{[e^{Zt}/(Z + 1)]}{(Z - 2)^2} \, dZ = 2\pi i \frac{[(d/dZ)[e^{Zt}/(Z + 1)]\big|_{Z = 2}]}{1!} = 2\pi i (\tfrac{1}{3} t e^{2t} - \tfrac{1}{9} e^{2t})$$

The final solution for $f(t)$ is then the sum of the residues for each of the poles,

$$f(t) = \frac{1}{2\pi i} \{2\pi i (\tfrac{1}{9} e^{-t}) + 2\pi i (\tfrac{1}{3} t e^{2t} - \tfrac{1}{9} e^{2t})\}$$

or

$$f(t) = \tfrac{1}{9} e^{-t} + \tfrac{1}{3} t e^{2t} - \tfrac{1}{9} e^{2t}$$

Example 2.7-2

Find the inverse Laplace transform of $\bar{f}(s) = s/[(s + 1)^3(s - 1)^2]$. There are singularities at $Z = -1$ (pole of order 3) and $Z = 1$ (pole of order 2). By the methods of Example 2.7-1 we can evaluate the inverse:

for $Z = -1$:

$$\oint_C \frac{[Ze^{Zt}/(Z-1)^2]}{(Z+1)^3}\,dZ = 2\pi i\frac{(d^2/dZ^2)[Ze^{Zt}/(Z-1)^2]\,|_{Z=-1}}{2!}$$

$$= 2\pi i[\tfrac{1}{16}e^{-t}(1-2t^2)]$$

for $Z = 1$:

$$\oint_C \frac{[Ze^{Zt}/(Z+1)^3]}{(Z-1)^2}\,dZ = 2\pi i\frac{(d/dZ)[Ze^{Zt}/(Z+1)^3]\,|_{Z=1}}{1!}$$

$$= 2\pi i[\tfrac{1}{16}e^{t}(2t-1)]$$

Thus

$$f(t) = \frac{1}{2\pi i}\{2\pi i[\tfrac{1}{16}e^{-t}(1-2t^2)] + 2\pi i[\tfrac{1}{16}e^{t}(2t-1)]\}$$

or

$$f(t) = \tfrac{1}{16}e^{-t}(1-2t^2) + \tfrac{1}{16}e^{t}(2t-1)$$

Example 2.7-3

Find the inverse Laplace transform of $\bar{f}(s) = 1/(s^2+1)^2$

$$\bar{f}(s) = \frac{1}{(s^2+1)^2} = \frac{1}{[(s+i)(s-i)]^2} = \frac{1}{(s+i)^2(s-i)^2}$$

This problem is solved exactly as shown in Example 2.7-1 and 2.7-2. There are singularities at $Z = +2$ (pole of order 2) and $Z = -i$ (pole of order 2).

for $Z = i$:

$$\oint_C \frac{[e^{Zt}/(Z+i)^2]}{(Z-i)^2}\,dZ = 2\pi i\frac{(d/dZ)[e^{Zt}/(Z+i)^2]\,|_{Z=i}}{1!}$$

$$= 2\pi i[-\tfrac{1}{4}te^{it} - \tfrac{1}{4}ie^{it}]$$

for $Z = -i$:

$$\oint_C \frac{[e^{Zt}/(Z-i)^2]}{(Z+i)^2}\,dZ = 2\pi i\frac{(d/dZ)[e^{Zt}/(Z-i)^2]\,|_{Z=-i}}{1!}$$

$$= 2\pi i[-\tfrac{1}{4}te^{-it} + \tfrac{1}{4}ie^{-it}]$$

Thus

$$f(t) = \frac{1}{2\pi i}\{2\pi i[-\tfrac{1}{4}te^{it} - \tfrac{1}{4}ie^{it}] + 2\pi i[-\tfrac{1}{4}te^{-it} + \tfrac{1}{4}ie^{-it}]\}$$

$$= \tfrac{1}{2}(\sin t - t\cos t)$$

Example 2.7-4

We consider next the case where there are an infinite number of singularities. Find the inverse Laplace transform of

$$\bar{f}(Z) = \frac{\cosh x \sqrt{Z}}{Z \cosh \sqrt{Z}}$$

Because of the presence of $\sqrt{Z}$ it would appear that $Z = 0$ is a multivalued point (called a branch point). If this were the case there would be a slight alteration in the inversion operation which we will cover in an example later in this section. We can show here, however, that $Z = 0$ is not multivalued and hence not a branch point. This is shown by expanding the functions of the integrand in an infinite series:

$$\bar{f}(Z) = \frac{\cosh x \sqrt{Z}}{Z \cosh \sqrt{Z}} = \frac{1 + (x \sqrt{Z})^2/2! + \cdots}{Z\{1 + (\sqrt{Z})^2/2! + \cdots\}}$$

$$= \frac{1 + x^2 Z/2! + \cdots}{Z\{1 + Z/2! + \cdots\}}$$

It becomes evident that there is no $\sqrt{Z}$ term actually involved and hence no multivalued points, only a simple pole at $Z = 0$.

Returning now to the original $\bar{f}(Z)$ integrand we note that there are infinitely many poles determined from $\cosh \sqrt{Z} = 0$ in the integrand. Expanding $\cosh \sqrt{Z}$ in terms of exponentials, we have

$$\cosh \sqrt{Z} = \frac{e^{\sqrt{Z}} + e^{-\sqrt{Z}}}{2} = 0$$

From this relationship we determine that

$$e^{2\sqrt{Z}} = -1 = e^{\pi i + 2k\pi i} \qquad k = 0, \pm 1, \pm 2, \ldots$$

from which we get the infinite number of singularities of $\cosh \sqrt{Z} = 0$, i.e.,

$$\sqrt{Z} = (k + \tfrac{1}{2})\pi i$$

or

$$Z = -(k + \tfrac{1}{2})^2 \pi^2 \qquad \text{(simple poles of order 1)}$$

Thus, in the original integrand to be inverted we have found simple poles at $Z = 0$ and $Z = Z_n$, where $Z_n = -(n - \tfrac{1}{2})^2 \pi^2$, $n = 1, 2, 3, \ldots$

for $Z = 0$:

$$\oint_C \frac{(e^{Zt} \cosh x \sqrt{Z}/\cosh \sqrt{Z})}{Z} dZ = 2\pi i [e^{Zt} \cosh x \sqrt{Z}/\cosh \sqrt{Z}]|_{Z=0}$$
$$= 2\pi i(1)$$

In evaluating the complex integral for Z_n, where there are an infinite number of singularities, we make use of the technique shown in Example 2.6-10 of Section 2.6:

for $Z = Z_n$:

$$\oint_C \frac{e^{Zt} \cosh x \sqrt{Z}}{Z \cosh \sqrt{Z}} dZ = 2\pi i \lim_{Z \to Z_n} \left\{ (Z - Z_n) \left[\frac{e^{Zt} \cosh x \sqrt{Z}}{Z \cosh \sqrt{Z}} \right] \right\}$$

$$= 2\pi i \lim_{Z \to Z_n} \left[\frac{(Z - Z_n)}{\cosh \sqrt{Z}} \right] \lim_{Z \to Z_n} \left[\frac{e^{Zt} \cosh x \sqrt{Z}}{Z} \right]$$

The first limit is evaluated by L'Hôpital's rule, as shown below.

$$\oint_C \frac{e^{Zt} \cosh x \sqrt{Z}}{Z \cosh \sqrt{Z}} dZ$$

$$= 2\pi i \lim_{Z \to Z_n} \left[\frac{(d/dZ)(Z - Z_n)}{(d/dZ) \cosh \sqrt{Z}} \right] \lim_{Z \to Z_n} \left[\frac{e^{Zt} \cosh x \sqrt{Z}}{Z} \right]$$

$$= 2\pi i \left[\frac{4(-1)^n}{\pi(2n-1)} e^{-(n-1/2)^2 \pi^2 t} \cos(n - \tfrac{1}{2})\pi x \right]$$

where $n = 1, 2, 3, \ldots$.

Thus, finally,

$$f(t) = \frac{1}{2\pi i} \left\{ 2\pi i(1) + 2\pi i \left[\frac{4(-1)^n}{\pi(2n-1)} \right] e^{-(n-1/2)^2 \pi^2 t} \cos(n - \tfrac{1}{2})\pi x \right\}$$

or

$$f(t) = 1 + \frac{4}{\pi} \sum_{n=1}^{\infty} \frac{(-1)^n}{(2n-1)} e^{-(2n-1)^2 \pi^2 t/4} \cos \frac{(2n-1)}{2} \pi x$$

Example 3.7-5

This is another example where the integrand has infinitely many singularities. Find the inverse Laplace transform of

$$\bar{f}(Z) = \frac{\sinh Zx}{Z^2 \cosh Za}$$

Note that we are actually working on

$$\mathscr{L}^{-1}_{s \to t} \bar{f}(s) = f(t) = \mathscr{L}^{-1} \frac{\sinh sx}{s^2 \cosh sa}$$

and have shown in this section that $\bar{f}(s)$ can be evaluated by integration in the complex (Z) plane. So we actually need to solve the integral involving $e^{Zt} \bar{f}(Z)$.

The singularities are at $Z = 0$ and at $Z = Z_n$, where Z_n satisfies $\cosh Za = 0$. For $Z = 0$ it is not yet certain whether this is a simple pole (of order 1) or a pole of order 2. This uncertainty is resolved by expanding $\bar{f}(Z)$ in an infinite series,

$$\frac{\sin Zx}{Z^2 \cosh Za} = \frac{Zx + (Zx)^3/3! + \cdots}{Z^2 \{1 + (Za)^2/2! + \cdots\}} = \frac{x + Z^2 x^3/3! + \cdots}{Z\{1 + Z^2 a^2/2! + \cdots\}}$$

where $Z = 0$ is determined to be a simple pole (of order 1). The infinite singularities corresponding to $\cosh Za = 0$ are obtained in a similar manner to that shown in Example 2.7-4, yielding $Z = Z_n = (n - \frac{1}{2})\pi i/a$, $n = 1, 2, 3, \ldots$. Proceeding as in the previous problem and utilizing L'Hôpital's rule, we find

for $Z = 0$:

$$\oint_C \frac{e^{Zt} \sinh Zx}{Z^2 \cosh Za} dZ = 2\pi i \lim_{Z \to 0} \left[\frac{(Z - 0)e^{Zt} \sinh Zx}{Z^2 \cosh Za} \right]$$

$$= 2\pi i \lim_{Z \to 0} \left[\frac{\sinh Zx}{Z} \right] \lim_{Z \to 0} \left[\frac{e^{Zt}}{\cosh Za} \right]$$

$$= 2\pi i \lim_{Z \to 0} \left[\frac{d/dZ \sinh Zx}{(d/dZ)Z} \right] \cdot 1$$

$$= 2\pi i (x)$$

for $Z = Z_n$:

$$\oint_C \frac{e^{Zt} \sinh Zx}{Z^2 \cosh Za} dZ = 2\pi i \lim_{Z \to Z_n} \left[\frac{(Z - Z_n)e^{Zt} \sinh Zx}{Z^2 \cosh Za} \right]$$

$$= 2\pi i \lim_{Z \to Z_n} \left[\frac{(Z - Z_n)}{\cosh Za} \right] \lim_{Z \to Z_n} \left[\frac{e^{Zt} \sinh Zx}{Z^2} \right]$$

$$= 2\pi i \lim_{Z \to Z_n} \left[\frac{(d/dZ)(Z - Z_n)}{d/dZ \cosh Za} \right] \lim_{Z \to Z_n} \left[\frac{e^{Zt} \sinh Zx}{Z^2} \right]$$

$$= 2\pi i \lim_{Z \to Z_n} \left[\frac{1}{a \sinh Za} \right] \lim_{Z \to Z_n} \left[\frac{e^{Zt} \sinh Zx}{Z^2} \right]$$

$$= 2\pi i \left[\frac{1}{ai \sin(n - \frac{1}{2})\pi} \right] \left[\frac{e^{(n-1/2)\pi i t/a} \, i \sin(n - \frac{1}{2})\pi x/a}{-(n - \frac{1}{2})^2 \pi^2/a^2} \right]$$

Thus, finally, we get the inverse Laplace transform

$$f(t) = \mathscr{L}^{-1}_{s \to t} \bar{f}(s) = \frac{1}{2\pi i} \int_{\gamma - i\infty}^{\gamma + i\infty} \bar{f}(s) e^{st} \, ds = \frac{1}{2\pi i} \oint_C \bar{f}(Z) e^{Zt} \, dZ$$

and by adding the contribution of the residues, $Z = 0$ and $Z = Z_n$, we get

$$f(t) = x + \frac{8a}{\pi^2} \sum_{n=1}^{\infty} \frac{(-1)^n}{(2n-1)^2} \sin \frac{(2n-1)\pi x}{2a} \cos \frac{(2n-1)\pi t}{2a}$$

Example 2.7-6

Find the inverse Laplace transform of $\bar{f}(Z) = (\cosh Zx)/(Z^3 \cosh Zb)$. This problem differs from the Example 2.7-5 in that we can show that $Z = 0$ is a pole of order 3 by expanding the integrand in an infinite series:

$$\frac{e^{Zt} \cosh Zx}{Z^3 \cosh Zx} = \frac{1}{Z^3} (1 + Zt + \cdots) \left(\frac{1 + Z^2 x^2/2! + \cdots}{1 + Z^2 b^2/2! + \cdots} \right)$$

This shows that $Z = 0$ is a pole of order 3. From the previous examples we can show that the poles of $\cosh Zb = 0$ are infinite in number and are given as $Z = Z_n = (2n-1)\pi i/2b$ with $n = 1, 2, 3, \ldots$. Proceeding to the solution as in the previous examples, we get

for $Z = 0$:

$$\oint_C \frac{e^{Zt} \cosh Zx}{Z^3 \cosh Zb} \, dZ = 2\pi i \left[\frac{(d^2/dZ^2)(e^{Zt} \cosh Zx/\cosh Zb) \, |_{Z=0}}{2!} \right]$$

for $Z = Z_n$:

$$\oint_C \frac{e^Z \cosh Zx}{Z^3 \cosh Zb} \, dZ$$

$$= 2\pi i \lim_{Z \to Z_n} \left[\frac{(Z - Z_n) e^{Zt} \cosh Zx}{Z^3 \cosh Zb} \right]$$

$$= 2\pi i \lim_{Z \to Z_n} \left[\frac{(Z - Z_n)}{\cosh Zb} \right] \lim_{Z \to Z_n} \left[\frac{e^{Zt} \cosh Zx}{Z^3} \right]$$

$$= 2\pi i \left[\frac{1}{b \sinh(2n-1)\pi i/2} \right] \left[\frac{e^{(2n-1)\pi i t/2b} \cosh(2n-1)\pi i x/2b}{[(2n-1)\pi i/2b]^3} \right]$$

With some simplification we can get the final solution by adding the above contributions:

$$f(t) = \tfrac{1}{2}(t^2 + x^2 - b^2)$$

$$- \frac{16b^2}{\pi^3} \sum_{n=1}^{\infty} \frac{(-1)^n}{(2n-1)^3} \cos \frac{(2n-1)\pi t}{2b} \cos \frac{(2n-1)\pi x}{2b}$$

Example 2.7-7

As a final example of the inverse Laplace transform operation by integration in the complex plane we evaluate

$$f(t) = \mathscr{L}_{s \to t}^{-1} \bar{f}(s) = \mathscr{L}_{s \to t}^{-1} \frac{e^{-a\sqrt{s}}}{s} = \frac{1}{2\pi i} \int_{\gamma-i\infty}^{\gamma+i\infty} \frac{e^{-a\sqrt{s}}}{s} e^{st} \, ds$$

At first glance it would seem that $s = 0$ is a singularity which is a pole of order 1 (simple pole). However, if we proceed in the usual manner, we get multiple values for $e^{-\sqrt{s}}$ and thus the "usual" methods are not applicable. Even expanding $\bar{f}(s)$ in an infinite series still does not resolve the multivalued property. We say that $s = 0$ is a "branch point" for reasons that will soon become more apparent.

The integration procedure is not as simple as before, but nevertheless it is still straightforward. It is necessary to devise a more complicated integration path than that given in Figure 9. The new path is given by Figure 10 and is designed to avoid (branch around) the multivalued branch point, $Z = 0$.

For integration around the closed path C indicated in Figure 10, we can write

$$\oint_C \frac{e^{-a\sqrt{Z}}}{Z} e^{Zt} \, dZ = \int_{AB} \frac{e^{-a\sqrt{Z}}}{Z} e^{Zt} \, dZ + \int_{BDE} \frac{e^{-a\sqrt{Z}}}{Z} e^{Zt} \, dZ$$

$$+ \int_{EH} \frac{e^{-a\sqrt{Z}}}{Z} e^{Zt} \, dZ + \int_{HJK} \frac{e^{-a\sqrt{Z}}}{Z} e^{Zt} \, dZ$$

$$+ \int_{KL} \frac{e^{-a\sqrt{Z}}}{Z} e^{Zt} \, dZ + \int_{LNA} \frac{e^{-a\sqrt{Z}}}{Z} e^{Zt} \, dZ$$

The closed path C is composed of line AB and arcs BDE and LNA of a circle of radius R with the center at the origin ($Z = 0$). The closed path also contains the arc HJK of a concentric circle of radius ε. Lines EH and KL are also part of the closed path and are located at distances of less than ε above and below the line $y = 0$.

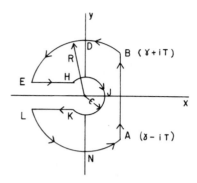

Fig. 10. Integration in the complex plane
around a branch point.

Since the only singularity, $Z = 0$, is not inside the closed path C, the
integral around the closed path, $\oint_C (e^{-a\sqrt{Z}}/Z)e^{Zt}\, dZ$, is zero by Cauchy's
theorem [equation (2.6-23)]. It is also possible to show that for $R \to \infty$,
the integrals involving paths BDE and LNA go to zero.[13] This is usually
true and can for all practical purposes be considered to be always true.
Thus, the above equation becomes

$$0 = \lim_{\substack{R \to \infty \\ \varepsilon \to 0}} \left\{ \int_{AB} \frac{e^{-a\sqrt{Z}}}{Z} e^{Zt}\, dZ + \int_{EH} \frac{e^{-a\sqrt{Z}}}{Z} e^{Zt}\, dZ \right.$$

$$\left. + \int_{HJK} \frac{e^{-a\sqrt{Z}}}{Z} e^{Zt}\, dZ + \int_{KL} \frac{e^{-a\sqrt{Z}}}{Z} e^{Zt}\, dZ \right\}$$

Rearranging this equation, we get

$$\lim_{\substack{R \to \infty \\ \varepsilon \to 0}} \int_{AB} \frac{e^{-a\sqrt{Z}}}{Z} e^{Zt}\, dZ = \int_{\gamma - i\infty}^{\gamma + i\infty} \frac{e^{-a\sqrt{Z}}}{Z} e^{Zt}\, dZ$$

$$= -\lim_{\substack{R \to \infty \\ \varepsilon \to 0}} \left\{ \int_{EH} \frac{e^{-a\sqrt{Z}}}{Z} e^{Zt}\, dZ \right.$$

$$\left. + \int_{HJK} \frac{e^{-a\sqrt{Z}}}{Z} e^{Zt}\, dZ + \int_{KL} \frac{e^{-a\sqrt{Z}}}{Z} e^{Zt}\, dZ \right\}$$

Now we need to evaluate separately each of the integrals along paths EH,
KJK, and KL. Path AB, with $R \to \infty$, $\varepsilon \to 0$, is the Bromwich path,
$\gamma - i\infty$, $\gamma + i\infty$, required for the inverse Laplace transform.

[13] R. V. Churchill, *op. cit.*, p. 131.

Path EH:

$$Z = xe^{\pi i}$$

(the angle is π and $e^{\pi i} = \cos \pi + i \sin \pi$) or

$$\sqrt{Z} = \sqrt{x} e^{\pi i/2} = \sqrt{x} i$$

which yields $Z = -x$. From the limits on Z for the integral over path *EH*, we note that as Z goes from $-R$ to $-\varepsilon$ (path *EH*), x goes from R to ε. Substituting the relationships found between Z and x over path *EH* into the integral, we get

$$\int_{EH} \frac{e^{-a\sqrt{Z}}}{Z} e^{Zt} \, dZ = \int_{-R}^{-\varepsilon} \frac{e^{-a\sqrt{Z}}}{Z} e^{Zt} \, dZ = \int_{R}^{\varepsilon} \frac{e^{-xt-ai\sqrt{x}}}{x} \, dx$$

Path KL:

Along path *KL*, in a manner similar to that developed for path *EH*, we get

$$Z = xe^{-\pi i}$$

(the angle is $-\pi$) or

$$\sqrt{Z} = \sqrt{x} e^{-i\pi/2} = -\sqrt{x} i$$

which yields $Z = -x$. As Z goes from $-\varepsilon$ to $-R$, x goes from ε to R. We thus get

$$\int_{KL} \frac{e^{-a\sqrt{Z}}}{Z} e^{Zt} \, dZ = \int_{-\varepsilon}^{-R} \frac{e^{-a\sqrt{Z}}}{Z} e^{Zt} \, dZ = \int_{\varepsilon}^{R} \frac{e^{-xt+ai\sqrt{x}}}{x} \, dx$$

Path HJK:

Along path *HJK*, $Z = \varepsilon e^{i\theta}$ (a circle of radius ε about the origin, going from the angle π to $-\pi$). When this form of Z is substituted into the original integral, we get

$$\int_{HJK} \frac{e^{-a\sqrt{Z}}}{Z} e^{Zt} \, dZ = \int_{\pi}^{-\pi} \frac{e^{\varepsilon e^{i\theta} t - a\sqrt{\varepsilon} e^{i\theta/2}}}{\varepsilon e^{i\theta}} i\varepsilon e^{i\theta} \, d\theta$$

$$= i \int_{\pi}^{-\pi} e^{\varepsilon e^{i\theta} t - a\sqrt{\varepsilon} e^{i\theta/2}} \, d\theta$$

If now the altered forms shown for paths *EH*, *KL*, and *HJK* are substituted into the original integral, the result is

$$\int_{\gamma-i\infty}^{\gamma+i\infty} \frac{e^{-a\sqrt{Z}}}{Z} e^{Zt}\, dZ$$

$$= -\lim_{\substack{R\to\infty \\ \varepsilon\to 0}} \left\{ \int_R^\varepsilon \frac{e^{-xt-ai\sqrt{x}}}{x}\, dx + \int_\varepsilon^R \frac{e^{-xt+ai\sqrt{x}}}{x}\, dx \right.$$

$$\left. + i \int_\pi^{-\pi} e^{\varepsilon e^{i\theta}-a\sqrt{\varepsilon}\,e^{i\theta/2}}\, d\theta \right\}$$

$$= \lim_{\substack{R\to\infty \\ \varepsilon\to 0}} \left\{ \int_\varepsilon^R \frac{e^{-xt}(e^{ai\sqrt{x}} - e^{-ai\sqrt{x}})}{x}\, dx \right.$$

$$\left. + i \int_\pi^{-\pi} e^{\varepsilon e^{i\theta}t - a\sqrt{\varepsilon}\,e^{i\theta/2}}\, d\theta \right\}$$

$$= -\lim_{\substack{R\to\infty \\ \varepsilon\to 0}} \left\{ 2i \int_\varepsilon^R \frac{e^{-xt}\sin a\sqrt{x}}{x}\, dx + i \int_\pi^{-\pi} e^{\varepsilon e^{i\theta}t - a\sqrt{\varepsilon}\,e^{i\theta/2}}\, d\theta \right\}$$

It can be shown that

$$\lim_{\substack{R\to\infty \\ \varepsilon\to 0}} \int_\pi^{-\pi} e^{\varepsilon e^{i\theta}t - a\sqrt{\varepsilon}\,e^{i\theta/2}}\, d\theta = \int_\pi^{-\pi} 1\, d\theta = -2\pi$$

where we have inverted the order of operation, i.e., the limit operation
was performed first. Using this result, the original equation is simplified:

$$\int_{\gamma-i\infty}^{\gamma+i\infty} \frac{e^{-a\sqrt{Z}}}{Z} e^{Zt}\, dZ = -\lim_{\substack{R\to\infty \\ \varepsilon\to 0}} \left\{ 2i \int_\varepsilon^R \frac{e^{-xt}\sin a\sqrt{x}}{x}\, dx - 2\pi i \right\}$$

$$= -2i \int_0^\infty \frac{e^{-xt}\sin a\sqrt{x}}{x}\, dx + 2\pi i$$

The inverse Laplace transform of the original function, $e^{-a\sqrt{s}}/s$, is finally

$$\mathscr{L}^{-1}_{s\to t} \frac{e^{-a\sqrt{s}}}{s} = \frac{1}{2\pi i} \int_{\gamma-i\infty}^{\gamma+i\infty} \frac{e^{-a\sqrt{s}}\,e^{st}}{s}\, ds = \frac{1}{2\pi i} \int_{\gamma-i\infty}^{\gamma+i\infty} \frac{e^{-a\sqrt{Z}}\,e^{Zt}}{Z}\, dZ$$

$$= 1 - \frac{1}{\pi} \int_0^\infty \frac{e^{-xt}\sin a\sqrt{x}}{x}\, dx$$

$$= 1 - \text{erf}\left(\frac{a}{2\sqrt{t}}\right) = \text{erfc}\left(\frac{a}{2\sqrt{t}}\right)$$

2.8. Solution of Partial Differential Equations

We can illustrate the application of Laplace transform techniques to partial differential equations by the following examples, presented with a minimum of comment.

Example 2.8-1

Solve by Laplace transform techniques the partial differential equation

$$\frac{\partial U}{\partial t} = k \frac{\partial^2 U}{\partial x^2}$$

with boundary and initial conditions

$$U(x, t) = 0 \qquad \text{at } t = 0$$
$$U(x, t) = U_0 \qquad \text{at } x = 0$$
$$U(x, t) = \text{finite} \qquad \text{at } x \to \infty$$

The Laplace transform of the governing equation, $\mathscr{L}_{t \to s}$, yields

$$s\bar{U}(x, s) - U(x, 0) = k \frac{d^2 \bar{U}(x, t)}{dx^2}$$

or

$$\frac{d^2 \bar{U}(x, s)}{dx^2} - \frac{s}{k} \bar{U}(x, s) = 0$$

The solution of this ordinary differential equation is

$$\bar{U}(x, s) = C_1 e^{\sqrt{s/k}\,x} + C_2 e^{-\sqrt{s/k}\,x}$$

We can infer $C_1 = 0$ since $U(x, t)$ [and hence $\bar{U}(x, s)$] must be finite for $x \to \infty$. From $U(x, t) = U_0$ at $x = 0$ we can get

$$\mathscr{L}_{t \to s}\{U(x, t) = U_0\} \qquad \text{or} \qquad \bar{U}(x, s) = \frac{U_0}{s} \qquad \text{at } x = 0$$

This allows the evaluation of C_2:

$$C_2 = \frac{U_0}{s}$$

so that we get

$$\bar{U}(x, s) = \frac{U_0}{s} e^{-\sqrt{s/k}\,x}$$

From the inversion tables the final solution is found:

$$U(x, t) = U_0 \operatorname{erfc}\left(\frac{x}{2\sqrt{kt}}\right)$$

Example 2.8-2

If the condition $U(x, t) = U_0$ at $x = 0$ in Example 2.8-1 is replaced by

$$U(x, t) = g(t) \qquad \text{at } x = 0$$

we proceed as before with the new condition (transformed):

$$\bar{U}(x, s) = \bar{g}(s) \qquad \text{at } x = 0$$

The Laplace transform of the original equation becomes

$$\bar{U}(x, s) = \bar{g}(s)e^{-\sqrt{s/k}\,x}$$

which inverts by the convolution integral to

$$U(x, t) = \int_0^t \frac{x}{2\sqrt{\pi k}} u^{-3/2} e^{-x^2/4ku} g(t - u)\, du$$

Note that

$$\mathscr{L}_{s \to t}^{-1} \bar{g}(s) = g(t)$$

and

$$\mathscr{L}_{s \to t}^{-1} e^{-\sqrt{s/k}\,x} = \frac{x}{2\sqrt{\pi k}} t^{-3/2} e^{-x^2/4kt}$$

If we substitute $v = x^2/4ku$ into the $U(x, t)$ equation, the final solution is

$$U(x, t) = \frac{2}{\sqrt{\pi}} \int_{x/2\sqrt{\pi kt}}^{\infty} e^{-v^2} g\left(t - \frac{x^2}{4kv^2}\right) dv$$

Example 2.8-3

Suppose we have the same governing equation

$$\frac{\partial U(x, t)}{\partial t} = k \frac{\partial^2 U(x, t)}{\partial x^2}$$

but the boundary conditions are a bit different from the previous examples, i.e.,

$$U(x, t) = U_0 \qquad \text{at } t = 0$$

$$\frac{\partial U(x, t)}{\partial x} = 0 \qquad \text{at } x = 0$$

$$U(x, t) = U_1 \qquad \text{at } x = l$$

We take the Laplace transform $\mathscr{L}_{t \to s}$ of the governing equation as usual,

$$s\bar{U}(x, s) - U(x, 0) = k \frac{d^2\bar{U}(x, s)}{dx^2}$$

or

$$\frac{d^2\bar{U}(x, s)}{dx^2} - \frac{s\bar{U}(x, s)}{k} = -\frac{U_0}{k}$$

which yields

$$\bar{U}(x, s) = C_1 \cosh \sqrt{\frac{s}{k}}\, x + C_2 \sinh \sqrt{\frac{s}{k}}\, x + \frac{U_0}{s}$$

The derivative boundary condition can be utilized:

$$\mathscr{L}_{t \to s}\left\{\frac{\partial U(x, t)}{\partial x} = 0\right\} \quad \text{yields} \quad \frac{\partial \bar{U}(x, s)}{\partial x} = 0 \qquad \text{at } x = 0$$

Using this transformed boundary condition in the $\bar{U}(x, s)$ solution we get $C_2 = 0$, and with the boundary condition at $x = l$ we get also

$$C_1 = (U_1 - U_0)/s \cosh \sqrt{s/k}\, l$$

Thus,

$$\bar{U}(x, s) = \frac{U_0}{s} + (U_1 - U_0) \frac{\cosh \sqrt{s/k}\, x}{s \cosh \sqrt{s/k}\, l}$$

We can invert this equation, $\mathscr{L}_{s \to t}^{-1}\bar{U}(x, s) = U(x, t)$, by simply looking up the result in the tables. However, it might be more instructional to show how to get the inverse Laplace transform by integration in the complex plane according to the inversion formula:

$$U(x, t) = \mathscr{L}_{s \to t}^{-1}\bar{U}(x, s) = \frac{1}{2\pi i} \int_{\gamma - i\infty}^{\gamma + i\infty} e^{st}\left[\frac{U_0}{s} + (U_1 - U_0) \frac{\cosh \sqrt{s/k}\, x}{s \cosh \sqrt{s/k}\, l}\right] ds$$

$$= \frac{1}{2\pi i} \oint_C e^{Zt}\left[\frac{U_0}{Z} + (U_1 - U_0) \frac{\cosh \sqrt{Z/k}\, x}{Z \cosh \sqrt{Z/k}\, l}\right] dZ$$

For integrand $\dfrac{e^{Zt}U_0}{Z}$:

There is a singularity at $Z = 0$ (simple pole)

for $Z = 0$:

$$\oint_C \frac{e^{Zt}U_0}{Z}\,dZ = 2\pi i U_0$$

For integrand $\dfrac{e^{Zt}(U_1 - U_0)\cosh\sqrt{Z/k}\,x}{Z\cosh\sqrt{Z/k}\,l}$:

There are singularities at $Z = 0$ (simple pole) and

$$Z = Z_n = \frac{-(2n-1)^2\pi^2 k}{4l^2} \qquad n = 1, 2, 3, \ldots$$

which was obtained from Example 2.7-4.

for $Z = 0$:

$$\oint_C \frac{e^{Zt}(U_1 - U_0)\cosh\sqrt{Z/k}\,x}{Z\cosh\sqrt{Z/k}\,l}\,dZ = 2\pi i(1)(U_1 - U_0)$$

for $Z = Z_n$:

$$\oint_C \frac{e^{Zt}(U_1 - U_0)\cosh\sqrt{Z/k}\,x}{Z\cosh\sqrt{Z/k}\,l}\,dZ$$

$$= 2\pi i \lim_{Z\to Z_n}\left\{\frac{(Z - Z_n)e^{Zt}(U_1 - U_0)\cosh\sqrt{Z/k}\,x}{Z\cosh\sqrt{Z/k}\,l}\right\}$$

$$= 2\pi i \lim_{Z\to Z_n}\left[\frac{(Z - Z_n)}{\cosh\sqrt{Z/k}\,l}\right]\lim_{Z\to Z_n}\left[\frac{e^{Zt}(U_1 - U_0)\cosh\sqrt{Z/k}\,x}{Z}\right]$$

$$= 2\pi i \lim_{Z\to Z_n}\left[\frac{1}{(\sinh\sqrt{Z/k}\,l)(1/2\sqrt{kZ})}\right]$$

$$\times \lim_{Z\to Z_n}\left[\frac{e^{Zt}(U_1 - U_0)\cosh\sqrt{Z/k}\,x}{Z}\right]$$

$$= 2\pi i\left[\frac{4(-1)^n}{(2n-1)\pi}\,e^{-(2n-1)^2\pi^2 kt/4l^2}\cos\frac{(2n-1)\pi x}{2l}\right]$$

using L'Hôpital's rule in the first limit term.

Adding the results for both integrands and all singularities, we finally get

$$U(x, t) = U_1 + \frac{4(U_1 - U_0)}{\pi} \sum_{n=1}^{\infty} \frac{(-1)^n}{(2n-1)} e^{-(2n-1)^2\pi^2kt/4l^2} \cos\frac{(2n-1)}{2l}\pi x$$

Example 2.8-3

Solve the following by Laplace transform techniques:

$$\frac{\partial U}{\partial t} = \left(\frac{\partial^2 U}{\partial r^2} + \frac{1}{r}\frac{\partial U}{\partial r}\right)$$

$$U(r, t) = 0 \qquad \text{for } r = 1$$

$$U(r, t) = T \qquad \text{for } t = 0$$

$$U(r, t) \text{ is finite for } r = 0$$

Taking the Laplace transform, $\mathcal{L}_{t \to s}$, we get

$$s\bar{U}(r, s) - U(r, 0) = \frac{d^2\bar{U}(r, s)}{dr^2} + \frac{1}{r}\frac{d\bar{U}(r, s)}{dr}$$

or

$$\frac{d^2\bar{U}(r, s)}{dr^2} + \frac{1}{r}\frac{d\bar{U}(r, s)}{dr} - s\bar{U}(r, s) = -T$$

This equation yields a Bessel function solution:

$$\bar{U}(r, s) = C_1 J_0(i\sqrt{s}\,r) + C_2 Y_0(i\sqrt{s}\,r) + \frac{T}{s}$$

Since $Y_0(i\sqrt{s}\,r)$ is not finite for $r = 0$, this would lead to the conclusion that $\bar{U}(r, s)$ and $U(r, t)$ were not finite for $r = 0$. Hence it is necessary that $C_2 = 0$, yielding

$$\bar{U}(r, s) = C_1 J_0(i\sqrt{s}\,r) + \frac{T}{s}$$

The constant C_1 is found from the first boundary condition on $r = 1$. We get $C_1 = -T/sJ_0(i\sqrt{s}\,)$ and

$$\bar{U}(r, s) = \frac{T}{s} - \frac{TJ_0(i\sqrt{s}\,r)}{sJ_0(i\sqrt{s}\,)}$$

This can be inverted by looking up the final result in the tables. However for heuristic reasons we again show the inversion by integration in the complex plane. The solution is given as

$$U(r, t) = \mathscr{L}_{s \to t}^{-1} \bar{U}(r, s) = \frac{1}{2\pi i} \int_{\gamma - i\infty}^{\gamma + i\infty} e^{st} \bar{U}(r, s) \, ds$$

$$= \frac{1}{2\pi i} \int_{\gamma - i\infty}^{\gamma + i\infty} e^{st} \left[\frac{T}{s} - \frac{T J_0(i \sqrt{s} \, r)}{s J_0(i \sqrt{s})} \right] ds$$

$$= \frac{1}{2\pi i} \oint_C e^{Zt} \left[\frac{T}{Z} - \frac{T J_0(i \sqrt{Z} r)}{Z J_0(i \sqrt{Z})} \right] dZ$$

$$= \frac{1}{2\pi i} \oint \frac{e^{Zt} T}{Z} \, dZ - \frac{1}{2\pi i} \oint_C \frac{e^{Zt} T J_0(i \sqrt{Z} r)}{Z J_0(i \sqrt{Z})} \, dZ$$

For integrand $\dfrac{e^{Zt} T}{Z}$

This integrand has a singularity at $Z = 0$

for $Z = 0$:

$$\oint_C \frac{e^{Zt} T}{Z} \, dZ = 2\pi i T$$

For integrand $\dfrac{e^{Zt} T J_0(i \sqrt{Z} r)}{Z J_0(i \sqrt{Z})}$

This integrand has singularities at $Z = 0$ and $Z = Z_n = -\lambda_n^2$, $n = 1, 2, 3, \ldots$, where $\lambda_n = i \sqrt{Z}$. Values for λ_n arise from $J_0(i \sqrt{Z}) = 0$

for $Z = 0$:

$$\oint_C \frac{e^{Zt} T J_0(i \sqrt{Z} r)}{Z J_0(i \sqrt{Z})} \, dZ = 2\pi i \lim_{Z \to 0} \left\{ \frac{(Z - 0) e^{Zt} T J_0(i \sqrt{Z} r)}{Z J_0(i \sqrt{Z})} \right\} = 2\pi i T$$

for $Z = Z_n$:

$$\oint_C \frac{e^{Zt} T J_0(i \sqrt{Z} r)}{Z J_0(i \sqrt{Z})} \, dZ = 2\pi i \lim_{Z \to Z_n} \left\{ \frac{(Z - Z_n) e^{Zt} T J_0(i \sqrt{Z} r)}{Z J_0(i \sqrt{Z})} \right\}$$

$$= 2\pi i \lim_{Z \to Z_n} \left[\frac{(Z - Z_n)}{J_0(i \sqrt{Z})} \right] \lim_{Z \to Z_n} \left[\frac{e^{Zt} T J_0(i \sqrt{Z} r)}{Z} \right]$$

and by L'Hôpital's rule

$$= 2\pi i \left[\lim_{Z \to Z_n} \frac{(d/dZ)(Z - Z_n)}{(d/dZ)J_0(i\sqrt{Z})} \right]$$

$$\times \lim_{Z \to Z_n} \left[\frac{e^{Zt}TJ_0(i\sqrt{Z}r)}{Z} \right]$$

$$= 2\pi i \left(-\frac{2e^{-\lambda_n^2 t}TJ_0(\lambda_n r)}{\lambda_n J_1(\lambda_n)} \right)$$

Summing the results for both integrands and all singularities,

$$U(r, t) = T - T \left\{ 1 - \sum_{n=1}^{\infty} \frac{2e^{-\lambda_n^2 t}J_0(\lambda_n r)}{\lambda_n J_1(\lambda_n)} \right\}$$

$$= 2T \sum_{n=1}^{\infty} \frac{e^{-\lambda_n^2 t}J_0(\lambda_n r)}{\lambda_n J_1(\lambda_n)}$$

Example 2.8-4

In some heat flow problems, a quantity of heat may be generated instantaneously at a point $x = a$ and is represented as

$$U(x, t) = Q\delta(t) \qquad \text{for } x = a$$

where $\delta(t)$ is the Dirac delta (impulse function) and Q is a constant. For use as a boundary condition when the Laplace transform technique is employed, it is necessary to transform this condition as follows:

$$\mathscr{L}_{t \to s}\{U(x, t) = Q\delta(t)\} \quad \text{yields} \quad \bar{U}(x, s) = Q$$

where $\mathscr{L}_{t \to s}\delta(t) = 1$.

2.9. Laplace Transforms and Nonlinear Equations

In all of the previous material in this chapter, the operations performed and equations solved were linear with respect to the dependent variable and derivatives of the dependent variable. Thus we transformed such functions as $y(t)$, dy/dt, d^2y/dt^2, $ty(t)$, etc. We did not operate upon forms such as y^2, $(dy/dt)^2$, $y(dy/dt)$, etc. From the definition of the Laplace transform, it becomes apparent why the nonlinear forms present extra problems:

$$\mathscr{L}_{t\to s}\, y(t) = \bar{y}(s) = \int_0^\infty y(t)e^{-st}\, dt$$

$$\mathscr{L}_{t\to s}\, y^2(t) = \overline{y^2}(s) = \int_0^\infty y^2(t)e^{-st}\, dt$$

It is not true that $[\bar{y}(s)]^2$ is identically equal to $[\overline{y^2}(s)]$; rather $\overline{y^2}(s)$ is a different dependent variable. For example, suppose we wish to solve

$$\frac{dy}{dt} + y^2 = 0 \qquad\qquad (2.9\text{-}1)$$

By a separation of variables technique we can get

$$y = \frac{1}{t+1} \qquad\qquad (2.9\text{-}2)$$

which is an exact solution of (2.9-1); here we have used as an initial condition $y = 1$ at $t = 0$. The solution of equation (2.9-1) by the Laplace transform technique, $\mathscr{L}_{t\to s}$, would result in

$$s\bar{y} - y(0) + \overline{y^2} = 0 \qquad\qquad (2.9\text{-}3)$$

where

$$\mathscr{L}_{t\to s}\, y(t) = \bar{y}(s) = \bar{y}$$

and

$$\mathscr{L}_{t\to s}\, y^2(t) = \overline{y^2}(s) = \overline{y^2}$$

Thus we find two dependent variables in equation (2.9-3) where originally there was one. What remains is to find the relationship between $\bar{y}$ and $\overline{y^2}$. In some cases it may be possible to linearize this relationship or approximate it by

$$\overline{y^2} \cong k\bar{y} \qquad\qquad (2.9\text{-}4)$$

or

$$\overline{y^2} \cong g(s)\bar{y} \qquad\qquad (2.9\text{-}5)$$

Using equation (2.9-4) in (2.9-3), we could get

$$s\bar{y} - 1 + k\bar{y} = 0 \qquad\qquad (2.9\text{-}6)$$

or

$$\bar{y} = \frac{1}{s+k} \qquad\qquad (2.9\text{-}7)$$

The inversion of $\bar{y}$ is accomplished easily to yield

$$y(t) = e^{-kt} \qquad (2.9\text{-}8)$$

This solution conforms to the exact solution $y = 1/(t + 1)$ at $t = 0$ and $t = \infty$. For intermediate values, the quality of the approximation depends upon the value of k.

If we used equation (2.9-5) as the approximation for $\overline{y^2}$, then equation (2.9-3) becomes a bit more complex than equation (2.9-6), and this is shown in

$$s\bar{y} - 1 + g(s)\bar{y} = 0 \qquad (2.9\text{-}9)$$

or

$$\bar{y} = \frac{1}{s + g(s)} \qquad (2.9\text{-}10)$$

It is not yet readily apparent how $g(s)$ is to be evaluated. Obviously $\bar{y}$ cannot be inverted until $g(s)$ is explicitly expressed.

Another approximation technique involves a series solution approach. From equation (2.9-1), define

$$y = y_0 + \alpha y_1 + \alpha^2 y_2 + \cdots \qquad (2.9\text{-}11)$$

and

$$\bar{y}(s) = \bar{y}_0(s) + \alpha \bar{y}_1(s) + \alpha^2 \bar{y}_2(s) + \cdots \qquad (2.9\text{-}12)$$

where $\bar{y}_0(s)$ is the Laplace transform of $y_0(t)$, $\bar{y}_1(s)$ corresponds to the transform of $y_1(t)$, etc. By substituting equations (2.9-11) and (2.9-12) into (2.9-3) and equating the coefficients of the α, α^2, α^3, ... terms, it is possible in principle to determine $y_1, y_2, y_3, \ldots$ in terms of α. From boundary conditions and other constraints it may be possible to determine α.

Finally, from a modification of the mean value theorem,[14] it is possible to write

$$\mathcal{L}_{t\to s}\, y^2(t) = \overline{y^2}(s) = \int_0^\infty y^2(t)e^{-st}\, dt \cong \langle y \rangle \int_0^\infty y e^{-st}\, dt \qquad (2.9\text{-}13)$$

where $\langle y \rangle$ is the mean value of the y function. Proceeding as before, we can produce

$$\bar{y} = \frac{1}{s + \langle y \rangle} \qquad (2.9\text{-}14)$$

[14] A. E. Taylor, *Advanced Calculus*, Ginn, New York (1955), p. 51.

which can be inverted to yield

$$y = e^{-\langle y \rangle t} \tag{2.9-15}$$

If equation (2.9-15) is now averaged,

$$\langle y \rangle = \frac{\int_{a_1}^{a_2} y \, dt}{\int_{a_1}^{a_2} dt}$$

the result, with some rearrangement, is

$$(a_2 - a_1)\langle y \rangle^2 = e^{-a_1 \langle y \rangle} - e^{-a_2 \langle y \rangle} \tag{2.9-16}$$

If now the value of $\langle y \rangle$ is determined from equation (2.9-16) over a section of the time range (a_1 to a_2), it is possible to reasonably approximate the behavior of the y function by equation (2.9-15) for the same time range. Proceeding in discrete time sections, it is then possible to determine the y function over the entire time range.

Another example of this "mean value" procedure is illustrated below. Given:

$$\left(\frac{dy}{dt} \right)^2 = t \tag{2.9-17}$$

with

$$y = 0 \qquad \text{at } t = 0$$

The rigorous solution by the separation of variables method is obtained sequentially:

$$\frac{dy}{dt} = t^{1/2} \tag{2.9-18}$$

$$dy = t^{1/2} \, dy \tag{2.9-19}$$

and

$$y = \tfrac{2}{3} t^{3/2} + c \tag{2.9-20}$$

With the initial condition, we get $c = 0$, so that the final solution is

$$y = \tfrac{2}{3} t^{3/2}$$

The approximate Laplace transform solution of equation (2.9-17), using the modification of the mean value theorem, is outlined below.

$$\mathcal{L}_{t \to p}\{\text{eq. (2.9-17)}\} = \left\langle \frac{dy}{dt} \right\rangle [p\bar{y} - y(0)] = \frac{1}{p^2} \qquad (2.9\text{-}21)$$

where

$$\mathcal{L}_{t \to p}\left(\frac{dy}{dt}\right)^2 = \int_0^\infty \left(\frac{dy}{dt}\right)^2 e^{-pt}\,dt = \left\langle \frac{dy}{dt} \right\rangle \int_0^\infty \frac{dy}{dt} e^{-pt}\,dt \qquad (2.9\text{-}22)$$

and $\langle dy/dt \rangle$ is the mean value of dy/dt over the range of the time variable. Equation (2.9-22) can be rearranged to yield

$$\bar{y} = \frac{1}{\langle dy/dt \rangle} \left(\frac{1}{p^3}\right) \qquad (2.9\text{-}23)$$

Inverting equation (2.9-23) gives

$$y = \frac{1}{\langle dy/dt \rangle} \frac{t^2}{2!} \qquad (2.9\text{-}24)$$

The derivative dy/dt from equation (2.9-24) is

$$\frac{dy}{dt} = \frac{1}{\langle dy/dt \rangle} t \qquad (2.9\text{-}25)$$

and the mean is

$$\left\langle \frac{dy}{dt} \right\rangle = \frac{1}{\langle dy/dt \rangle} \frac{\displaystyle\int_{a_1}^{a_2} t\,dt}{\displaystyle\int_{a_1}^{a_2} dt} \qquad (2.9\text{-}26)$$

or

$$\left\langle \frac{dy}{dt} \right\rangle^2 = \frac{\frac{1}{2}(a_2{}^2 - a_1{}^2)}{(a_2 - a_1)} = \frac{\frac{1}{2}(a_2 + a_1)(a_2 - a_1)}{(a_2 - a_1)} = \frac{1}{2}(a_2 + a_1) \qquad (2.9\text{-}27)$$

The value of $\langle dy/dt \rangle$ is, therefore, from equation (2.9-27)

$$\left\langle \frac{dy}{dt} \right\rangle = \sqrt{\frac{(a_2 + a_1)}{2}} \qquad (2.9\text{-}28)$$

Thus we finally get from equations (2.9-24) and (2.9-28) the approximate solution to the original problem [equation (2.9-17)]:

$$y = \frac{1}{\sqrt{(a_2 + a_1)/2}} \frac{t^2}{2!} = \frac{t^2}{\sqrt{2(a_2 + a_1)}} \qquad (2.9\text{-}29)$$

Equation (2.9-29) is applicable sectionally over the time interval between a_1 and a_2.

Assignments in Chapter 2

2.1. Verify equations (2.3-5), (2.3-18), (2.3-24), (2.3-36), (2.4-10), (2.4-13), and (2.5-34).

2.2. Given the governing equation

$$\frac{\partial^2 y}{\partial t^2} = a^2 \frac{\partial^2 y}{\partial x^2} \qquad x > 0 \qquad t > 0$$

with initial and boundary conditions

$$y(x, 0) = 0 \qquad \frac{\partial y}{\partial t}\bigg|_{t=0} = 0 \qquad y(0, t) = A_0 \sin wt$$

(a) Show that by a Laplace transform technique, $\mathscr{L}_{t \to s}$, the solution can be found to be

$$y(x, t) = \begin{cases} A_0 \sin w(t - x/a), & t > x/a \\ 0 & t < x/a \end{cases}$$

(b) Sketch the solution found in (a). What physical phenomena might part (a) describe?

2.3. Given the governing equation

$$\frac{\partial^2 y}{\partial t^2} + b^2 \frac{\partial^4 y}{\partial x^4}$$

$$y(x, 0) = 0 \qquad \frac{\partial y}{\partial t}\bigg|_{t=0} = 0 \qquad y(0, t) = h \qquad \frac{\partial^2 y}{\partial x^2}\bigg|_{x=0} = 0$$

(a) By taking the Laplace transform once, $\mathscr{L}_{t \to s}$, and then solving the subsequent equation, show that we can get

$$\bar{y}(x, s) = e^{+\sqrt{s/2b}x}(C_1 \cos \sqrt{s/2b}x + C_2 \sin \sqrt{s/2b}x)$$

$$+ e^{-\sqrt{s/b}x}(C_3 \cos \sqrt{s/2b}x + (C_4 \sin \sqrt{s/2b}x)$$

(b) By integration in the complex plane show that the final solution is

$$y(x, t) = h\left\{ 1 - \frac{2}{\pi} \int_0^\infty \frac{e^{-2bv^2 t} \sin vx \cosh vx}{v} \, dv \right\}$$

2.4. Given the governing equation

$$\frac{\partial C}{\partial t} = k \frac{\partial^2 C}{\partial x^2}$$

with

$$C(x, 0) = C_0 \qquad \frac{\partial C}{\partial x}\bigg|_{x=0} = -\alpha C(0, t)$$

(a) Describe a transport phenomenon which is determined by these equations.
(b) Solve by Laplace transform techniques. Find the inverse Laplace transform from the tables.
(c) Solve by Laplace transform techniques. Find the inverse by performing the integration in the complex plane.

2.5.

(a) Describe the mass transport, momentum transport, and heat transport governed by

$$\frac{\partial U}{\partial t} = k\left(\frac{\partial^2 U}{\partial r^2} + \frac{1}{r}\frac{\partial U}{\partial r}\right)$$

$$U(1, t) = 0 \qquad U(r, 0) = T$$

(b) Solve the equations in part (a) by Laplace transform, $\mathscr{L}_{t\to s}$. Find $\mathscr{L}_{s\to t}^{-1}\bar{U}(r, s) = U(r, t)$ by consulting the appropriate tables.
(c) Find the inverse Laplace transform by integration in the complex plane.

2.6. In Problem 2.4, suppose the boundary and initial conditions are

$$C(x, 0) = 0 \qquad C(a, t) = Q\delta(t)$$

where Q is a constant and $\delta(t)$ is the Dirac delta function.

(a) Describe a phenomenon for which these conditions apply.
(b) Solve by Laplace transform techniques.

2.7. Many times the governing equation that arises in the analysis of a transport problem is called Bessel's differential equation of order zero:

$$x\frac{d^2y}{dx^2} + \frac{dy}{dx} + xy = 0$$

(a) Show that by taking the Laplace transform, $\mathscr{L}_{x\to s}$, we can get

$$(s^2 + 1)\frac{d\bar{y}(s)}{ds} + s\bar{y}(s) = 0$$

(b) Show that we can get

$$\bar{y}(s) = \frac{C}{\sqrt{s^2 + 1}}$$

(c) Invert (b) by looking up the solution in the tables, to get

$$y(x) = CJ_0(x)$$

2.8. Given the governing equation

$$\frac{\partial u}{\partial t} = \alpha\left(\frac{\partial^2 u}{\partial r^2} + \frac{1}{r}\frac{\partial u}{\partial r}\right)$$

which may crop up as a description of unsteady-state diffusion or heat flow phenomena. The dependent variable $u(r, t)$ may be concentration, temperature, or velocity. There may be associated initial and boundary conditions,

$$u(r, 0) = 0$$

$$u(a, t) = \begin{cases} 0, & t < 0 \\ U_0, & t > 0 \end{cases}$$

$$u(0, t) \text{ is finite}$$

(a) Take the Laplace transform, $\mathscr{L}_{t \to s}$, to get

$$\frac{d^2\bar{u}(r, s)}{dr^2} + \frac{1}{r}\frac{d\bar{u}(r, s)}{dr} - \frac{s}{\alpha}\bar{u}(r, s) = 0$$

(b) Show that the solution of part (a) with the initial and boundary condition yields

$$\bar{u}(r, s) = \frac{U_0 J_0(\lambda r)}{s J_0(\lambda a)}$$

where $\lambda = i\sqrt{s/\alpha}$

(c) Invert part (b) by looking up the inversion form in the tables and also by integration in the complex plane to get

$$u(r, t) = U_0\left[1 - \frac{2}{a}\sum_{n=1}^{\infty}\frac{J_0(\lambda_n r)}{\lambda_n J_1(\lambda_n a)}e^{-\alpha\lambda_n^2 t}\right]$$

2.9. One model for blood oxygenation in the red cell is represented as[15]

$$\frac{\partial C_{HbO_2}}{\partial t} = k'C_{Hb}C_{O_2} - kC_{HbO_2}$$

[15] *Blood Oxygenation*, D. Hershey, ed., Plenum Press, New York, 1970, pp. 72–106.

where

$C_{\mathrm{HbO_2}}$ = concentration of oxyhemoglobin in the red cell

C_{Hb} = concentration of hemoglobin in the red cell

$C_{\mathrm{O_2}}$ = concentration of oxygen in the red cell

k', k = rate constants for the hemoglobin-oxygen chemical reaction,

$$\mathrm{Hb} + \mathrm{O_2} \underset{k}{\overset{k'}{\rightleftharpoons}} \mathrm{HbO_2}$$

$C_{\mathrm{Hb_0}} = C_{\mathrm{Hb}} + C_{\mathrm{HbO_2}}$ = initial hemoglobin concentration.

Solve this equation by an approximate Laplace transform technique. For example, $\mathscr{L}_{t \to s} C_{\mathrm{Hb}} C_{\mathrm{O_2}} = \bar{C}_{\mathrm{O_2}} \mathscr{L}_{t \to s} C_{\mathrm{Hb}}$, where $\bar{C}_{\mathrm{O_2}}$ is a mean value. This is a linearization of the original equation.

For Further Reading

Laplace Transforms, by M. R. Spiegel, Schaum Pub. Co., New York, 1965.

Table of Laplace Transforms, by G. E. Roberts and H. Kaufman, W. B. Saunders Co., Philadelphia, 1966.

Introduction to Complex Variables and Applications, by R. V. Churchill, McGraw-Hill, New York, 1948.

Mathematics of Engineering Systems, by F. H. Raven, McGraw-Hill, New York, 1966.

Integral Transforms in Mathematical Physics, by C. J. Tranter, Methuen & Co., New York, 1951.

Blood Oxygenation, Daniel Hershey, ed., Plenum Press, New York, 1970.

Advanced Calculus, by Angus E. Taylor, Ginn and Company, New York, 1955.

Part II

TRANSPORT ANALYSIS
IN CONTINUOUS PROCESSES

Chapter 3

Derivation of the Momentum Transport Equations

In the analysis of fluid flow problems, the equation of continuity (material balance) is usually the initial principle invoked. The information gleaned from the equation of continuity is then applied to the equation of motion. From the equation of motion and the fluid properties, the velocity profiles, average velocities, and energy loss can be calculated.

This chapter will begin with derivations of the equation of continuity and the equation of motion. The results are generalized to vector and tensor notation and then tabulated for rectangular, cylindrical, and spherical coordinates. Non-Newtonian flow concepts are introduced, and the applicability of the equations of motion to non-Newtonian fluids is shown. Creeping flow is briefly discussed and the chapter concludes with a short discussion of energy dissipation in fluid flow.

3.1. The Equation of Continuity (Material Balance)

In making a material balance, we invoke equation (1.1-9) for a mass balance, yielding

$$\text{summation of mass} \,|_{\text{in}} - \text{summation of mass} \,|_{\text{out}} + \left\{ \begin{array}{c} \text{generation} \\ \text{or depletion} \\ \text{by chemical reaction} \end{array} \right\}$$

$$= \text{rate of accumulation} \tag{3.1-1}$$

Figure 11 sketches the differential element analyzed and the mass fluxes operating. The mass fluxes are expressed as ϱv_x, ϱv_y, and ϱv_z, where ϱ is the density and v_x, v_y, and v_z are the fluid velocities in the direction indicated by the subscript. The units of a flux could typically be $\text{lb/ft}^2 \cdot \text{sec}$ using the system of pound force (lbf), pound mass (lbm), feet, and seconds. Invoking

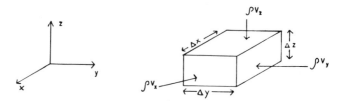

Fig. 11. Mass fluxes through a differential element, $\Delta x \Delta y \Delta z$.

equation (3.1-1) (for no generation or depletion of mass) we get the following equation for the differential element shown in Figure 11:

$$[(\varrho v_x)(\Delta y \Delta z)]\big|_x - [(\varrho v_x)(\Delta y \Delta z)]\big|_{x+\Delta x}$$
$$+ [(\varrho v_y)(\Delta x \Delta z)]\big|_y - [(\varrho v_y)(\Delta x \Delta z)]\big|_{y+\Delta y}$$
$$+ [(\varrho v_z)(\Delta x \Delta y)]\big|_z - [(\varrho v_z)(\Delta x \Delta y)]\big|_{z+\Delta z}$$
$$= \frac{\partial}{\partial t}(\varrho \Delta x \Delta y \Delta z) \tag{3.1-2}$$

The symbol $[\ ]\big|_x$ reads "evaluated at x." We also use the mathematical concept "in at x" and "out at $x + \Delta x$." By dividing equation (3.1-2) by $\Delta x \Delta y \Delta z$ and using the calculus limiting process [equation (1.2-2)], the result is the equation of continuity,

$$\frac{\partial \varrho}{\partial t} = -\left(\frac{\partial}{\partial x}\varrho v_x + \frac{\partial}{\partial y}\varrho v_y + \frac{\partial}{\partial z}\varrho v_z\right) \tag{3.1-3}$$

TABLE I

The Equation of Continuity [Equation (3.1-4)]

Rectangular coordinates (x, y, z)

$$\frac{\partial \varrho}{\partial t} = -\left[\frac{\partial}{\partial x}(\varrho v_x) + \frac{\partial}{\partial y}(\varrho v_y) + \frac{\partial}{\partial z}(\varrho v_z)\right] \tag{A}$$

Cylindrical coordinates (r, θ, z)

$$\frac{\partial \varrho}{\partial t} = -\left[\frac{1}{r}\frac{\partial}{\partial r}(\varrho r v_r) + \frac{1}{r}\frac{\partial}{\partial \theta}(\varrho v_\theta) + \frac{\partial}{\partial z}(\varrho v_z)\right] \tag{B}$$

Spherical coordinates (r, θ, φ)

$$\frac{\partial \varrho}{\partial t} = -\left[\frac{1}{r^2}\frac{\partial}{\partial r}(\varrho r^2 v_r) + \frac{1}{r \sin \theta}\frac{\partial}{\partial \theta}(\varrho v_\theta \sin \theta) + \frac{1}{r \sin \theta}\frac{\partial}{\partial \varphi}(\varrho v_\varphi)\right] \tag{C}$$

which is represented in vector notation as

$$\frac{\partial \varrho}{\partial t} = -(\nabla \cdot \varrho \bar{v}) \tag{3.1-4}$$

Table I shows the equation of continuity, equation (3.1-4), in rectangular, cylindrical, and spherical coordinates. For an incompressible fluid, where ϱ is constant, equation (3.1-4) becomes

$$\nabla \cdot \bar{v} = 0 \tag{3.1-5}$$

3.2. The Equations of Motion (Rate of Momentum Balance)

In formulating an equation of motion, it is useful to recall the basic physics statement "impulse is equivalent to momentum." Impulse is given as a force multiplied by the contact time, so that the impulse–momentum statement can be written as

$$(\text{force}) \cdot (\text{time of contact}) \eqsim mv \tag{3.2-1}$$

where the symbol $\eqsim$ reads "is equivalent to" and mv (momentum) is mass multiplied by velocity. In the units of pounds force (lbf), pounds mass (lbm), feet (ft), and seconds (sec), equation (3.2-1) reads

$$F(\text{lbf}) \cdot t(\text{sec}) = \frac{m(\text{lbm}) \cdot v(\text{ft/sec})}{g_c} \tag{3.2-1a}$$

where $g_c = 32.17$ (lbm $\cdot$ ft/sec)/lbf. Note that in equation (3.2-1a) we have introduced g_c in order to balance the units used. Momentum in these units is written as mv/g_c. Thus, in making a rate of momentum balance, the symbolic operation is $d/dt \, (mv/g_c)$, which reads "the rate of change (with respect to time) of momentum." From equation (3.2-1a) the rate of change of momentum yields units of (lbf $\cdot$ sec)/sec, or lbf. We conclude therefore that a rate of momentum balance is actually a force balance.

A rate of momentum balance, corresponding to equation (1.1-9), is given by

rate of momentum $|_{\text{in}}$ $-$ rate of momentum $|_{\text{out}}$ $+$ body forces

$$= \text{rate of momentum accumulation} \tag{3.2-2}$$

where the term "body forces" usually implies the contribution due to

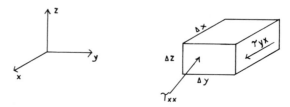

Fig. 12. Some shear stresses acting on a differential
element.

gravitational effects. Sometimes these body forces may incorporate electrical or electromagnetic effects. Figure 12 sketches the differential element
on which the momentum balance is applied. The terms, τ_{xx}, τ_{xy}, etc., are
the shear stress components of the shear stress tensor

$$\bar{\bar{\tau}} = \begin{pmatrix} \tau_{11} & \tau_{12} & \tau_{13} \\ \tau_{21} & \tau_{22} & \tau_{23} \\ \tau_{31} & \tau_{32} & \tau_{33} \end{pmatrix} \tag{3.2-3}$$

where in rectangular coordinates 1 corresponds to x, 2 corresponds to y,
and 3 corresponds to z. For cylindrical coordinates (r, θ, z) and spherical
coordinates (r, θ, φ) the same notation procedure is followed.

The shear stresses arise as internal drag forces associated with fluids
that have a resistance to flow (viscosity). Thus as sketched in Figure 13 the
shear stress component is envisioned as being aligned between hypothetical
layers of fluid, causing a "rubbing" of one layer with respect to the adjacent
layers. The more rapidly moving layer imparts some of its momentum to
the slower layer, i.e., momentum flows into the more slowly flowing layer.
The subscripts for each shear stress component are designed to place the τ
arrow in the proper plane. Thus τ_{yx} is the shear stress component in a plane
which is perpendicular to the y-axis and in the x-direction (the $\Delta x \Delta z$ plane).
By definition τ_{xx} designates a different type of stress component. The

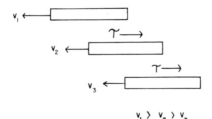

Fig. 13. Laminar model of fluid flow.

stress is in the x-direction but is perpendicular to the $\Delta y \Delta z$ plane. For most fluids $\tau_{xx} = \tau_{yy} = \tau_{zz} = 0$ but viscoelastic fluids or fluids flowing in unusual geometric configurations may exert such normal shear stresses. We shall see later in this chapter how to compute the shear stress components.

We now return to equation (3.2-2) and Figure 12; it is likely that there are a number of factors besides the shear stress which contribute momentum transport in the differential element. There is a volumetric flow rate through the $\Delta y \Delta z$ plane in the x-direction, $(v_x \Delta y \Delta z)|_x$, where the symbol $|_x$ again indicates "located at x." A similar expression is written at location $x + \Delta x$. Associated with this v_x flow is an intensive property, momentum per unit volume, $\varrho v_x / g_c$. The product of these two terms, (volumetric flow/time) (momentum/volumetric flow), is (momentum/time), or rate of momentum transport. The rate of convective momentum entering the volume element (at x) diminished by the rate of momentum leaving (at $x + \Delta x$) may be expressed as

net rate of momentum entering by convection (v_x) in the x-direction

$$= \left[(v_x \Delta y \Delta z)\left(\frac{\varrho v_x}{g_c} \right) \right]\Big|_x - \left[(v_x \Delta y \Delta z)\left(\frac{\varrho v_x}{g_c} \right) \right]\Big|_{x+\Delta x} \qquad (3.2\text{-}4)$$

For v_y and v_z the product of volumetric flow rate and the associated momentum per unit volume for the x-direction is given by

net rate of momentum entering by convection (v_y) in the x-direction

$$= \left[(v_y \Delta x \Delta z)\left(\frac{\varrho v_x}{g_c} \right) \right]\Big|_y - \left[(v_y \Delta x \Delta z)\left(\frac{\varrho v_x}{g_c} \right) \right]\Big|_{y+\Delta y} \qquad (3.2\text{-}5)$$

and

net rate of momentum entering by convection (v_z) in the x-direction

$$= \left[(v_z \Delta x \Delta y)\left(\frac{\varrho v_x}{g_c} \right) \right]\Big|_z - \left[(v_z \Delta x \Delta y)\left(\frac{\varrho v_x}{g_c} \right) \right]\Big|_{z+\Delta z} \qquad (3.2\text{-}6)$$

Another contribution to x-momentum is the pressure term,

net rate of momentum entering by a pressure gradient in the x-direction

$$= (P \Delta y \Delta z)|_x - (P \Delta y \Delta z)|_{x+\Delta x} \qquad (3.2\text{-}7)$$

Finally, the body force (gravitational) contribution is given by

rate of momentum (body force) entering by gravity in the x-direction

$$= \varrho \Delta x \Delta y \Delta z \, \frac{g_x}{g_c} \tag{3.2-8}$$

Note that in equations (3.2-4) through (3.2-8) the equivalence of "rate of momentum" and "force" is expressed. All of these x-momentum contributions are now added to the contribution by the viscous shear stress effects. The following equation sums up these viscous contributions:

net rate of momentum entering the differential element by viscous effects

in the x direction $= (\tau_{yx} \Delta x \Delta z)\big|_y - (\tau_{yx} \Delta x \Delta z)\big|_{y+\Delta y} + (\tau_{zx} \Delta x \Delta y)\big|_z$

$$- (\tau_{zx} \Delta x \Delta y)\big|_{z+\Delta z} + (\tau_{xx} \Delta y \Delta z)\big|_x - (\tau_{xx} \Delta y \Delta z)\big|_{x+\Delta x} \tag{3.2-9}$$

In equation (3.2-9) the shear stress (in lbf/ft^2) is multiplied by the area (in ft^2) over which it acts, yielding units of lbf (or rate of momentum).

The rate of accumulation of x-direction momentum within the volume element is all that remains to be derived:

rate of accumulation of x-direction momentum

$$= \frac{\partial}{\partial t} \left[(\Delta x \Delta y \Delta z) \left(\frac{\varrho v_x}{g_c} \right) \right] \tag{3.2-10}$$

where $\varrho v_x / g_c$ is momentum per unit volume and $\Delta x \Delta y \Delta z$ is the volume of the element. If now equations (3.2-4) through (3.2-10) are summed as indicated by equation (3.2-2), the result is the equation of motion for the x-component of momentum transport. Upon dividing this equation by $\Delta x \Delta y \Delta z$ and taking the limit as Δx, Δy, and Δz approach zero, we obtain

$$\frac{\partial}{\partial t} \left(\frac{\varrho v_x}{g_c} \right) = - \left(\frac{\partial}{\partial x} \frac{\varrho v_x v_x}{g_c} + \frac{\partial}{\partial y} \frac{\varrho v_y v_x}{g_c} + \frac{\partial}{\partial z} \frac{\varrho v_z v_x}{g_c} \right)$$

$$- \left(\frac{\partial}{\partial x} \tau_{xx} + \frac{\partial}{\partial y} \tau_{yx} + \frac{\partial}{\partial z} \tau_{zx} \right)$$

$$- \frac{\partial P}{\partial x} + \frac{\varrho g_x}{g_c} \tag{3.2-11}$$

In a similar manner the y and z components of the equation of motion can be derived. These equations can be combined and generalized by utilization of tensor and vector notation. The combined result is given by

$$\frac{\partial}{\partial t}(\varrho \bar{v}) = -[\nabla \cdot \varrho \bar{v}\bar{v}] - \overline{\nabla P} - \nabla \cdot \bar{\bar{\tau}} + \varrho \bar{g} \quad (3.2\text{-}12)$$

| accumulation contribution (unsteady state) | convective transport | pressure contribution | viscous transport | body force (gravitational) |

In equation (3.2-12), the g_c terms have been omitted. It should always be kept in mind that the units of each term should be checked for consistency. Each term in equation (3.2-12) has the units of rate of momentum transport (force) per unit volume. Equation (3.2-12) is applicable to all types of fluids. Tables II, III, and IV display equation (3.2-12) in rectangular, cylindrical, and spherical coordinates, with the assumption that the density, ϱ, is constant, i.e., that the fluid is incompressible and $\nabla \cdot \bar{v} = 0$.

With equation (3.2-12) and Tables II, III, and IV we have equations of motion for viscous fluids with components of the shear stress tensor,

TABLE II

The Equation of Motion in Rectangular Coordinates [Equation (3.2-12)]

(Incompressible fluid, ϱ = constant, $\nabla \cdot \bar{v} = 0$)

x-component

$$\varrho\left(\frac{\partial v_x}{\partial t} + v_x\frac{\partial v_x}{\partial x} + v_y\frac{\partial v_x}{\partial y} + v_z\frac{\partial v_x}{\partial z}\right)$$

$$= -\frac{\partial P}{\partial x} - \left(\frac{\partial \tau_{xx}}{\partial x} + \frac{\partial \tau_{yx}}{\partial y} + \frac{\partial \tau_{zx}}{\partial z}\right) + \varrho g_x \quad (A)$$

y-component

$$\varrho\left(\frac{\partial v_y}{\partial t} + v_x\frac{\partial v_y}{\partial x} + v_y\frac{\partial v_y}{\partial y} + v_z\frac{\partial v_y}{\partial z}\right)$$

$$= -\frac{\partial P}{\partial y} - \left(\frac{\partial \tau_{xy}}{\partial x} + \frac{\partial \tau_{yy}}{\partial y} + \frac{\partial \tau_{zy}}{\partial z}\right) + \varrho g_y \quad (B)$$

z-component

$$\varrho\left(\frac{\partial v_z}{\partial t} + v_x\frac{\partial v_z}{\partial x} + v_y\frac{\partial v_z}{\partial y} + v_z\frac{\partial v_z}{\partial z}\right)$$

$$= -\frac{\partial P}{\partial z} - \left(\frac{\partial \tau_{xz}}{\partial x} + \frac{\partial \tau_{yz}}{\partial y} + \frac{\partial \tau_{zz}}{\partial z}\right) + \varrho g_z \quad (C)$$

TABLE III

The Equation of Motion in Cylindrical Coordinates [Equation (3.2-12)]

(Incompressible fluid, ϱ = constant, $\nabla \cdot \bar{v} = 0$)

r-component

$$\varrho\left(\frac{\partial v_r}{\partial t} + v_r \frac{\partial v_r}{\partial r} + \frac{v_\theta}{r} \frac{\partial v_r}{\partial \theta} - \frac{v_\theta^2}{r} + v_z \frac{\partial v_r}{\partial z}\right)$$

$$= -\frac{\partial P}{\partial r} - \left(\frac{1}{r} \frac{\partial}{\partial r}(r\tau_{rr}) + \frac{1}{r} \frac{\partial \tau_{r\theta}}{\partial \theta} - \frac{\tau_{\theta\theta}}{r} + \frac{\partial \tau_{rz}}{\partial z}\right) + \varrho g_r \quad \text{(A)}$$

θ-component

$$\varrho\left(\frac{\partial v_\theta}{\partial t} + v_r \frac{\partial v_\theta}{\partial r} + \frac{v_\theta}{r} \frac{\partial v_\theta}{\partial \theta} + \frac{v_r v_\theta}{r} + v_z \frac{\partial v_\theta}{\partial z}\right)$$

$$= -\frac{1}{r} \frac{\partial P}{\partial \theta} - \left(\frac{1}{r^2} \frac{\partial}{\partial r}(r^2\tau_{r\theta}) + \frac{1}{r} \frac{\partial \tau_{\theta\theta}}{\partial \theta} + \frac{\partial \tau_{\theta z}}{\partial z}\right) + \varrho g_\theta \quad \text{(B)}$$

z-component

$$\varrho\left(\frac{\partial v_z}{\partial t} + v_r \frac{\partial v_z}{\partial r} + \frac{v_\theta}{r} \frac{\partial v_z}{\partial \theta} + v_z \frac{\partial v_z}{\partial z}\right)$$

$$= -\frac{\partial P}{\partial z} - \left(\frac{1}{r} \frac{\partial}{\partial r}(r\tau_{rz}) + \frac{1}{r} \frac{\partial \tau_{\theta z}}{\partial \theta} + \frac{\partial \tau_{zz}}{\partial z}\right) + \varrho g_z \quad \text{(C)}$$

τ_{ij}. It will be helpful at this point to illustrate how to characterize the viscous flow properties of a fluid. Referring to Figure 14, which depicts a fluid flowing in the x-direction only, the shear stress τ_{yx} is related to the velocity gradient (shear rate). The velocity gradient is written for this case as $\partial v_x/\partial y$. Certain fluids, called Newtonian fluids, conform to a rheological equation of the form

$$\tau_{yx} = -\mu \frac{\partial v_x}{\partial y} \quad (3.2\text{-}13)$$

where μ is called viscosity; for a Newtonian fluid the viscosity is a constant, independent of the applied τ_{yx} and $\partial v_x/\partial y$. The units of μ or μ/g_c are adjusted to be in conformity with those of the shear stress τ_{yx} and the shear rate $\partial v_x/\partial y$. In the lbf, lbm, ft, sec units, τ_{yx} has units of lbf/ft² and $\partial v_x/\partial y$ has units of sec⁻¹.

TABLE IV

The Equation of Motion in Spherical Coordinates [Equation (3.2-12)]

(Incompressible fluid, $\varrho = $ constant, $\nabla \cdot \bar{v} = 0$)

---

r-component

$$\varrho\left(\frac{\partial v_r}{\partial t} + v_r\frac{\partial v_r}{\partial r} + \frac{v_\theta}{r}\frac{\partial v_r}{\partial \theta} + \frac{v_\varphi}{r\sin\theta}\frac{\partial v_r}{\partial \varphi} - \frac{v_\theta^2 + v_\varphi^2}{r}\right)$$

$$= -\frac{\partial P}{\partial r} - \left(\frac{1}{r^2}\frac{\partial}{\partial r}(r^2\tau_{rr}) + \frac{1}{r\sin\theta}\frac{\partial}{\partial \theta}(\tau_{r\theta}\sin\theta)\right.$$

$$+ \frac{1}{r\sin\theta}\frac{\partial \tau_{r\varphi}}{\partial \varphi} - \left.\frac{(\tau_{\theta\theta} + \tau_{\varphi\varphi})}{r}\right) + g_r \qquad (A)$$

θ-component

$$\varrho\left(\frac{\partial v_\theta}{\partial t} + v_r\frac{\partial v_\theta}{\partial r} + \frac{v_\theta}{r}\frac{\partial v_\theta}{\partial \theta} + \frac{v_\varphi}{r\sin\theta}\frac{\partial v_\theta}{\partial \varphi} + \frac{v_r v_\theta}{r} - \frac{v_\varphi^2 \cot\theta}{r}\right)$$

$$= -\frac{1}{r}\frac{\partial P}{\partial \theta} - \left(\frac{1}{r^2}\frac{\partial}{\partial r}(r^2\tau_{r\theta}) + \frac{1}{r\sin\theta}\frac{\partial}{\partial \theta}(\tau_{\theta\theta}\sin\theta)\right.$$

$$+ \frac{1}{r\sin\theta}\frac{\partial \tau_{\theta\varphi}}{\partial \varphi} + \frac{\tau_{r\theta}}{r} - \left.\frac{\cot\theta}{r}\tau_{\varphi\varphi}\right) + \varrho g_\theta \qquad (B)$$

φ-component

$$\varrho\left(\frac{\partial v_\varphi}{\partial t} + v_r\frac{\partial v_\varphi}{\partial r} + \frac{v_\theta}{r}\frac{\partial v_\varphi}{\partial \theta} + \frac{v_\varphi}{r\sin\theta}\frac{\partial v_\varphi}{\partial \varphi} + \frac{v_\varphi v_\theta}{r} + \frac{v_\theta v_\varphi}{r}\cot\theta\right)$$

$$= -\frac{1}{r\sin\theta}\frac{\partial P}{\partial \varphi} - \left(\frac{1}{r^2}\frac{\partial}{\partial r}(r^2\tau_{r\varphi}) + \frac{1}{r}\frac{\partial \tau_{\theta\varphi}}{\partial \theta} + \frac{1}{r\sin\theta}\frac{\partial \tau_{\varphi\varphi}}{\partial \varphi}\right.$$

$$+ \frac{\tau_{r\varphi}}{r} + \left.\frac{2\cot\theta}{r}\tau_{\theta\varphi}\right) + \varrho g_\varphi \qquad (C)$$

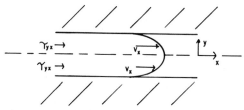

Fig. 14. Two-dimensional fluid flow between flat
plates.

TABLE V

Shear Stress–Shear Rate Components in Rectangular Coordinates—Newtonian Fluid

$$\tau_{xx} = -\mu \left[2 \frac{\partial v_x}{\partial x} - \frac{2}{3} \nabla \cdot \bar{v} \right] \tag{A}$$

$$\tau_{yy} = -\mu \left[2 \frac{\partial v_y}{\partial y} - \frac{2}{3} \nabla \cdot \bar{v} \right] \tag{B}$$

$$\tau_{zz} = -\mu \left[2 \frac{\partial v_z}{\partial z} - \frac{2}{3} \nabla \cdot \bar{v} \right] \tag{C}$$

$$\tau_{yz} = \tau_{zy} = -\mu \left[\frac{\partial v_y}{\partial z} + \frac{\partial v_z}{\partial y} \right] \tag{D}$$

$$\tau_{xy} = \tau_{yx} = -\mu \left[\frac{\partial v_x}{\partial y} + \frac{\partial v_y}{\partial x} \right] \tag{E}$$

$$\tau_{zx} = \tau_{xz} = -\mu \left[\frac{\partial v_z}{\partial x} + \frac{\partial v_x}{\partial z} \right] \tag{F}$$

TABLE VI

Shear Stress–Shear Rate Components in Cylindrical Coordinates—Newtonian Fluid

$$\tau_{rr} = -\mu \left[2 \frac{\partial v_r}{\partial r} - \frac{2}{3} \nabla \cdot \bar{v} \right] \tag{A}$$

$$\tau_{\theta\theta} = -\mu \left[2 \left(\frac{1}{r} \frac{\partial v_\theta}{\partial \theta} + \frac{v_r}{r} \right) - \frac{2}{3} \nabla \cdot \bar{v} \right] \tag{B}$$

$$\tau_{zz} = -\mu \left[2 \frac{\partial v_z}{\partial z} - \frac{2}{3} \nabla \cdot \bar{v} \right] \tag{C}$$

$$\tau_{r\theta} = \tau_{\theta r} = -\mu \left[r \frac{\partial}{\partial r} \left(\frac{v_\theta}{r} \right) + \frac{1}{r} \frac{\partial v_r}{\partial \theta} \right] \tag{D}$$

$$\tau_{\theta z} = \tau_{z\theta} = -\mu \left[\frac{\partial v_\theta}{\partial z} + \frac{1}{r} \frac{\partial v_z}{\partial \theta} \right] \tag{E}$$

$$\tau_{zr} = \tau_{rz} = -\mu \left[\frac{\partial v_z}{\partial r} + \frac{\partial v_r}{\partial z} \right] \tag{F}$$

TABLE VII

Shear Stress–Shear Rate Components in Spherical Coordinates—Newtonian Fluid

$$\tau_{rr} = -\mu\left[2\frac{\partial v_r}{\partial r} - \frac{2}{3}\nabla \cdot \bar{v}\right] \tag{A}$$

$$\tau_{\theta\theta} = -\mu\left[2\left(\frac{1}{r}\frac{\partial v_\theta}{\partial \theta} + \frac{v_r}{r}\right) - \frac{2}{3}\nabla \cdot \bar{v}\right] \tag{B}$$

$$\tau_{\varphi\varphi} = -\mu\left[2\left(\frac{1}{r\sin\theta}\frac{\partial v_\varphi}{\partial \varphi} + \frac{v_r}{r} + \frac{v_\theta \cot\theta}{r}\right) - \frac{2}{3}\nabla \cdot \bar{v}\right] \tag{C}$$

$$\tau_{r\theta} = \tau_{\theta r} = -\mu\left[r\frac{\partial}{\partial r}\left(\frac{v_\theta}{r}\right) + \frac{1}{r}\frac{\partial v_r}{\partial \theta}\right] \tag{D}$$

$$\tau_{\theta\varphi} = \tau_{\varphi\theta} = -\mu\left[\frac{\sin\theta}{r}\frac{\partial}{\partial \theta}\left(\frac{v_\varphi}{\sin\theta}\right) + \frac{1}{r\sin\theta}\frac{\partial v_\theta}{\partial \varphi}\right] \tag{E}$$

$$\tau_{\varphi r} = \tau_{r\varphi} = -\mu\left[\frac{1}{r\sin\theta}\frac{\partial v_r}{\partial \varphi} + r\frac{\partial}{\partial r}\left(\frac{v_\varphi}{r}\right)\right] \tag{F}$$

TABLE VIII

Equation of Motion for Newtonian Fluids—Rectangular Coordinates
(Navier–Stokes Equations)

x-component

$$\varrho\left(\frac{\partial v_x}{\partial t} + v_x\frac{\partial v_x}{\partial x} + v_y\frac{\partial v_x}{\partial y} + v_z\frac{\partial v_x}{\partial z}\right)$$

$$= -\frac{\partial P}{\partial x} + \mu\left(\frac{\partial^2 v_x}{\partial x^2} + \frac{\partial^2 v_x}{\partial y^2} + \frac{\partial^2 v_x}{\partial z^2}\right) + \varrho g_x \tag{A}$$

y-component

$$\varrho\left(\frac{\partial v_y}{\partial t} + v_x\frac{\partial v_y}{\partial x} + v_y\frac{\partial v_y}{\partial y} + v_z\frac{\partial v_y}{\partial z}\right)$$

$$= -\frac{\partial P}{\partial y} + \mu\left(\frac{\partial^2 v_y}{\partial x^2} + \frac{\partial^2 v_y}{\partial y^2} + \frac{\partial^2 v_y}{\partial z^2}\right) + \varrho g_y \tag{B}$$

z-component

$$\varrho\left(\frac{\partial v_z}{\partial t} + v_x\frac{\partial v_z}{\partial x} + v_y\frac{\partial v_z}{\partial y} + v_z\frac{\partial v_z}{\partial z}\right)$$

$$= -\frac{\partial P}{\partial z} + \mu\left(\frac{\partial^2 v_z}{\partial x^2} + \frac{\partial^2 v_z}{\partial y^2} + \frac{\partial^2 v_z}{\partial z^2}\right) + \varrho g_z \tag{C}$$

In considering more general Newtonian fluid cases than that posed in Figure 14, all the components of shear stress and shear rate for Newtonian fluids must be used. In the more general situation the Newtonian shear stress and shear rate relationships are given in Tables V, VI, and VII, where the terms in brackets on the right-hand side of the equations are the shear rate components.

When the Newtonian relationships of Tables V, VI, and VII are substituted into the equations of motion in Tables II, III, and IV, the results are flow equations applicable to Newtonian fluids (the Navier–Stokes equations). These Navier–Stokes equations are summarized in Tables VIII, IX, and X.

TABLE IX

**Equation of Motion for Newtonian Fluids—Cylindrical Coordinates
(Navier–Stokes Equations)**

---

r-component

$$\varrho\left(\frac{\partial v_r}{\partial t} + v_r\frac{\partial v_r}{\partial r} + \frac{v_\theta}{r}\frac{\partial v_r}{\partial \theta} - \frac{v_\theta^2}{r} + v_z\frac{\partial v_r}{\partial z}\right)$$

$$= -\frac{\partial P}{\partial r} + \mu\left[\frac{\partial}{\partial r}\left(\frac{1}{r}\frac{\partial}{\partial r}(rv_r)\right) + \frac{1}{r^2}\frac{\partial^2 v_r}{\partial \theta^2} - \frac{2}{r^2}\frac{\partial v_\theta}{\partial \theta} + \frac{\partial^2 v_r}{\partial z^2}\right] + \varrho g_r \qquad \text{(A)}$$

θ-component

$$\varrho\left(\frac{\partial v_\theta}{\partial t} + v_r\frac{\partial v_\theta}{\partial r} + \frac{v_\theta}{r}\frac{\partial v_\theta}{\partial \theta} + \frac{v_r v_\theta}{r} + v_z\frac{\partial v_\theta}{\partial z}\right)$$

$$= -\frac{1}{r}\frac{\partial P}{\partial \theta} + \mu\left[\frac{\partial}{\partial r}\left(\frac{1}{r}\frac{\partial}{\partial r}(rv_\theta)\right) + \frac{1}{r^2}\frac{\partial^2 v_\theta}{\partial \theta^2} + \frac{2}{r^2}\frac{\partial v_r}{\partial \theta} + \frac{\partial^2 v_\theta}{\partial z^2}\right] + \varrho g_\theta \qquad \text{(B)}$$

z-component

$$\varrho\left(\frac{\partial v_z}{\partial t} + v_r\frac{\partial v_z}{\partial r} + \frac{v_\theta}{r}\frac{v_z}{\partial \theta} + v_z\frac{\partial v_z}{\partial z}\right)$$

$$= -\frac{\partial P}{\partial z} + \mu\left[\frac{1}{r}\frac{\partial}{\partial r}\left(r\frac{\partial v_z}{\partial r}\right) + \frac{1}{r^2}\frac{\partial^2 v_z}{\partial \theta^2} + \frac{\partial^2 v_z}{\partial z^2}\right] + \varrho g_z \qquad \text{(C)}$$

TABLE X

Equation of Motion for Newtonian Fluids—Spherical Coordinates
(Navier–Stokes Equations)

---

r-component

$$\varrho\left(\frac{\partial v_r}{\partial t} + v_r\frac{\partial v_r}{\partial r} + \frac{v_\theta}{r}\frac{\partial v_r}{\partial \theta} + \frac{v_\varphi}{r\sin\theta}\frac{\partial v_r}{\partial \varphi} - \frac{v_\theta{}^2 + v_\varphi{}^2}{r}\right)$$

$$= -\frac{\partial P}{\partial r} + \mu\left[\nabla^2 v_r - \frac{2}{r^2}v_r - \frac{2}{r^2}\frac{\partial v_\theta}{\partial \theta} - \frac{2}{r^2}v_\theta\cot\theta - \frac{2}{r^2\sin\theta}\frac{\partial v_\varphi}{\partial \varphi}\right]$$

$$+ \varrho g_r \tag{A}$$

θ-component

$$\varrho\left(\frac{\partial v_\theta}{\partial t} + v_r\frac{\partial v_\theta}{\partial r} + \frac{v_\theta}{r}\frac{\partial v_\theta}{\partial \theta} + \frac{v_\varphi}{r\sin\theta}\frac{\partial v_\theta}{\partial \varphi} + \frac{v_r v_\theta}{r} - \frac{v_\varphi{}^2\cot\theta}{r}\right)$$

$$= -\frac{1}{r}\frac{\partial P}{\partial \theta} + \mu\left[\nabla^2 v_\theta + \frac{2}{r^2}\frac{\partial v_r}{\partial \theta} - \frac{v_\theta}{r^2\sin^2\theta} - \frac{2\cos\theta}{r^2\sin^2\theta}\frac{\partial v_\varphi}{\partial \varphi}\right] + \varrho g_\theta \tag{B}$$

φ-component

$$\varrho\left(\frac{\partial v_\varphi}{\partial t} + v_r\frac{\partial v_\varphi}{\partial r} + \frac{v_\theta}{r}\frac{\partial v_\varphi}{\partial \theta} + \frac{v_\varphi}{r\sin\theta}\frac{\partial v_\varphi}{\partial \varphi} + \frac{v_\varphi v_r}{r} + \frac{v_\theta v_\varphi}{r}\cot\theta\right)$$

$$= -\frac{1}{r\sin\theta}\frac{\partial P}{\partial \varphi} + \mu\left[\nabla^2 v_\varphi - \frac{v_\varphi}{r^2\sin^2\theta} + \frac{2}{r^2\sin\theta}\frac{\partial v_r}{\partial \varphi} + \frac{2\cos\theta}{r^2\sin^2\theta}\frac{\partial v_\theta}{\partial \varphi}\right]$$

$$+ \varrho g_\varphi \tag{C}$$

3.3. Non-Newtonian Fluid Behavior Applied to the Equations of Motion

Only when the shear stress–shear rate relationship is inserted into the equations of motion have we become specific about a particular fluid flow phenomenon. Thus equation (3.2-12) and Tables II, III, and IV apply to all fluids, while only Newtonian fluids conform to the equations in Tables VIII, IX, and X.

There are two non-Newtonian shear stress–shear rate rheological models that are used frequently. In the one-directional flow situation they may be expressed in rectangular coordinates as shown below in equations (3.3-1), (3.3-5), and (3.3-6).

The Power-Law Model (Ostwald–de Waele)

$$\tau_{yx} = -m \left| \frac{dv_x}{dy} \right|^{n-1} \frac{dv_x}{dy} \qquad (3.3\text{-}1)$$

where m and n are power-law parameters. We can remove the absolute value sign by the usual mathematical operations:

$$\left| \frac{dv_x}{dy} \right| = \frac{dv_x}{dy} \qquad \text{for } \frac{dv_x}{dy} > 0 \qquad (3.3\text{-}2)$$

$$\left| \frac{dv_x}{dy} \right| = -\frac{dv_x}{dy} \qquad \text{for } \frac{dv_x}{dy} < 0 \qquad (3.3\text{-}3)$$

The choice of either equation (3.3-2) or (3.3-3) depends upon the velocity profile. For example, for a fluid flowing as shown in Figure 14, it is observed that v_x decreases as y increases in the upper half of the conduit, yielding $dv_x/dy < 0$, requiring the use of equation (3.3-3) in (3.3-1). For this part of the conduit equation (3.3-1) becomes

$$\begin{aligned}
\tau_{yx} &= -m\left(-\frac{dv_x}{dy} \right)^{n-1}\left(\frac{dv_x}{dy} \right) \\
&= m\left(-\frac{dv_x}{dy} \right)^{n-1}\left(-\frac{dv_x}{dy} \right) \\
&= m\left(-\frac{dv_x}{dy} \right)^{n} \qquad (3.3\text{-}4)
\end{aligned}$$

The Bingham (Plastic) Model

Another common type of non-Newtonian fluid is the Bingham (plastic) model. For one-dimensional flow in rectangular coordinates the shear stress–shear rate relationship is given by

$$\tau_{yx} = -\mu_0 \frac{dv_x}{dy} + \tau_0 \qquad \text{if } |\tau_{yx}| > \tau_0 \qquad (3.3\text{-}5)$$

$$\frac{dv_x}{dy} = 0 \qquad \text{if } |\tau_{yx}| < \tau_0 \qquad (3.3\text{-}6)$$

The Bingham fluid which is initially at rest will not flow until a shear stress is imposed upon the system which exceeds τ_0, the yield stress. Once the fluid is in motion, there will be a variation of the shear stress across the flow channel. If we can establish that there exists a flow region in the

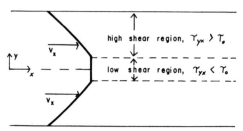

Fig. 15. Velocity profile for a Bingham plastic
flowing in a pipe.

channel where τ_{yx} is less than the yield stress τ_0, then equation (3.3-6) applies in that region. Equation (3.3-6) implies that the velocity does not change in this region and we have a "flat" velocity profile, a region of so-called "plug" flow. In the rest of the channel, where τ_{yx} is greater than τ_0, equation (3.3-5) is applicable. In the high-shear region, where τ_{yx} is greater than τ_0, the fluid behaves very much like a Newtonian fluid. Figure 15 shows a typical velocity profile for a Bingham fluid flowing in a pipe.

For unidirectional flow as discussed here a plot of τ_{yx} and dv_x/dy for the three models is shown in Figure 16.

3.4. Generalized Representation of Newtonian and Non-Newtonian Flow

Recognizing that flow in one direction is merely a special case of the more general three-dimensional case, empirical rheological equations corresponding to equations (3.2-13), (3.3-1), (3.3-5), and (3.3-6) have been developed. These are given below by equations (3.4-1), (3.4-2), and (3.4-3).

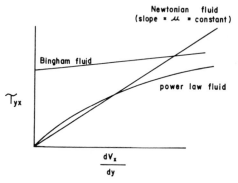

Fig. 16. Shear stress τ_{yx} vs. shear rate, dv_x/dy
for various fluid models.

Power-Law (Ostwald–de Waele) Model

$$\bar{\bar{\tau}} = -\{m \mid \sqrt{\tfrac{1}{2}\bar{\bar{\varDelta}} : \bar{\bar{\varDelta}}} \mid^{n-1}\}\bar{\bar{\varDelta}} \tag{3.4-1}$$

Bingham (Plastic) Model

$$\bar{\bar{\tau}} = -\left\{\mu_0 + \frac{\tau_0}{\sqrt{\tfrac{1}{2}\bar{\bar{\varDelta}} : \bar{\bar{\varDelta}}}}\right\}\bar{\bar{\varDelta}} \quad \text{for } \tfrac{1}{2}(\bar{\bar{\tau}} : \bar{\bar{\tau}}) > \tau_0^2 \tag{3.4-2}$$

and

$$\bar{\bar{\varDelta}} = 0 \quad \text{for } \tfrac{1}{2}(\bar{\bar{\tau}} : \bar{\bar{\tau}}) < \tau_0^2 \tag{3.4-3}$$

The shear stress tensor, $\bar{\bar{\tau}}$, and the shear rate tensor, $\bar{\bar{\varDelta}}$, can be displayed as shown:

$$\bar{\bar{\tau}} = \begin{pmatrix} \tau_{11} & \tau_{12} & \tau_{13} \\ \tau_{21} & \tau_{22} & \tau_{23} \\ \tau_{31} & \tau_{32} & \tau_{33} \end{pmatrix} \tag{3.4-4}$$

$$\bar{\bar{\varDelta}} = \begin{pmatrix} \varDelta_{11} & \varDelta_{12} & \varDelta_{13} \\ \varDelta_{21} & \varDelta_{22} & \varDelta_{23} \\ \varDelta_{31} & \varDelta_{32} & \varDelta_{33} \end{pmatrix} \tag{3.4-5}$$

where $1, 2, 3$ correspond to x, y, z in rectangular coordinates, r, θ, z in cylindrical coordinates, and r, θ, φ in spherical coordinates.

Equations (3.4-1), (3.4-2), and (3.4-3) have been developed by arguments that begin with a generalized Newtonian behavior,

$$\bar{\bar{\tau}} = -\mu\bar{\bar{\varDelta}} \tag{3.4-6}$$

where $\bar{\bar{\varDelta}}$ is the shear rate and μ is the Newtonian viscosity and is constant. Thus, for a Newtonian fluid, the components of equation (3.4-6) are

$$\tau_{ij} = -\mu\varDelta_{ij} \tag{3.4-7}$$

The ij components of τ_{ij} and $\varDelta_{ij}$ can simply be looked up in Tables V, VI, and VII.

For the power-law or Bingham plastic non-Newtonian cases, it is necessary to determine $\tfrac{1}{2}\bar{\bar{\varDelta}} : \bar{\bar{\varDelta}}$ in the proper coordinate system.[1] Table XI

[1] R. B. Bird, W. E. Stewart, and E. N. Lightfoot, *Transport Phenomena*, John Wiley, New York (1960), p. 103.

TABLE XI
$\frac{1}{2}\bar{\bar{\Delta}} : \bar{\bar{\Delta}}$ in Rectangular, Cylindrical, and Spherical Coordinates

Rectangular coordinates

$$\frac{1}{2}\bar{\bar{\Delta}} : \bar{\bar{\Delta}} = 2\left[\left(\frac{\partial v_x}{\partial x}\right)^2 + \left(\frac{\partial v_y}{\partial y}\right)^2 + \left(\frac{\partial v_z}{\partial z}\right)^2\right] + \left[\frac{\partial v_y}{\partial x} + \frac{\partial v_x}{\partial y}\right]^2$$

$$+ \left[\frac{\partial v_z}{\partial y} + \frac{\partial v_y}{\partial z}\right]^2 + \left[\frac{\partial v_x}{\partial z} + \frac{\partial v_z}{\partial x}\right]^2$$

Cylindrical coordinates

$$\frac{1}{2}\bar{\bar{\Delta}} : \bar{\bar{\Delta}} = 2\left[\left(\frac{\partial v_r}{\partial r}\right)^2 + \left(\frac{1}{r}\frac{\partial v_\theta}{\partial \theta} + \frac{v_r}{r}\right)^2 + \left(\frac{\partial v_z}{\partial z}\right)^2\right]$$

$$+ \left[r\frac{\partial}{\partial r}\left(\frac{v_\theta}{r}\right) + \frac{1}{r}\frac{\partial v_r}{\partial \theta}\right]^2 + \left[\frac{1}{r}\frac{\partial v_z}{\partial \theta} + \frac{\partial v_\theta}{\partial z}\right]^2 + \left[\frac{\partial v_r}{\partial z} + \frac{\partial v_z}{\partial r}\right]^2$$

Spherical coordinates

$$\frac{1}{2}\bar{\bar{\Delta}} : \bar{\bar{\Delta}} = 2\left[\left(\frac{\partial v_r}{\partial r}\right)^2 + \left(\frac{1}{r}\frac{\partial v_\theta}{\partial \theta} + \frac{v_r}{r}\right)^2 + \left(\frac{1}{r\sin\theta}\frac{\partial v_\varphi}{\partial \varphi} + \frac{v_r}{r} + \frac{v_\theta \cot\theta}{r}\right)^2\right]$$

$$+ \left[r\frac{\partial}{\partial r}\left(\frac{v_\theta}{r}\right) + \frac{1}{r}\frac{\partial v_r}{\partial \theta}\right]^2 + \left[\frac{\sin\theta}{r}\frac{\partial}{\partial \theta}\left(\frac{v_\varphi}{\sin\theta}\right) + \frac{1}{r\sin\theta}\frac{\partial v_\theta}{\partial \varphi}\right]^2$$

$$+ \left[\frac{1}{r\sin\theta}\frac{\partial v_r}{\partial \varphi} + r\frac{\partial}{\partial r}\left(\frac{v_\varphi}{r}\right)\right]^2$$

shows this in rectangular, cylindrical, and spherical coordinates. Thus the procedure for arriving at the working forms of equations (3.4-1), (3.4-2), and (3.4-3) is analogous to the Newtonian case. The terms in the brackets of equations (3.4-1), (3.4-2), and (3.4-3) determine the viscous character-istics. The equations are empirical, having been arrived at partly because the viscous terms in the brackets are invariant under coordinate transforma-tions. Since the viscous terms represent the intrinsic fluid properties, it seems reasonable that they not be variable when the coordinate system is changed.

3.5. Alternative Forms of the Equations of Motion

Equation (3.2-12) can be alternatively written as[2]

$$\frac{\partial}{\partial t}(\varrho \bar{v}) = -[\nabla \cdot \bar{\bar{\varphi}}] + \sum_{\alpha=1}^{n} \varrho_{\alpha} \bar{g}_{\alpha} \qquad (3.5\text{-}1)$$

$$\underbrace{\phantom{\frac{\partial}{\partial t}(\varrho \bar{v})}}_{\text{accumulation}} \quad \underbrace{\phantom{[\nabla \cdot \bar{\bar{\varphi}}]}}_{\substack{\text{transport} \\ \text{through the} \\ \text{surface}}} \quad \underbrace{\phantom{\sum \varrho_{\alpha}}}_{\text{generation}}$$

where

$$\bar{\bar{\varphi}} = \varrho \bar{v}\bar{v} + \bar{\bar{\tau}} + P\bar{\bar{\delta}}$$

P = pressure

ϱ = density

$\bar{\bar{\delta}}$ = unit tensor

$\bar{g}_{\alpha}$ = gravitational acceleration of component α

$\bar{\bar{\tau}}$ = shear stress tensor

We can also equate the external forces exerted by the surroundings to the rate at which momentum is being created within the fluid element. Thus we obtain[3]

$$\frac{\partial}{\partial t}(\varrho \bar{v}) \;+\; \nabla \cdot (\varrho \bar{v}\bar{v}) \;=\; \nabla \cdot \bar{\bar{\pi}} \;+\; \varrho \bar{F} \qquad (3.5\text{-}2)$$

| rate of increase of momentum, per unit volume | rate of momentum loss by convection through the surface, per unit volume | stresses on the surface, per unit volume | external body force on the element, per unit volume (gravity, for example) |

$$\underbrace{\phantom{\text{rate of creation of momentum,}}}_{\substack{\text{rate of creation of momentum,} \\ \text{per unit volume}}} \qquad \underbrace{\phantom{\text{external forces,}}}_{\substack{\text{external forces,} \\ \text{per unit volume}}}$$

where

$\bar{\bar{\pi}}$ = pressure or stress tensor

$$= -P\bar{\bar{\delta}} + \varkappa(\nabla \cdot \bar{v})\bar{\bar{\delta}} + \mu\bar{\bar{\varDelta}} \quad \text{(Newtonian fluids)} \qquad (3.5\text{-}3)$$

P = hydrostatic pressure the fluid would be supporting if it were at rest

$\bar{\bar{\delta}}$ = unit tensor

$\varkappa$ = bulk or volume viscosity

μ = shear viscosity

$\bar{\bar{\varDelta}}$ = rate of deformation tensor (shear rate)

[2] D. M. Himmelblau and K. B. Bischoff, *Process Analysis and Simulation*, John Wiley, New York (1968), p. 13.

[3] J. Happel and H. Brenner, *Low Reynolds Number Hydrodynamics*, Prentice-Hall, Englewood Cliffs, N. J. (1965), p. 24.

The bulk viscosity, $\varkappa$, relates shear stress to volumetric deformation in the same way that shear viscosity relates shear stress to the shear rate. The bulk viscosity is important in the case of fluids subjected to rapidly varying forces such as ultrasonic vibrations. The components of the pressure tensor for an incompressible fluid, with $\varkappa = 0$ and $\nabla \cdot \bar{v} = 0$, are given for rectangular coordinates by

$$\pi_{xx} = -P + 2\mu \frac{\partial v_x}{\partial x} \tag{3.5-4}$$

$$\pi_{yy} = -P + 2\mu \frac{\partial v_y}{\partial y} \tag{3.5-5}$$

$$\pi_{zz} = -P + 2\mu \frac{\partial v_z}{\partial z} \tag{3.5-6}$$

$$\pi_{xy} = \pi_{yx} = \mu\left(\frac{\partial v_x}{\partial y} + \frac{\partial v_y}{\partial x}\right) \tag{3.5-7}$$

$$\pi_{yz} = \pi_{zy} = \mu\left(\frac{\partial v_y}{\partial z} + \frac{\partial v_z}{\partial y}\right) \tag{3.5-8}$$

$$\pi_{zx} = \pi_{xz} = \mu\left(\frac{\partial v_z}{\partial x} + \frac{\partial v_x}{\partial z}\right) \tag{3.5-9}$$

An "ideal" fluid is defined as having the property $\mu = 0$ in the equation of motion. Irrotational motion is characterized by $\nabla \times \bar{v} = 0$, which is sometimes also called potential flow. In the case of a viscous fluid flowing past a solid, the tangential fluid velocity is experimentally observed to be zero (no slip condition). The velocity of the fluid normal to the solid surface is the same as that of the boundary. Thus at a stationary solid surface the vector boundary condition is $\bar{v} = 0$. The assumption of no fluid slippage at the wall is valid unless the mean free path of the fluid molecules becomes large compared with the dimensions of the boundary surfaces. The Navier–Stokes equations are based on the assumption that the fluid may be treated as a continuum. Where there are gases at low pressures or for flow in tubes of very small diameter, the mean free path of the fluid may be large compared to the other dimensions and intermolecular collisions could be rare. (Momentum transport depends upon the collisions of molecules with the bounding surfaces.) For these situations, we define a Knudsen number, $\varkappa = L/d$, where L is the molecular mean free path and d is the macroscopic length (or diameter) of the container. For $\varkappa$ less than 0.01, the equations of motion apply.[4] For $\varkappa$ between 0.01 and 0.1 it is still possible to employ the equations of motion, but a correction must be made to allow for slippage between the fluid and the solid boundary. Supersonic aerodynamics and high-vacuum flows through small capillaries can have slippage at the boundaries and require, in many cases, a molecular rather than continuum analytical approach.

For a flow to be stable, it is necessary that small disturbances in the flow decay with time. If these perturbations tend to be amplified with time, the flows tend to instability, and turbulence normally results. At high Reynolds numbers where the flow is turbulent we sometimes can neglect the viscous terms in equation (3.2-12) in comparison with the inertial convective terms. However, the resulting simplified equation furnishes no information on the viscous drag experienced by bodies immersed in the fluid or on the flow resistance in pipes. Because the omission of viscous terms reduces the order of the equation of motion, this equation also cannot be made to satisfy the no-slip condition at the solid boundaries. For the high-flow, high-Reynolds-number situations it is possible to subdivide the flow field into an external region where the flow is irrotational (neglect the viscous effects) and a thin layer near the solid boundary where viscous effects are not negligible. The velocities are then matched for the boundary layer and the external region.

Another modification of the equation of motion involves the analysis of very slow or creeping flow phenomena. If at steady state the inertial convective terms are neglected when compared with the viscous terms, we get from (3.2-12) for a Newtonian fluid

$$\nabla^2 \bar{v} = \frac{1}{\mu} \nabla P \qquad\qquad (3.5\text{-}10)$$

where the body forces have been neglected.

Assignments in Chapter 3

3.1. Verify equation (3.1-3) and (3.2-11).

3.2. Write out the τ_{rz} and $\tau_{r\theta}$ components of equation (3.4-1) in cylindrical coordinates and apply it to laminar flow in a pipe.

[4] *Ibid.*, p. 50.

3.3. For steady flow of a Newtonian fluid between two infinite parallel plates spaced $2h$ units apart, show that the governing flow equation is

$$\frac{d^2v_z}{d\xi^2} = +\frac{\Delta P}{\mu L}$$

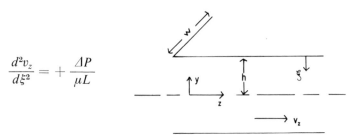

where ξ is the distance from one of the plates. Show further that we can get

$$v_z = \frac{\Delta P}{\mu L}\,\xi\left(\frac{\xi}{2} - h\right)$$

and

$$Q = -\frac{2}{3}\,\frac{w\,\Delta p}{\mu L}\,h^3$$

where Q is the volumetric flow rate.

For Further Reading

Transport Phenomena, by R. B. Bird, W. E. Stewart, and E. N. Lightfoot, John Wiley, New York, 1960.

Process Analysis and Simulation, by D. M. Himmelblau and K. B. Bischoff, John Wiley, New York, 1968.

Low Reynolds Number Hydrodynamics, by J. Happel and H. Brenner, Prentice-Hall, Englewood Cliffs, N. J., 1965.

Chapter 4

Transport Analysis in Fluid Flow Phenomena

In this chapter velocity profiles and volumetric flow rate equations are developed from the equations of motion introduced in Chapter 3. From these equations a discussion of macroscopic flow phenomena is presented, including an analysis of friction factors in laminar and turbulent flow. Then non-Newtonian fluid behavior is discussed, from the point of view of Reynolds number–friction factor plots. Some examples of viscoelastic properties are then briefly presented, along with two-phase solid–liquid suspension flow. The chapter concludes with an analysis of some fluid flow phenomena by boundary layer theory.

4.1. Flow of Fluids in Thin Films

Figure 17 sketches the flow of a fluid in a thin film. From the equation of continuity in rectangular coordinates, equation A of Table I, we get, for an incompressible fluid at steady state,

$$\frac{\partial v_z}{\partial z} = 0 \qquad (4.1-1)$$

From equation C of Table II, a steady-state one-dimensional flow situation

Fig. 17. Flow of a fluid in a thin film down an inclined plane.

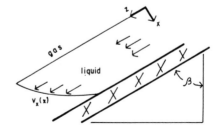

111

is governed by the momentum flux expression

$$\frac{d\tau_{xz}}{dx} = \varrho \frac{g}{g_c} \cos \beta \qquad (4.1\text{-}2)$$

where g_z, the component of the vertical gravity force, is equal to $(g/g_c) \cos \beta$. In arriving at equation (4.1-2) we have used the information obtained from the equation of continuity, equation (4.1-1). (Note that we first apply the equation of continuity and glean information which is substituted into the equation of motion.) Equation (4.1-2) is general, at this point applicable to Newtonian as well as non-Newtonian fluids. Equation (4.1-2) could also have been derived by setting up a momentum balance [equation (3.2-2)] around the differential element shown in Figure 18. The result is

$$\underbrace{\left[(LW\tau_{xz})\bigg|_{x} - (LW\tau_{xz})\bigg|_{x+\Delta x}\right]}_{\text{I}} + \underbrace{\left[(W\Delta x v_z)\left(\frac{\varrho v_z}{g_c}\right)\bigg|_{z=0} - (W\Delta x v_z)\left(\frac{\varrho v_z}{g_c}\right)\bigg|_{z=L}\right]}_{\text{II}}$$

$$\underbrace{+ (LW\Delta x)\left(\varrho \frac{g}{g_c} \cos \beta\right)}_{\text{III}} = 0 \qquad (4.1\text{-}3)$$

where I represents the net momentum transport into the differential element by viscous forces and II refers to the convective momentum transport terms. In II it will be recalled from Chapter 3 that $\varrho v_z/g_c$ is the momentum per unit volume of the flowing fluid (lbf sec/ft³). The last term, III, is the body (gravity) force. Since the velocity v_z is the same at $z = 0$ and $z = L$ ($\partial v_z/dz = 0$ from the equation of continuity), we find that II = 0. If equation (4.1-3) is divided by $LW\Delta x$ and if Δx goes to zero in the manner shown in Chapter 1, the differential of τ_{xz} is formed, resulting finally in equation (4.1-2). The boundary conditions for this falling film phenomenon are given by

$$x = 0 \qquad \tau_{xz} = 0 \qquad (4.1\text{-}4)$$

$$x = \delta \qquad v_z = 0 \qquad (4.1\text{-}5)$$

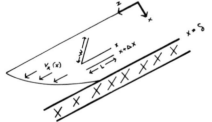

Fig. 18. Differential element in a falling liquid film.

Equation (4.1-4) states the assumption that at gas–liquid interfaces the shear stress τ_{xz} (momentum flux) is negligibly small. Equation (4.1-5) expresses the condition that the fluid is assumed to cling to the solid surface i.e., the liquid at the liquid–solid interface has the same velocity as the solid surface. Integrating equation (4.1-2) and using equation (4.1-4) to evaluate the integration constant yields

$$\tau_{xz} = \varrho \, \frac{g}{g_c} \, x \cos \beta \qquad (4.1\text{-}6)$$

which shows that the shear stress profile is linear. If we now wish to determine the velocity profile, it is necessary to substitute into equation (4.1-6) the rheological expression for τ_{xz}. For a Newtonian fluid, we go to equation F of Table V for τ_{xz}, to get $\tau_{xz} = -\mu(\partial v_z/\partial x)$. This expression, combined with equations (4.1-6) and (4.1-5), finally gives the velocity profile for a Newtonian fluid:

$$v_z = \frac{\varrho(g/g_c)\delta^2 \cos \beta}{2\mu} \left[1 - \left(\frac{x}{\delta} \right)^2 \right] \qquad (4.1\text{-}7)$$

The term g/g_c is inserted for emphasis, to serve as a reminder that the units in each of the equations must balance. Thus, depending on the units of the viscosity μ, equation (4.1-7) may or may not contain g_c. [Equation (4.1-7) could also have been obtained for a Newtonian fluid by going directly to Table VIII.] The maximum velocity is found at $x = 0$, and the average velocity, found by the mean value theorem, is given by

$$\langle v_z \rangle = \frac{\displaystyle\int_0^w \int_0^\delta v_z \, dx \, dy}{\displaystyle\int_0^w \int_0^\delta dx \, dy} = \frac{\varrho(g/g_c)\delta^2 \cos \beta}{3\mu} \qquad (4.1\text{-}8)$$

Note that this is a "volumetric" average, that is, the point function v_z is weighted by the differential area $dx\,dy$, so that the volumetric flow rate is generated. The utility of this average velocity, $\langle v_z \rangle$, is reflected by the ease with which one can produce the volumetric flow rate, $Q = \langle v_z \rangle (W\delta)$, where $W\delta$ is the total cross-sectional area. The volumetric flow rate equation for a Newtonian fluid is therefore given by

$$Q = \frac{\varrho(g/g_c)W\delta^3 \cos \beta}{3\mu} \qquad (4.1\text{-}9)$$

The drag F of this Newtonian fluid over the solid surface can be calculated from

$$F = \int_0^L \int_0^W \tau_{xz} \big|_{x=\delta} \, dydz$$

$$= \int_0^L \int_0^W \mu \frac{dv_z}{dx} \bigg|_{x=\delta} \, dydz$$

$$= \varrho \frac{g}{g_c} \, \delta L W \cos \beta \qquad (4.1\text{-}10)$$

Equation (4.1-7) was used in determining dv_z/dx. A Reynolds number is defined for film flow, $N_{Re} = 4\delta\langle v_z\rangle\varrho/\mu$, and criteria have been established for determining whether the foregoing laminar flow analysis is applicable[1]:

Laminar flow without rippling $N_{Re} = 4\text{-}25$
Laminar flow with rippling $N_{Re} = (4\text{-}25)$ to $(1000\text{-}2000)$
Turbulent flow $N_{Re} = $ greater than 2000

If a power-law fluid were flowing in a thin film, equation (4.1-2) would remain the same. However instead of inserting the Newtonian expression for τ_{xz}, we use the power-law expression from equation (3.3-1) to give

$$\tau_{xz} = m\left(-\frac{dv_z}{dx}\right)^n \qquad (4.1\text{-}11)$$

where $dv_z/dx < 0$. It is now a straightforward procedure to rederive equations (4.1-7) through (4.1-10) for a power-law fluid.

4.2. Flow in Circular-Shaped Conduits

For flow of fluids in conduits with circular geometry such as pipes and wetted-wall columns, the governing equation is obtained from Table III and is given by

$$\frac{1}{r} \frac{d}{dr} (r\tau_{rz}) = \frac{P_0 - P_L}{L} + \varrho \frac{g}{g_c} \qquad (4.2\text{-}1)$$

where we used $dv_z/dz = 0$, obtained from the equation of continuity in Table I. Figure 19 sketches the general flow situation.

[1] R. B. Bird, W. E. Stewart, and E. N. Lightfoot, *Transport Phenomena*, John Wiley, New York (1960), p. 41.

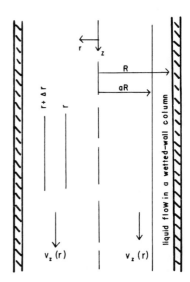

Fig. 19. Flow of fluids in circular conduits.

In equation (4.2-1) we have written $\partial P/\partial z = (P_0 - P_L)/L$ and $\varrho g_z = \varrho g/g_c$. If the fluid fills the pipe, the boundary conditions are given by

$$r = 0 \qquad \tau_{rz} \text{ and } v_z \text{ are finite} \qquad (4.2\text{-}2)$$

$$r = R \qquad v_z = 0 \qquad (4.2\text{-}3)$$

If the fluid wets only the surface of the conduit but does not fill the cross section, we refer to the equipment as a wetted-wall column. For the wetted-wall column the governing equation is the same as equation (4.2-1) but the boundary conditions are now given by

$$r = aR \qquad \tau_{rz} = 0 \qquad (4.2\text{-}4)$$

$$r = R \qquad v_z = 0 \qquad (4.2\text{-}5)$$

where the location aR in Figure 19 is the gas–liquid interface. Note that the wetted-wall situation does not utilize equation (4.2-2) since for the wetted-wall column the region of interest is $aR \le r \le R$ and the value of r never goes to zero.

Equation (4.2-1) could also have been derived from a momentum analysis of the differential element shown in Figure 19. From equation (3.2-2) we write the various terms that contribute to a "rate of momentum" balance

$$\underbrace{[(\tau_{rz}2\pi rL)\big|_r - (\tau_{rz}2\pi rL)\big|_{r+\Delta r}]}_{\text{I}}$$

$$+ \underbrace{\left[(2\pi r\Delta rv_z)\left(\frac{\varrho v_z}{g_c}\right)\bigg|_{z=0} - (2\pi r\Delta rv_z)\left(\frac{\varrho v_z}{g_c}\right)\bigg|_{z=L}\right]}_{\text{II}}$$

$$+ \underbrace{2\pi r\Delta rL\varrho\,\frac{g}{g_c}}_{\text{III}} + \underbrace{2\pi r\Delta r(P_0 - P_L)}_{\text{IV}} = 0 \qquad (4.2\text{-}6)$$

where

I = net rate of momentum transport by viscous effects across the
cylindrical surface (in force units)

II = net rate of momentum transport by convective effects (in force
units)

III = body (gravity) force

IV = net force associated with pressure acting on the differential
element

By dividing equation (4.2-6) by $2\pi L\Delta r$ and noting that $v_z\big|_{z=0} = v_z\big|_{z=L}$
equation (4.2-6) can be written as

$$\lim_{\Delta r\to 0}\frac{r\tau_{rz}\big|_{r+\Delta r} - r\tau_{rz}\big|_r}{\Delta r} = \left(\frac{P_0 - P_L}{L} + \varrho\,\frac{g}{g_c}\right)r \qquad (4.2\text{-}7)$$

which yields equation (4.2-1) when the limiting process is performed. If the
governing equation (4.2-1) is integrated, the result is

$$\tau_{rz} = \left(\frac{P_0 - P_L}{L} + \varrho\,\frac{g}{g_c}\right)\frac{r}{2} + \frac{C_1}{r} \qquad (4.2\text{-}8)$$

If the flowing fluid fills the pipe, then equations (4.2-2) and (4.2-3) are
applicable. Specifically, invoking equation (4.2-2) in equation (4.2-8) implies
$C_1 = 0$, leading to

$$\tau_{rz} = \left(\frac{P_0 - P_L}{L} + \varrho\,\frac{g}{g_c}\right)\frac{r}{2} \qquad \text{(vertical pipe flow)} \qquad (4.2\text{-}9)$$

For vertical wetted-wall columns, equation (4.2-4) used in (4.2-8) yields

$$\tau_{rz} = \left(\frac{P_0 - P_L}{L} + \varrho\,\frac{g}{g_c}\right)\left[\frac{r^2 - a^2R^2}{2r}\right] \qquad \text{(vertical wetted-wall column flow)}$$
$$(4.2\text{-}10)$$

Horizontal Pipe Flow

When considering flow in horizontal pipes, there is no gravitational component in the flow direction and thus equation (4.2-9) becomes

$$\tau_{rz} = \frac{\Delta P}{L}\frac{r}{2} \quad \text{(horizontal pipe flow)} \tag{4.2-11}$$

Equation (4.2-11) is general as far as the liquid properties are concerned. It is equally applicable to Newtonian and non-Newtonian fluids. For a Newtonian fluid, we can get from Table VI the proper τ_{rz} expression:

$$\tau_{rz} = -\mu\frac{dv_z}{dr} \tag{4.2-12}$$

where the units of viscosity μ balance the units in the equation. If equation (4.2-12) is substituted into equation (4.2-11) and the boundary conditions (4.2-3) are applied, the result is the velocity profile for the steady flow of a Newtonian fluid in a horizontal pipe:

$$v_z = \frac{(P_0 - P_L)R^2}{4\mu L}\left[1 - \left(\frac{r}{R}\right)^2\right] \tag{4.2-13}$$

The maximum velocity is easily seen to be $v_{z_{\max}} = (P_0 - P_L)R^2/4\mu L$. By the mean value theorem discussed in Chapter 1 the average velocity is computed:

$$\langle v_z\rangle = \frac{\int_0^{2\pi}\int_0^R v_z r\,dr\,d\theta}{\int_0^{2\pi}\int_0^R r\,dr\,d\theta} = \frac{(P_0 - P_L)R^2}{8\mu L} \tag{4.2-14}$$

where the differential cross-sectional area is $r\,dr\,d\theta$. The volumetric flow rate Q is obtained by either

$$Q = \int_0^{2\pi}\int_0^R v_z r\,dr\,d\theta \tag{4.2-15}$$

or

$$Q = \langle v_z\rangle(\pi R^2) \tag{4.2-16}$$

In either case, combining equations (4.2-13) through (4.2-16) yields the Hagen–Poiseuille equation

$$Q = \frac{(P_0 - P_L)\pi R^4}{8\mu L} \tag{4.2-17}$$

The fluid drag, F, for horizontal flow of a Newtonian fluid in a circular pipe is computed from

$$F = \tau_{rz}|_{r=R}(2\pi RL)$$

$$= \left(-\mu \frac{dv_z}{dr}\bigg|_{r=R}\right)(2\pi RL)$$

$$= \pi R^2(P_0 - P_L) \qquad (4.2\text{-}18)$$

It should be emphasized that all of these results apply in laminar flow where the Reynolds number is generally less than 2100.

Power-Law Fluids

A similar analysis for a power-law fluid would yield the following results for horizontal steady pipe flow:

$$v_z = \left[\frac{(P_0 - P_L)R}{2mL}\right]^{1/n}\left(\frac{R}{1/n + 1}\right)\left[1 - \left(\frac{r}{R}\right)^{1/n+1}\right] \qquad (4.2\text{-}19)$$

$$Q = \pi\left[\frac{(P_0 - P_L)R}{2mL}\right]^{1/n}\frac{R^3}{1/n + 3} \qquad (4.2\text{-}20)$$

Bingham Fluids

For a Bingham fluid, there is a slight variation in the derivation of the shear stress equation and velocity profiles. In Chapter 3 it was shown that a Bingham fluid has two zones of differing behavior, as sketched in Figure 11. The governing shear stress equation for this non-Newtonian fluid is still given by equation (4.2-11), applicable to Newtonian and non-Newtonian fluids. Thus it is applicable to Bingham fluids also. It remains for us to substitute the Bingham model into equation (4.2-11) for τ_{rz} in order to generate velocity profiles. The region to be analyzed is the outer region in Figure 11, near the wall where the shear stress can be greater than the yield stress τ_0. We can find τ_{rz} for this region from equation (3.4-2). The result is

$$\tau_{rz} = -\left\{\mu_0 - \frac{\tau_0}{\sqrt{\frac{1}{2}\bar{\bar{\Delta}} : \bar{\bar{\Delta}}}}\right\}\Delta_{rz} \qquad (4.2\text{-}21)$$

where Δ_{rz} is obtained from Table VI and $\frac{1}{2}\bar{\bar{\Delta}} : \bar{\bar{\Delta}}$ is obtained from Table XI. Equation (4.2-21) becomes

$$\tau_{rz} = \tau_0 - \mu_0 \frac{dv_z}{dr} \qquad (4.2\text{-}22)$$

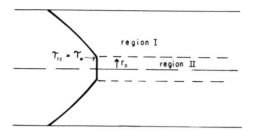

Fig. 20. Velocity profile for a Bingham fluid in
a horizontal pipe.

If equation (4.2-22) is substituted into (4.2-11), with the boundary condition
$v_z = 0$ at $r = R$, we can get the velocity profile for the outer region of a
circular pipe (near the wall). For a Bingham plastic this result is given by

$$v_z^I = \frac{(\Delta P)R^2}{4\mu_0 L} \left[1 - \left(\frac{r}{R}\right)^2 \right] - \frac{\tau_0 R}{\mu_0} \left[1 - \left(\frac{r}{R}\right) \right] \qquad r \geq r_0 \qquad (4.2\text{-}23)$$

where

$$\tau_0 = \frac{\Delta P}{L} \frac{r_0}{2}$$

$$r_0 = \text{radius} \qquad \text{where } \tau_{rz} = \tau_0$$

and the superscript I indicates the outer region, as shown in Figure 20.
Since it is assumed that there is continuity of velocity throughout regions
I and II, equation (4.2-23) can be used to find the velocity of the inner
"plug flow" region II at $r = r_0$. Upon substituting $r = r_0$ into equation
(4.2-23), the velocity of the inner zone is given by

$$v_z^{II} = \frac{(\Delta P)R^2}{4\mu_0 L} \left(1 - \frac{r_0}{R} \right)^2 \qquad r \leq r_0 \qquad (4.2\text{-}24)$$

The volumetric flow rate Q is calculated as usual, except that Q is composed
of two summed quantities, related to the flows through the two zones.
Thus the result we seek[2] is

$$Q = \int_0^{2\pi} \int_0^R v_z r \, dr \, d\theta$$

$$= \int_0^{2\pi} \int_0^{r_0} v_z^{II} r \, dr \, d\theta + \int_0^{2\pi} \int_{r_0}^R v_z^I r \, dr \, d\theta$$

$$= \frac{\pi(\Delta P)R^4}{8\mu_0 L} \left[1 - \frac{4}{3}\left(\frac{\tau_0}{\tau_R}\right) + \frac{1}{3}\left(\frac{\tau_0}{\tau_R}\right)^4 \right] \qquad (4.2\text{-}25)$$

[2] *Ibid.*, p. 50.

Flow Through an Annulus (Newtonian Fluids)

For flow through an annulus, as sketched in Figure 21, the situation is altered only in that there are two boundary conditions where $v_z = 0$:

$$v_z = 0 \qquad \text{at } r = R \qquad\qquad (4.2\text{-}26)$$

$$v_z = 0 \qquad \text{at } r = \varkappa R \qquad\qquad (4.2\text{-}27)$$

For flow in an annulus there is no possibility that the radial dimension r can go to zero. Thus equation (4.2-8) is applicable, rather than (4.2-11), in calculating the shear stress profile. More simply, however, we can get the flow equation directly from Table III (non-Newtonian fluids) or Table VI (Newtonian fluids). If boundary condition equations (4.2-26) and (4.2-27) are used in conjunction with the governing equation of motion in Table VI, we can derive the velocity profile for a Newtonian fluid in an annulus (in steady laminar flow):

$$v_z = \frac{(\Delta P)R^2}{4\mu L}\left[1 - \left(\frac{r}{R}\right)^2 + \frac{(1 - \varkappa^2)}{\ln(1/\varkappa)}\ln\frac{r}{R}\right] \qquad (4.2\text{-}28)$$

where $\Delta P = (P_0 - P_L)/L$ if the conduit is horizontal and $\Delta P/L = (P_0 - P_L)/L + \varrho g/g_c$ for vertical systems.

The maximum velocity in the annulus could be found by finding the value of r that satisfies $dv_z/dr = 0$ and then substituting this value of r back into equation (4.2-28). The average velocity is computed as usual by the mean value theorem[3]:

$$\langle v_z \rangle = \frac{\displaystyle\int_0^{2\pi}\int_{\varkappa R}^{R} rdrd\theta}{\displaystyle\int_0^{2\pi}\int_{\varkappa R}^{R} rdrd\theta} = \frac{(\Delta P)R^2}{8\mu L}\left(\frac{1 - \varkappa^4}{1 - \varkappa^2} - \frac{1 - \varkappa^2}{\ln(1/\varkappa)}\right) \qquad (4.2\text{-}29)$$

The volumetric flow rate Q can be obtained as usual:

$$Q = \int_0^{2\pi}\int_{\varkappa R}^{R} v_z rdrd\theta \qquad\qquad (4.2\text{-}30)$$

or

$$Q = \langle v_z \rangle \cdot (\text{flow area}) = \langle v_z \rangle[\pi R^2(1 - \varkappa^2)] \qquad (4.2\text{-}31)$$

[3] *Ibid.*, p. 53.

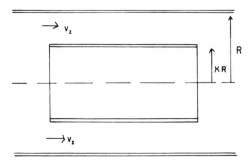

Fig. 21. Flow through an annulus.

From either equation (4.2-30) or (4.2-31), the final result is

$$Q = \frac{\pi(\Delta P)R^4}{8\mu L}\left[(1 - \varkappa^4) - \frac{(1 - \varkappa^2)^2}{\ln(1/\varkappa)}\right] \qquad (4.2\text{-}32)$$

For flow through an annulus, the fluid drag is the sum of the drag on both wetted solid surfaces and is given by

$$F = \tau_{rz}\big|_{r=\varkappa R}(2\pi RL) + \tau_{rz}\big|_{r=R}(2\pi RL) \qquad (4.2\text{-}33)$$

Liquid–Liquid Immiscible Flow (Rectangular Coordinates)

For two immiscible fluids flowing horizontally in contact with each other, as sketched in Figure 22, the problem is one of keeping track of the separate fluid properties. From Table II the governing equation is obtained:

$$\frac{d\tau_{xz}}{dx} = \frac{\Delta P}{L} \qquad (4.2\text{-}34)$$

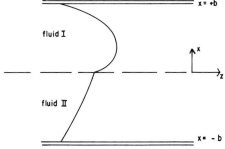

Fig. 22. Horizontal flow of two adjacent immiscible fluids.

In this situation we invoke a new boundary condition, namely, the assumption that there is, at the liquid–liquid interface, continuity of both velocity and shear stress:

$$x = 0 \qquad \tau_{xz}{}^I = \tau_{xz}{}^{II} \qquad\qquad\qquad (4.2\text{-}35)$$

$$x = 0 \qquad v_z{}^I = v_z{}^{II} \qquad\qquad\qquad (4.2\text{-}36)$$

The superscripts I and II refer to the adjacent fluids. Thus, from equation (4.2-34), for each region, the equations

$$\tau_{xz} = \frac{\Delta P}{L} x + C_1 \qquad\qquad\qquad (4.2\text{-}37)$$

or

$$\tau_{xz}{}^I = \frac{\Delta P}{L} x + C_1{}^I \qquad\qquad\qquad (4.2\text{-}38)$$

$$\tau_{xz}{}^{II} = \frac{\Delta P}{L} x + C_1{}^{II} \qquad\qquad\qquad (4.2\text{-}39)$$

are applicable. From equation (4.2-35) we get $C_1{}^I = C_1{}^{II} = C$. If each fluid is Newtonian with viscosity μ^I and μ^{II}, then from Table V we obtain

$$\tau_{xz} = -\mu \frac{dv_z}{dx} \qquad\qquad\qquad (4.2\text{-}40)$$

which, when combined with equations (4.2-38) and (4.2-39), yields

$$-\mu^I \frac{dv_z{}^I}{dx} = \frac{\Delta P}{L} x + C \qquad\qquad\qquad (4.2\text{-}41)$$

$$-\mu^{II} \frac{dv_z{}^{II}}{dx} = \frac{\Delta P}{L} x + C \qquad\qquad\qquad (4.2\text{-}42)$$

or

$$v_z{}^I = \frac{-\Delta P}{2\mu^I L} x^2 - \frac{C}{\mu^I} x + C_2{}^I \qquad\qquad\qquad (4.2\text{-}43)$$

$$v_z{}^{II} = \frac{-\Delta P}{2\mu^{II} L} x^2 - \frac{C}{\mu^{II}} x + C_2{}^{II} \qquad\qquad\qquad (4.2\text{-}44)$$

The three boundary conditions needed to solve for the integration constants are equation (4.2-36) and

$$x = -b \qquad v_z{}^I = 0 \qquad\qquad\qquad (4.2\text{-}45)$$

$$x = +b \qquad v_z{}^{II} = 0 \qquad\qquad\qquad (4.2\text{-}46)$$

Combining these boundary condition equations with equations (4.2-43) and (4.2-44) yields the final results for velocity profiles and shear stress profiles[4]:

$$v_z{}^I = \frac{(\Delta P)b^2}{2\mu^I L} \left[\left(\frac{2\mu^I}{\mu^I + \mu^{II}} \right) + \left(\frac{\mu^I - \mu^{II}}{\mu^I + \mu^{II}} \right)\left(\frac{x}{b} \right) - \left(\frac{x}{b} \right)^2 \right] \qquad (4.2\text{-}47)$$

$$v_z{}^{II} = \frac{(\Delta P)b^2}{2\mu^{II} L} \left[\left(\frac{2\mu^{II}}{\mu^I + \mu^{II}} \right) + \left(\frac{\mu^I - \mu^{II}}{\mu^I + \mu^{II}} \right)\left(\frac{x}{b} \right) - \left(\frac{x}{b} \right)^2 \right] \qquad (4.2\text{-}48)$$

$$\tau_{xz} = \frac{(\Delta P)b}{L} \left[\left(\frac{x}{b} \right) - \tfrac{1}{2}\left(\frac{\mu^I - \mu^{II}}{\mu^I + \mu^{II}} \right) \right] \qquad (4.2\text{-}49)$$

There is only one shear stress equation for both fluids, which is consistent with the previous specification that the "τ" equation did not distinguish the type of fluid involved. The τ_{xz} profile is linear, which infers that $\tau_{xz} = 0$ can only occur once in the cross section. The maximum velocity is characterized by $dv_z/dx = 0$, which implies $\tau_{xz} = 0$ from $\tau_{xz} = -\mu\, dv_z/dx$. Thus the location of the maximum velocity is obtained from equation (4.2-49). While there is continuity of velocity at $x = 0$ (the liquid–liquid interface) the slopes of the velocity profiles of $x = 0$ are not necessarily the same. We show this by noting first that at $x = 0$, $v_z{}^I = v_z{}^{II}$, and $\tau_{xz}{}^I = \tau_{xz}{}^{II}$. Then we write

$$\tau_{xz}{}^I = \mu^I\, \frac{dv_z{}^I}{dz} = \tau_{xz}{}^{II} = \mu^{II}\, \frac{dv_z{}^{II}}{dz} \qquad (4.2\text{-}50)$$

If μ^I is not equal to μ^{II}, then it is possible that $dv_z{}^I/dz$ is not equal to $dv_z{}^{II}/dz$ at $x = 0$, though they should be of the same sign.

4.3. Flow Equations Used in Viscometry

The equations of motion developed in Chapter 3 are useful in viscometry, which is the study of fluid flow properties and viscosity. To extract shear stress and shear rate data from flow experiments it is important to have a flow configuration (equipment) that is analyzable mathematically. One of the most common viscometers is the concentric cylinder (rotating spindle) sketched in Figure 23. This viscometer is known as the Couette–Hatschek or Stormer type. In Figure 23 the inner cylinder is stationary while the outer cup is rotated. The stationary cylinder may be attached to a calibrated spring system that will measure the torque caused by the fluid

---

[4] *Ibid.*, p. 56.

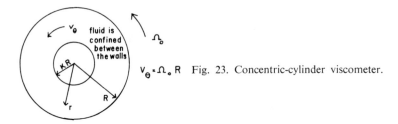

Fig. 23. Concentric-cylinder viscometer.

drag. With torque and angular velocity measured, it is possible with the derived equations of motion to extract the viscosity. The governing equations for the viscometer shown in Figure 23 are obtained from Table IX:

$$-\varrho \frac{v_\theta^2}{r} = -\frac{dP}{dr} + \frac{\varrho g_r}{g_c} \tag{4.3-1}$$

$$0 = \frac{d}{dr}\left[\frac{1}{r}\frac{d}{dr}(rv_\theta)\right] + \frac{\varrho g_\theta}{g_c} \tag{4.3-2}$$

$$0 = -\frac{dP}{dz} + \varrho\frac{g_z}{g_c} \tag{4.3-3}$$

These equations are applicable for Newtonian fluids. Equation (4.3-1) is simply an expression for centripetal force acting on the fluid. While the equation is true, this information is not directly useful. Equation (4.3-3) is a statement of the hydraulic pressure head due to the height of the fluid. Equation (4.3-2) is the governing equation we are seeking. The boundary conditions are

$$r = \varkappa R \qquad v_\theta = 0 \tag{4.3-4}$$

$$r = R \qquad v_\theta = \Omega_0 R \tag{4.3-5}$$

Equations (4.3-4) and (4.3-5) both express the condition that the fluid clings to the solid surfaces and achieves the velocity of the solid surfaces. If equation (4.3-2) is integrated with respect to r, neglecting the gravity term, the result is

$$v_\theta = \Omega_0 R\left(\frac{\varkappa R}{r} - \frac{r}{\varkappa R}\right)\Big/\left(\varkappa - \frac{2}{\varkappa}\right) \tag{4.3-6}$$

where the boundary conditions were used to evaluate the integration constants. From Table VI we find that the only shear stress term that is not zero is $\tau_{r\theta}$, which becomes

$$\tau_{r\theta} = -\mu\left[r\frac{d}{dr}\left(\frac{v_\theta}{r}\right)\right] \tag{4.3-7}$$

If equation (4.3-6) is combined with (4.3-7), the result is

$$\tau_{r\theta} = -\mu\left[2\Omega_0 R^2\left(\frac{1}{r^2}\right)\left(\frac{\varkappa^2}{2 - \varkappa^2}\right)\right] \qquad (4.3\text{-}8)$$

The shear stress $\tau_{r\theta}$ is seen to vary across the thickness of the viscometer. If an average value is to be used, it becomes desirable to reduce the separation distance between the cup and spindle. Under experimental conditions $\tau_{r\theta}$ is calculated from equation (4.3-8). (The term in brackets is the shear rate.) A viscometer such as this usually measures torque, and it is a simple matter to convert shear stress to torque (T) readings:

$$T = (2\pi RL)(\tau_{r\theta}|_{r=R})(R) \qquad (4.3\text{-}9)$$

where the shear stress is multiplied by the contacting surface, $2\pi RL$, to get the drag force. This force is then multiplied by the lever arm, R, to get torque.

Another common viscometer is the plate-and-cone type, sketched in Figure 24. Essentially this device consists of a flat plate, above which is a conical section with the cone tip barely contacting the plate. The test fluid is placed in the space between the plate and the cone. Either the plate or the cone may be rotated. In Figure 24 the cone is depicted as the rotating component. The angle between the cone and plate is kept as small as possible, of the order of one-half of a degree, to simplify the analysis. One of the advantages of this viscometer is that the shear stress is approximately constant throughout the fluid. If the fluid is rotating in laminar flow, then

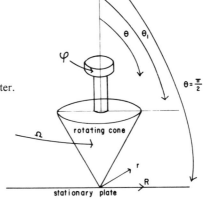

Fig. 24. Plate-and-cone viscometer.

the following governing equations are obtained from Table IV:

$$-\varrho \frac{v_\phi^2}{r} = -\frac{\partial P}{\partial r} \qquad (4.3\text{-}10)$$

$$-\varrho \cot \theta \frac{v_\phi^2}{r} = -\frac{1}{r} \frac{\partial P}{\partial \theta} \qquad (4.3\text{-}11)$$

$$0 = -\left(\frac{1}{r^2} \frac{\partial}{\partial r} (r^2 \tau_{r\phi}) + \frac{1}{r} \frac{\partial \tau_{\theta\phi}}{\partial \theta} + \frac{\tau_{r\phi}}{r} + 2 \cot \theta \frac{\tau_{\theta\phi}}{r} \right) \qquad (4.3\text{-}12)$$

The use of these equations does not imply Newtonian fluids, rather the equations are applicable to non-Newtonian fluids also. All that has been assumed is steady state, incompressible fluid behavior, and $v_r = v_\theta = 0$ with $v_\phi = v_\phi(r, \theta)$. From the equation of continuity in Table I, we obtain $\partial v_\phi / \partial \phi = 0$, which is used to obtain equations (4.3-10), (4.3-11), and (4.3-12). From Table VII we find that the shear rate components $\Delta_{rr} = \Delta_{\theta\theta} = \Delta_{\phi\phi} = 0$, which implies $\tau_{rr} = \tau_{\theta\theta} = \tau_{\phi\phi} = 0$. If v_ϕ is small, then v_ϕ^2 can be considered negligibly small, so that equations (4.3-10) and (4.3-11) can be considered trivial. Further, if it is assumed that $v_\phi / r = f(\theta)$, then from Table VII we find that $\Delta_{r\phi} = 0$, and hence $\tau_{r\phi} = 0$. This simplifies equation (4.3-12) to

$$\frac{d\tau_{\theta\phi}}{d\theta} = -2\tau_{\theta\phi} \cot \theta \qquad (4.3\text{-}13)$$

There are a number of boundary conditions that are applicable here, one of which is

$$\text{Torque on the flat plate,} \quad T = \int_0^{2\pi} \int_0^R \tau_{\theta\phi} \big|_{\theta=\pi/2} \, r^2 dr d\phi \qquad (4.3\text{-}14)$$

where $r dr d\phi$ is the differential area over which $\tau_{\theta\phi}$ acts and r is the lever arm. Torque is equal to the drag force ($\tau_{\theta\phi} r dr d\phi$) multiplied by the lever arm. In addition to the boundary condition of equation (4.3-14), two other velocity boundary conditions are

$$v_\phi = 0 \qquad \text{at } \theta = \frac{\pi}{2} \qquad (4.3\text{-}15)$$

$$v_\phi = \Omega R \sin \theta \qquad \text{at } \theta = \theta_1 \qquad (4.3\text{-}16)$$

If equation (4.3-13) is integrated, the result is

$$\tau_{\theta\phi} = \frac{C_1}{\sin^2 \theta} \qquad (4.3\text{-}17)$$

which, when combined with equation (4.3-14), allows C_1 to be eliminated, yielding

$$\tau_{\theta\phi} = \frac{3T}{2\pi R^3 \sin^2 \theta} \qquad (4.3\text{-}18)$$

Since θ is of the order of $90°$, $\sin^2 \theta$ is approximately equal to unity. Equation (4.3-18) thus is simplified to

$$\tau_{\theta\phi} \cong \frac{3T}{2\pi R^3} \qquad (4.3\text{-}19)$$

which suggests that the shear stress is uniform throughout the fluid and can be calculated easily by reading the torque meter and measuring the radius of the flat plate. To obtain velocity profiles for a Newtonian fluid, we substitute $\tau_{\theta\phi}$ from Table VII into equation (4.3-18). The result is

$$-\mu \frac{\sin \theta}{r} \frac{\partial}{\partial \theta} \left(\frac{V_\phi}{\sin \theta} \right) = \frac{3T}{2\pi R^3 \sin^2 \theta} \qquad (4.3\text{-}20)$$

which can be integrated to obtain[5]

$$\frac{v_\phi}{r} = \frac{3T}{4\pi R^3 \mu} \left[\cot \theta + \tfrac{1}{2} \left(\ln \frac{1 + \cos \theta}{1 - \cos \theta} \right) \sin \theta \right] + C_2 \qquad (4.3\text{-}21)$$

With equation (4.3-15) applied to (4.3-21), we find $C_2 = 0$. Similarly, combining equation (4.3-16) with (4.3-21), the working viscometer equation (4.3-22) is obtained, which enables us to calculate the Newtonian viscosity when torque T and angular velocity Ω are measured

$$\Omega \sin \theta_1 = \frac{3T}{4\pi R^3 \mu} \left[\cot \theta_1 + \tfrac{1}{2} \left(\ln \frac{1 + \cos \theta_1}{1 - \cos \theta_1} \right) \sin \theta_1 \right] \qquad (4.3\text{-}22)$$

Suppose a power-law fluid is confined in the concentric-cylinder viscometer shown in Figure 23. From Table VI, for this flow configuration, all the shear rate components Δ_{ij} are zero except $\Delta_{r\theta}$. Thus the only shear stress component τ_{ij} that is nonzero is $\tau_{r\theta}$. Now we go to Table III to find the governing equation of motion:

$$0 = \frac{1}{r^2} \frac{d}{dr} (r^2 \tau_{r\theta}) \qquad (4.3\text{-}23)$$

which upon integration yields

$$\tau_{r\theta} = \frac{C_1}{r^2} \qquad (4.3\text{-}24)$$

[5] *Ibid.*, p. 100.

We still have not invoked the non-Newtonian behavior. From equation (3.4-1) $\bar{\bar{\Delta}} : \bar{\bar{\Delta}}$ is required. This is found from Table XI to be

$$\tfrac{1}{2}\bar{\bar{\Delta}} : \bar{\bar{\Delta}} = \Delta_{r\theta}^2 = \left[r \frac{r}{dr} \left(\frac{v_\theta}{r} \right) \right]^2 \tag{4.3-25}$$

Thus from equation (3.4-1) and (4.3-25) we can get the $\tau_{r\theta}$ component:

$$\tau_{r\theta} = -\left\{ m \left[r \frac{d}{dr} \left(\frac{v_\theta}{r} \right) \right]^{n-1} \right\} r \frac{d}{dr} \left(\frac{v_\theta}{r} \right) \tag{4.3-26}$$

where

$$\left| r \frac{d}{dr} \left(\frac{v_\theta}{r} \right) \right| = + r \frac{d}{dr} \left(\frac{v_\theta}{r} \right)$$

since $r\, d/dr(v_\theta/r)$ is positive. Before combining equations (4.3-24) and (4.3-26), it will be helpful to evaluate the constant C_1 in equation (4.3-24). In terms of torque T, a boundary condition can be written as

$$T = \tau_{r\theta} \big|_{r=R} (2\pi R L)(R) \tag{4.3-27}$$

From equation (4.3-24) and (4.3-27), we evaluate $C_1 = T/2\pi L$, and hence equation (4.3-24) is now

$$\tau_{r\theta} = \frac{T}{2\pi L r^2} \tag{4.3-28}$$

The other boundary conditions remain as before:

$$v_\theta = 0 \qquad \text{at } r = \varkappa R \tag{4.3-29}$$

$$v_\theta = \Omega_0 R \qquad \text{at } r = R \tag{4.3-30}$$

It is now a straightforward procedure to combine equations (4.3-26), (4.3-28), (4.3-29), and (4.3-30) to arrive at a relationship between Ω_0 and T which allows the determination of the power-law coefficients m and n.

4.4. Periodic and Unsteady Flow Phenomena

Velocity profiles, pressure drop, and other fluid flow concepts are altered by the nonsteady movement of fluids. For example, suppose an incompressible fluid (Newtonian) is in the proximity of a flat wall that is suddenly set in motion. Figure 25 shows such a situation in rectangular

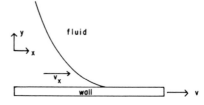

Fig. 25. Fluid in the vicinity of a wall suddenly set in motion.

coordinates. From Tables I and V, we obtain the governing equation of motion

$$\frac{\partial v_x}{\partial t} = \nu \frac{\partial^2 v_x}{\partial y^2} \tag{4.4-1}$$

in a one-dimensional laminar flow situation. The boundary and initial conditions for this setup are given by

$$v_x = 0 \quad \text{at } t \leq 0 \quad \text{for all } y \tag{4.4-2}$$

$$v_x = V \quad \text{at } y = 0 \quad \text{for } t > 0 \tag{4.4-3}$$

$$v_x = 0 \quad \text{at } y = \infty \quad \text{for } t > 0 \tag{4.4-4}$$

There are a number of methods available for the solution of equation (4.4-1) subject to equations (4.4-2), (4.4-3), and (4.4-4). One solution, by a similarity transformation, is shown here. A Laplace transform solution technique was presented in Chapter 2.

With the similarity technique we assume that the independent variable v_x can be transformed as shown by

$$\frac{v_x}{v} = \phi(\eta) \tag{4.4-5}$$

and

$$\eta = \frac{y}{(4\nu t)^{1/2}} \tag{4.4-6}$$

By partial differentiation of v_x with respect to t and y (using the "chain rule") we can reduce equation (4.4-1) to an ordinary differential equation with one independent variable. The details of the differentiation are given by

$$\frac{\partial v_x}{\partial t} = \frac{\partial v_x}{\partial \eta} \frac{\partial \eta}{\partial t} = \left(V \frac{\partial \phi}{\partial \eta} \right) [y(4\nu t)^{-3/2}(4\nu)] = -\frac{1}{2} \frac{\eta}{t} \frac{d\phi}{d\eta} \tag{4.4-7}$$

$$\frac{\partial v_x}{\partial y} = \frac{\partial v_x}{\partial \eta} \frac{\partial \eta}{\partial y} = \left(V \frac{\partial \phi}{\partial \eta} \right) (4\nu t)^{-1/2} \tag{4.4-8}$$

$$\frac{\partial^2 v_x}{\partial y^2} = \frac{\partial}{\partial y} \left(\frac{\partial v_x}{\partial y} \right) = \frac{\partial}{\partial \eta} \left[\frac{\partial v_x}{\partial y} \right] \frac{\partial \eta}{\partial y} = \frac{\eta^2}{y^2} \frac{d^2\phi}{d\eta^2} \tag{4.4-9}$$

By substituting equations (4.4-7), (4.4-8), and (4.4-9) into (4.4-1), we obtain

$$\frac{d^2\phi}{d\eta^2} + 2\eta\frac{d\phi}{d\eta} = 0 \qquad (4.4\text{-}10)$$

Equation (4.4-10) can be broken down into a simpler form by the transformation $P = d\phi/d\eta$ and $dP/d\eta = d^2\phi/d\eta^2$. With these forms put into equation (4.4-10), we can produce the solution to equation (4.4-10) as shown below:

$$P = C_1 e^{-\eta^2} \qquad (4.4\text{-}11)$$

and

$$\phi = C_1 \int_0^\eta e^{-\eta^2}d\eta + C_2 \qquad (4.4\text{-}12)$$

In order to evaluate constants C_1 and C_2 in equation (4.4-12), it is necessary to convert the boundary and initial conditions, equations (4.4-2), (4.4-3), and (4.4-4), into the ϕ, η form. This is done by using equations (4.4-5) and (4.4-6), yielding

$$\eta = 0 \qquad \phi = 1 \qquad (4.4\text{-}13)$$

$$\eta = \infty \qquad \phi = 0 \qquad (4.4\text{-}14)$$

Application of these two new boundary conditions allows the determination of C_1 and C_2, producing

$$\phi = 1 - \frac{2}{\sqrt{\pi}} \int_0^\eta e^{-\eta^2}d\eta \qquad (4.4\text{-}15)$$

and, finally,

$$\frac{v_x}{V} = 1 - \text{erf}\,\frac{y}{\sqrt{4vt}} \qquad (4.4\text{-}16)$$

where "erf" is the error function, a tabulated function (as are sine and cosine functions.) The error function is sometimes tabulated under headings such as Gaussian distribution or probability integrals.

If we are interested in startup conditions (the transient flow situation) we can go to Tables I and III, but this time we do not disregard the terms containing derivatives with respect to time. If the governing equation can be solved with the proper boundary and initial conditions, then we have the answer to the question of how the velocity profile is changing with time and how long it takes the velocity to become "fully developed." For the flow of a Newtonian fluid in a horizontal circular tube, starting from rest, the governing equation is obtained from Tables I and IX:

$$\varrho \frac{\partial v_z}{\partial t} = \frac{P_0 - P_L}{L} + \mu \frac{1}{r} \frac{\partial}{\partial r}\left(r \frac{\partial v_z}{\partial r}\right) \tag{4.4-17}$$

The boundary and initial conditions are given by

$$t = 0 \qquad v_z = 0 \tag{4.4-18}$$

$$r = 0 \qquad v_z = \text{finite} \tag{4.4-19}$$

$$r = R \qquad v_z = 0 \tag{4.4-20}$$

Equations (4.4-17) through (4.4-20) yield a solution by a number of techniques such as Laplace transforms, separation of variables, and similarity transforms. The solution is given by[6]

$$\phi = (1 - \xi^2) - 8 \sum_{n=1}^{\infty} \frac{J_0(\alpha_n \xi)}{\alpha_n^3 J_1(\alpha_n)} e^{-\alpha_n^2 \tau} \tag{4.4-21}$$

where

$$\phi = \frac{v_z}{(P_0 - P_L)R^2/4\mu L} \tag{4.4-22}$$

$$\xi = \frac{r}{R} \tag{4.4-23}$$

$$\tau = \mu t / \varrho R^2 \tag{4.4-24}$$

Note that ϕ is a dimensionless velocity, made dimensionless by dividing v_z by the maximum steady-flow velocity [set $dv_z/dt = 0$ in equation (4.4-17) and let $r = 0$ in the solution to this equation]. The dimensionless time, equation (4.4-24), is determined by working with the units of the various terms in the governing equation until a suitable dimensionless form evolves. Figure 26 shows how the velocity profile develops with time until steady conditions set in. This figure is a plot of equation (4.4-21).

Another nonsteady type of flow is pulsing flow, where the upstream (input) pressure is sinusoidal and we wish to know how the flow changes in this periodically varying problem. This type of analysis has some important applications in physiology (blood flow) and engineering (augmented heat and mass transfer operations). For sinusoidal pulsing flow of a Newtonian

[6] *Ibid.*, p. 129.

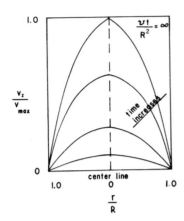

Fig. 26. Developing velocity profile from startup.

fluid in a rigid pipe, equation (4.4-17) is still applicable except we use $\partial P/\partial z$ instead of $(P_0 - P_L)/L$. The pressure downstream, for reasons discussed below, is

$$P = \left(P_0 - \frac{f\varrho\langle v_z\rangle^2 z}{Rg_c}\right)(1 + \sin \omega t) \qquad (4.4\text{-}25)$$

Assume that the pressure loss ΔP is in phase with the upstream pressure $P_0(1 + \sin \omega t)$. Thus, we write for ΔP,

$$\Delta P = \left(\frac{f\varrho\langle v_z\rangle^2 z}{Rg_c}\right)(1 + \sin \omega t) \qquad (4.4\text{-}26)$$

where

$$f = \text{Fanning friction factor}$$

$$\varrho = \text{density}$$

$$\langle v_z\rangle = \text{average velocity}$$

$$z = \text{distance along the pipe}$$

$$R = \text{radius of the pipe}$$

$$\omega = \text{angular velocity}$$

Note that as $\omega \to 0$ (steady flow), the pressure drop given in equation (4.4-26) is the expression called the Fanning equation, defining the Fanning friction factor. With equation (4.4-25) substituted into equation (4.4-17) and with appropriate boundary conditions similar to equations (4.4-18), (4.4-19), and (4.4-20), we can solve by Laplace transform techniques to yield the velocity profile[7]

$$v_z = 2\alpha k \sum_{n=1}^{\infty} \frac{J_0(r\mu_n/R)}{\mu_n J_1(\mu_n)} \left[\frac{R^2}{\alpha\mu_n^2} - \frac{\alpha R^2 \mu_n^2 \cos \omega\tau + R^4 \omega \sin \omega\tau}{\omega^2 R^4 + \alpha^2 \mu_n^4} \right.$$
$$\left. - \left(\frac{R^2}{\alpha\mu_n^2} - \frac{\alpha\mu_n^2 R^2}{\omega^2 R^4 + \alpha^2 \mu_n^4} \right) \right] \exp\left\{ -\frac{\alpha\mu_n^2}{R^2} \tau \right\} \qquad (4.4\text{-}27)$$

where

$$\tau = t - \frac{3\pi/2}{\omega}$$

μ_n = roots of the zero order Bessel function

α = kinematic viscosity

$k = f\langle v_z \rangle^2 / R\alpha$

$$\langle v_z \rangle = \frac{\displaystyle\int_0^{2\pi/\omega} \int_0^R v_z 2\pi r\, dr\, d\tau}{\displaystyle\int_0^{2\pi/\omega} \int_0^R 2\pi r\, dr\, d\tau} = \text{average velocity}$$

The average velocity $\langle v_z \rangle$ is obtained by the mean value theorem, where the velocity is averaged over one cycle of time, $t_c = 2\pi/\omega$, as well as over the radius. We apply the mean value theorem to equation (4.4-27) to produce $\langle v_z \rangle$ and from this result extract an explicit expression for the friction factor[8]:

$$f = \left(\frac{\pi}{16s} \right) \left(\frac{16}{N_{\text{Re}}} \right) \qquad (4.4\text{-}28)$$

where

$$s = \sum_{n=1}^{\infty} \left[\frac{2\pi}{\mu_n^4} + \frac{1}{\lambda\mu_n^4 + (\mu_n^6/\lambda)} - \frac{\lambda}{\mu_n^6} \right] (1 - e^{-2\pi\nu\mu_n^6/R^2}) \qquad (4.4\text{-}29)$$

and

$$\lambda = R^2 \omega / \nu$$

Thus for sinusoidally pulsing flow the friction factor equation is quite similar to the steady-flow expression, except for a theoretical factor which can be easily computed without the need for experimental data. Figure 27 shows some theoretical curves based on equation (4.4-28) and some experimental results superimposed on these curves.[9] Figure 28 correlates the critical Reynolds number with λ.[10]

---

[7] D. Hershey and G. Song, *A.I.Ch.E. Journal*, **13**, 491–496 (1967).

[8] *Ibid.*

[9] *Ibid.*, p. 493.

[10] D. Hershey and C. S. Im, *A.I.Ch.E. Journal*, **14**, 808 (1968).

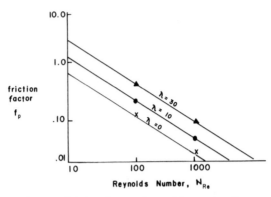

Fig. 27. Friction factors for pulsing water flow.

4.5. Flow in Various Geometrical Configurations

In some flow situations it is useful to develop a model which proposes that when a fluid flows in the vicinity of a wall, the region near the wall is in laminar flow, whereas the bulk fluid has a uniform velocity, v_∞. It is assumed that in the "boundary layer" of thickness δ there is a velocity profile. This situation is sketched in Figure 29. From Table I we get the continuity relationship for this situation:

$$\frac{\partial v_x}{\partial x} + \frac{\partial v_y}{\partial y} = 0 \tag{4.5-1}$$

The governing equation for the boundary layer is obtained from Table VIII (Newtonian fluid) and is

$$v_x \frac{\partial v_x}{\partial x} + v_y \frac{\partial v_x}{\partial y} = \nu \left(\frac{\partial^2 v_x}{\partial x^2} + \frac{\partial^2 v_x}{\partial y^2} \right) \tag{4.5-2}$$

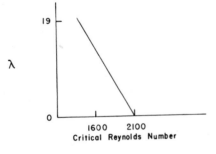

Fig. 28. Critical Reynolds number vs λ for sinusoidally pulsing flow.

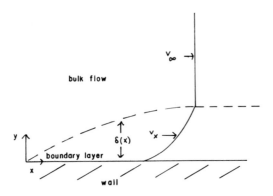

Fig. 29. Flow of a fluid near a wall.

The boundary conditions can be written as

$$v_x = 0 \qquad \text{at } y = 0 \qquad\qquad (4.5\text{-}3)$$

$$v_x = v_\infty \qquad \text{at } y = \infty \qquad\qquad (4.5\text{-}4)$$

$$v_x = v_\infty \qquad \text{at } x = 0 \qquad\qquad (4.5\text{-}5)$$

The y-component of motion has been neglected here since v_y is considered negligibly small. Because of the $v_x\, \partial v_x/\partial x$ and $v_y\, \partial v_x/\partial y$ terms, equation (4.5-2) is a nonlinear partial differential equation and hence is difficult to solve by straightforward techniques. There are some boundary layer methods of solution involving similarity transformations. The velocity profile in the boundary layer which is the solution of equations (4.5-1) and (4.5-2) can be written as[11]

$$\frac{v_x}{v_\infty} = \tfrac{3}{2}\left(\frac{y}{4.64\sqrt{vx/v_\infty}}\right) - \tfrac{1}{2}\left(\frac{y}{4.64\sqrt{vx/v_\infty}}\right)^3 \qquad (4.5\text{-}6)$$

and the thickness is

$$\delta(x) = 4.64\sqrt{vx/v_\infty} \qquad\qquad (4.5\text{-}7)$$

Another flow situation, between two flat plates, is shown in Figure 30. The flow equations from Tables I and VIII are given by

$$\varrho\,\frac{\partial v_x}{\partial t} = -\frac{dP}{dx} + \mu\left(\frac{\partial^2 v_x}{\partial y^2}\right) \qquad (4.5\text{-}8)$$

[11] R. B. Bird, W. E. Stewart, and E. N. Lightfoot, *op. cit.*, p. 145.

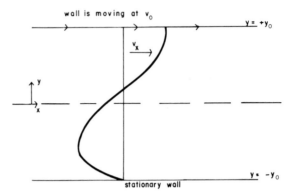

Fig. 30. Flow of a fluid between flat plates with a moving
wall.

with boundary conditions

$$v_x = 0 \qquad \text{at } y = -y_0 \qquad (4.5\text{-}9)$$

$$v_x = v_0 \qquad \text{at } y = +y_0 \qquad (4.5\text{-}10)$$

This phenomenon, with a moving boundary, is called a Couette problem.
For a steady-state situation, $\partial v_x/\partial t = 0$. If in equation (4.5-8) dP/dx is a
constant, then by separating the variables and integrating we can get

$$v_x = \frac{v_0}{2}\left(1 + \frac{y}{y_0}\right) - \frac{y_0^2}{2\mu}\frac{dP}{dx}\left(1 - \frac{y^2}{y_0^2}\right) \qquad (4.5\text{-}11)$$

For $dP/dx = 0$ we see from equation (4.5-11) that the velocity profile is
linear. If dP/dx is negative, then the velocity v_x is positive. However where
dP/dx is positive, the velocity can become negative and hence a backflow
can exist. The point of maximum flow reversal is found from the condition
$\partial v_x/\partial y = 0$ or

$$y = \frac{-\mu v_0}{2y_0 dP/dx} \qquad (4.5\text{-}12)$$

For flow through a pipe of elliptical cross section of major and minor
semiaxes a and b, the velocity profile equation is given by [12]

$$v_x = \frac{(\Delta P)a^2 b^2}{2\mu L(a^2 + b^2)}\left(1 - \frac{y^2}{a^2} - \frac{z^2}{b^2}\right) \qquad (4.5\text{-}13)$$

[12] J. Happel and H. Brenner, *Low Reynolds Number Hydrodynamics*, Prentice-Hall,
Englewood Cliffs, N.J. (1965), p. 38.

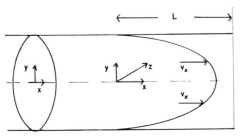

Fig. 31. Flow in an elliptical pipe.

where $v_x = 0$ is satisfied on the circumference of the ellipse, $y^2/a^2 + z^2/b^2 = 1$. Figure 31 illustrates the geometry. The volumetric flow is given by

$$Q = \frac{\pi \Delta P}{4\mu L} \frac{a^3 b^3}{a^2 + b^2} \qquad (4.5\text{-}14)$$

For a pipe with a rectangular cross section, with sides a and b in the directions y and z respectively, the velocity profile equation is[13]

$$v_x = -\frac{\Delta P}{2\mu L} y(y - a)$$

$$+ \sum_{m=1}^{\infty} \sin\left(\frac{m\pi y}{a}\right)\left(A_m \cosh \frac{m\pi z}{a} + B_m \sinh \frac{m\pi z}{a}\right) \qquad (4.5\text{-}15)$$

Figure 32 shows the flow with boundary conditions $v_x = 0$ at $y = 0$ and $y = a$. We determine A_m and B_m using the boundary conditions

$$v_x = 0 \qquad \text{at } z = 0 \qquad (4.5\text{-}16)$$

$$v_x = 0 \qquad \text{at } z = b \qquad (4.5\text{-}17)$$

From equations (4.5-15) and (4.5-16), we get

$$0 = -\frac{\Delta P}{2\mu L} y(y - a) + \sum_{m=1}^{\infty} A_m \sin \frac{m\pi y}{a} \qquad (4.5\text{-}18)$$

By multiplying equation (4.5-18) by $\sin(m\pi y/a)$ and invoking the orthogonality property of the sine function, we can evaluate A_m:

$$A_m = \frac{\Delta P}{2\mu a L} \int_0^a y(y - a) \sin \frac{m\pi y}{a} \, dy$$

$$= \frac{a^2 \Delta P}{\mu m^3 \pi^3 L} (\cos m\pi - 1) \qquad (4.5\text{-}19)$$

[13] *Ibid.*, p. 38.

With equation (4.5-17) and $b = \eta a$, we get from equation (4.5-15)

$$0 = -\frac{\Delta P}{2\mu L}y(y - a) + \sum_{m=1}^{\infty} \sin\left(\frac{m\pi y}{a}\right)(A_m \cosh m\eta\pi + B_m \sinh m\eta\pi)$$

$$(4.5\text{-}20)$$

which gives

$$B_m = -A_m\frac{(\cosh m\pi - 1)}{\sinh m\eta\pi} \tag{4.5-21}$$

Now it is possible to derive the volumetric flow rate from the following equation:

$$Q = 2\int_0^{h/2}\int_0^{a/2} v_x dy dz$$

$$= \frac{\Delta P}{24\mu L}ab(a^2 + b^2) - \frac{8\Delta P}{\pi^5\mu L}\sum_{n=1}^{\infty}\frac{1}{(2n - 1)^5}\left[a^4 \tanh\left(\frac{2n - 1}{2a}\pi b\right)\right.$$

$$\left. + b^4 \tanh\left(\frac{2n - 1}{2b}\pi a\right)\right] \tag{4.5-22}$$

For square ducts, where $a = b$, equation (4.5-22) becomes

$$Q = \frac{(\Delta P)a^4}{12\mu L}\left[1 - \frac{192}{\pi^5}\sum_{n=1}^{\infty}\frac{1}{(2n - 1)^5}\tanh\left(\frac{2n - 1}{2}\pi\right)\right] \quad (4.5\text{-}23)$$

If we sum the series in equation (4.5-23), simplification yields the final result

$$Q = \frac{0.03514\Delta Pa^4}{\mu L} \tag{4.5-24}$$

For an equilateral triangular conduit with sides of length b, with the origin at the center of the cross section and with the y-axis parallel to one of the

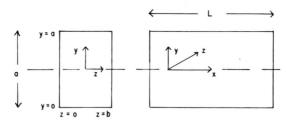

Fig. 32. Flow in a rectangular duct.

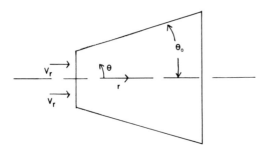

Fig. 33. Flow through a conical duct.

sides, we get for the velocity profile[14]

$$v_x = \frac{\sqrt{3}\,\Delta P}{6\mu bL}\left(z - \frac{b}{2\sqrt{3}}\right)\left(z + \sqrt{3}\,y - \frac{b}{\sqrt{3}}\right)\left(z - \sqrt{3}\,y - \frac{b}{\sqrt{3}}\right)$$

(4.5-25)

with the volumetric flow rate

$$Q = \frac{\sqrt{3}\,\Delta P b^4}{320\mu L}$$

(4.5-26)

The configuration used in the analysis of flow in conical sections, employing spherical coordinates r, θ, ϕ, is shown in Figure 33. The velocity profile is given by[15]

$$v_r = \frac{3Q}{2\pi r^2}\,\frac{\xi^2 - \xi_0^2}{(2\xi_0)(1 - \xi_0)^2}$$

(4.5-27)

where $\xi = \cos\theta$ and Q is the volumetric flow rate. For small values of θ_0, the velocity profile is simplified:

$$v_r \simeq \frac{2Q}{\pi(r\theta_0)^4}\,[(r\theta_0)^2 - (r\theta)^2]$$

(4.5-28)

With

$$a = r\sin\theta_0 \simeq r\theta_0$$

$$\bar{w} = r\sin\theta \simeq r\theta$$

$$z = r\cos\theta \simeq r$$

[14] Ibid., p. 39.
[15] Ibid., p. 140.

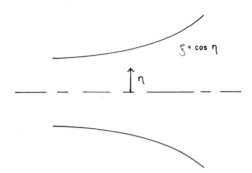

Fig. 34. Flow in a Venturi section.

we can get, for small values of θ_0, a compact equation for velocity, applicable for very slow flow situations:

$$v_z = \frac{2Q}{\pi a^4} (a^2 - \overline{w}^2) \tag{4.5-29}$$

A significant aspect of the solution for creeping flow (very slow flow) in a conical section is the lack of distinction between converging and diverging flows. This is not true when the inertial effects are taken into account. In the Navier–Stokes equations the inertial term is $\varrho \overline{v} \, \nabla v$, and this is neglected in very slow flow situations. For slow flow in a Venturi section, the situation is sketched in Figure 34, in spherical coordinates. The velocity profile is given by[16]

$$\phi = \frac{Q}{2\pi} \; \frac{\xi(\xi^2 - 3\xi_0^2) - (1 - 3\xi_0^2)}{(1 + 2\xi_0)(1 - \xi_0)^2} \tag{4.5-30}$$

where ϕ is the stream function defined by

$$v_r = -\frac{1}{r^2 \sin \theta} \frac{\partial \phi}{\partial \theta} \tag{4.5-31}$$

and

$$v_\theta = \frac{1}{r \sin \theta} \frac{\partial \phi}{\partial r} \tag{4.5-32}$$

4.6. Macroscopic Flow, Friction Factors, and Turbulent Flow

Most of the material presented in this chapter is applicable only for laminar flow where the Reynolds number is less than about 2100. For Reynolds numbers greater than 2100 a knowledge of instantaneous values

[16] *Ibid.*, p. 150.

of fluctuating pressure and flow is required. The fluid, in this nonlaminar (turbulent) flow, is moving in some disorganized, random regime. In turbulent flow, one could time-average everything and apply the equations of motion. With this approach we introduce additional terms to account for fluctuations and study the turbulence statistically and empirically. The velocity profiles tend, in turbulent flow, toward the uniform or flat shape sketched in Figure 35. Three regions of flow behavior can be considered as noted in Figure 35.

We are obviously quite interested in the criteria established to determine whether a flow is laminar, transitional, or turbulent. One can do the classic experiments of injecting dye solutions into the flowing stream and observe the motion of the stream. As the velocity of the main flow is increased the dye stream begins a wave or sinuous motion, signifying the beginning of the transitional regime. Still higher flows cause the dye to break up into what are called turbulent eddies. If the pipe is smooth and no disturbances are introduced at the entrance region, Reynolds numbers of 30,000 are possible before the laminar regime is altered. This introduces the concept of the critical Reynolds number, above which there is no longer laminar flow. In general the transition from laminar flow is said to occur when the generation of two- or three-dimensional waves is amplified causing turbulent zones to appear and propagate. In seeking convenient ways to pinpoint the departure from laminar flow, the empirical friction factor is useful for predicting the onset of the transitional behavior. Usually the friction factor is defined in terms of the drag that a fluid exerts on the walls of a pipe and is written as

$$F = Akf \qquad (4.6\text{-}1)$$

where F is the fluid drag force, A the area of contact, f the friction factor,

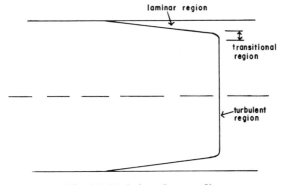

Fig. 35. Turbulent flow profile.

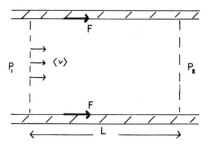

Fig. 36. Flow situation defining the friction factor.

and k the correlating factor. For flow in circular pipes equation (4.6-1) becomes

$$F = (2\pi RL)(\tfrac{1}{2}\varrho\langle v\rangle^2)(f) \qquad (4.6\text{-}2)$$

where

$$A = 2\pi RL$$

$$k = \tfrac{1}{2}\varrho\langle v\rangle^2 = \text{kinetic energy of the fluid}$$

$$\langle v\rangle = \text{average velocity of the fluid}$$

For high-flow conditions the drag force F is balanced by the pressure drop, as sketched in Figure 36. Thus $F = (\Delta P)(\pi R^2)$, and if this relationship is combined with equation (4.6-2), the result is the Fanning equation

$$f = \frac{R\Delta P}{L\varrho\langle v\rangle^2} \qquad (4.6\text{-}3)$$

Figure 37 shows how the friction factor behaves typically. For turbulent flow an empirical relationship, referred to as the Blasius equation, is given by

$$f = \frac{0.0791}{N_{\mathrm{Re}}^{1/4}} \qquad (4.6\text{-}4)$$

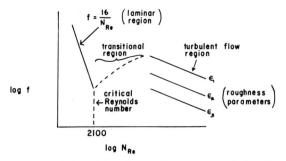

Fig. 37. Friction factor–Reynolds number relationship.

Another empirical turbulent flow relationship is

$$\frac{1}{\sqrt{f}} = 4.0 \log\left(\frac{N_{\text{Re}}}{\sqrt{f}}\right) - 0.40 \qquad (4.6\text{-}5)$$

In Figure 37 the turbulent flow parameter ε is related to the roughness of the pipe. Sometimes it is helpful to define a "mean hydralic radius," R_H, by

$$R_H = \frac{s}{z} \qquad (4.6\text{-}6)$$

where s is the cross section of the pipe and z is the wetted perimeter of the pipe. Thus from equation (4.6-3) and (4.6-6) the Fanning equation is rewritten,

$$f = \frac{2\Delta P R_H}{L\rho\langle v\rangle^2} \qquad (4.6\text{-}7)$$

where for circular pipes $R_H = \pi R^2/2\pi R = R/2$. For turbulent flow, where it is not convenient to work with velocity profiles, it is useful to write averaged (macroscopic) equations for momentum flow. Usually in the macroscopic approaches we set up the analysis as shown in Figure 38. A mass balance between sections 1 and 2 yields

$$\rho_1\langle v_1\rangle S_1 - \rho_2\langle v_2\rangle S_2 = \frac{dm}{dt} \qquad (4.6\text{-}8)$$

where

$$\text{summation in} - \text{summation out} = \text{rate of accumulation}$$

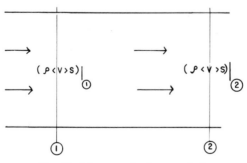

Fig. 38. Macroscopic flow analysis.

and

$\varrho_1 = $ density at plane 1

$\langle v_1 \rangle = $ average velocity at plane 1

$S_1 = $ cross-sectional area at plane 1

$m = $ mass in the section bounded by planes 1 and 2

If a macroscopic momentum (force) balance is made on this same section, the result is

$$\left(\frac{\varrho_1 v_1}{g_c} \right)(v_1 S_1) - \left(\frac{\varrho_2 v_2}{g_c} \right)(v_2 S_2) + P_1 S_1 - P_2 S_2 - F + m \frac{g}{g_c} = \frac{d\mathfrak{P}}{dt}$$

(4.6-9)

where

$\dfrac{\varrho_1 v_1}{g_c} = \dfrac{\text{momentum}}{\text{volume}}$ at section 1

$v_1 S_1 = $ volumetric flow rate at section 1

$P_1 = $ fluid pressure at section 1

$S_1 = $ cross-sectional area at section 1

$F = $ frictional drag force of the fluid on the pipe wall

$m \dfrac{g}{g_c} = $ gravity force

$\mathfrak{P} = $ momentum of the fluid system between sections 1 and 2

Note that the frictional drag F incorporates the less well defined momentum flux terms arising from viscous effects (shear stress). Thus F is a correction term making equation (4.6-9) more exact, though it now contains an empirical factor. Equation (4.6-9) may also be written as

$$\frac{d\mathfrak{P}}{dt} = -\varDelta \left[\frac{\langle v^2 \rangle}{\langle v \rangle} w + PS \right] - F + m \frac{g}{g_c}$$

(4.6-10)

where

$\varDelta = [\;]|_2 - [\;]|_1$, the bracketed quantity at location 2 minus the bracketed quantity at location 1

$w = \varrho \langle v \rangle S = $ mass flow rate

For steady state, $d\mathfrak{B}/dt = 0$ and equation (4.6-10) becomes

$$F = -\varDelta\left[\frac{\langle v^2\rangle}{\langle v\rangle}\, w + PS\right] + m\,\frac{g}{g_c} \qquad (4.6\text{-}11)$$

The ratio $\langle v^2\rangle/\langle v\rangle$ can be evaluated if the velocity profiles are known theoretically (laminar flow) or from experimental data. For most turbulent flow cases, where the velocity profile is essentially flat, we can write $\langle v^2\rangle/\langle v\rangle \simeq \langle v\rangle$.

By an analogous approach it is possible to derive a mechanical energy equation[17]

$$\frac{d}{dt}\,(K + \phi + A) = -\varDelta\left[\left(\frac{1}{2}\,\frac{\langle v^3\rangle}{\langle v\rangle} + \phi + G\right)w\right] - W - E_v \qquad (4.6\text{-}12)$$

where

$$K = \text{kinetic energy}$$

$$\phi = \text{potential energy}$$

$$A = \text{Helmholtz free energy}$$

$$W = \text{rate of work done by the fluid on the surroundings}$$

$$E_v = \text{friction loss}$$

$$G = \text{Gibbs free energy}$$

For steady state, $d(K + \phi + A)/dt = 0$ and $w_1 = w_2$. For isothermal systems

$$\varDelta G = \int_{P_1}^{P_2} \frac{1}{\varrho}\, dP$$

With these substitutions equation (4.6-12) becomes the Bernoulli equation

$$\varDelta\,\frac{1}{2}\,\frac{\langle v^3\rangle}{\langle v\rangle} + \varDelta\phi + \int_{P_1}^{P_2} \frac{1}{\varrho}\, dP + W + E_v = 0 \qquad (4.6\text{-}13)$$

In turbulent flow $\langle v^3\rangle/\langle v\rangle \simeq \langle v\rangle^2$, making computations much more convenient. Also the potential energy term $\varDelta\phi$ can be replaced by $(g/g_c)\varDelta z$, (the gravity term associated with changes of elevation). For isothermal ideal gases, $1/\varrho$ can be replaced by $RT/(\text{M.W.})(P)$, where R is the gas constant, T the temperature, and M.W. the molecular weight. In evaluating

[17] R. B. Bird, W. E. Stewart, and E. N. Lightfoot, op. cit., p. 212.

TABLE XII[a]

Empirical Values of e_v for Use in Equation (4.6-14)

Sudden contraction	$0.45(1 - \beta)$
Sudden expansion	$\left(\dfrac{1}{\beta} - 1\right)^2$
Orifice (sharp-edged)	$2.7(1 - \beta)(1 - \beta^2)\dfrac{1}{\beta^2}$
90° elbows	0.4–0.9
45° elbows	0.3–0.4
Globe valve (open)	6–10
Gate valve (open)	0.2
$\beta = \dfrac{\text{smaller cross-sectional area}}{\text{larger cross-sectional area}}$	

[a] R. B. Bird, W. E. Stewart, and E. N. Lightfoot, *Transport Phenomena*, John Wiley, New York (1960), p. 217.

the friction loss term E_v in equation (4.6-13), the energy loss is proportional to the kinetic energy:

$$E_v = \tfrac{1}{2}\langle v\rangle^2 e_v \tag{4.6-14}$$

where e_v is evaluated for different types of friction loss. For example, $e_v = (L/R_H)f$ for frictional loss due to fluid contact with the pipe wall. For circular pipes the mean hydraulic radius R_H is $R/2$, so that $e_v = 2Lf/R$. For other forms of energy loss such as that at pipe fittings and valves, or pipe expansions, the appropriate e_v is empirically determined. Table XII summarizes some e_v values. When determining E_v in equation (4.6-14), we sum all the e_v terms that contribute to the friction loss, such as globe valves, orifices, fluid friction, etc.

4.7. Non-Newtonian Macroscopic Properties

In the mathematical analysis of fluid flow phenomena it is necessary to impute to the fluid some internal (viscometric) properties. Generally for each fluid there is available an empirical flow behavior curve as shown in Figure 16. The majority of non-Newtonian fluids are characterized as pseudoplastic (power-law) in behavior. Power-law behavior was described in Section 3.3. From equation (3.3-1) it is possible to define an apparent viscosity μ_A for a power-law fluid:

$$\mu_A = \frac{\text{shear stress}}{\text{shear rate}} = \frac{\tau_{xy}}{dv_x/dy} = -m \left| \frac{dv_x}{dy} \right|^{n-1} \qquad (4.7\text{-}1)$$

Since n is less than unity for power-law fluids, equation (4.7-1) shows that the apparent viscosity decreases with increasing shear rate. This suggests that there may be a shear-related orientation process at work, particularly for flow of suspensions. Examples of pseudoplastic fluids are rubber solutions, polymer solutions, greases, soap, mayonnaise, paper pulp, paints, blood. Another class of non-Newtonian fluids is called dilatant, characterized by an increase in the apparent viscosity with increasing shear rate. This type of behavior is not very common and the classic examples are suspensions in which the solids concentration is high (in low-viscosity fluids). Other examples are some starch solutions, wet beach sand, and quicksand. It is suggested that for dilatant materials the suspending liquid is originally just sufficient to fill the voids. When the system is stressed there is an expansion in the volume (volume dilatancy) and a decreased "lubrication" effect. There are also time-dependent fluids, whose behavior depends upon their previous temporal history. Some of these fluids are called thixotropic, and exhibit reversible hysteresis properties as shown in Figure 39. Examples of thixotropic fluids are margarine, paints, and some heavy polymers. Rheopectic fluids, such as some clay solutions and gypsum suspensions, are thixotropic fluids that seem to thicken or dilate when stressed.

In working with these complex non-Newtonian fluids, and in attempting to characterize them, many times the only data available are shear stress at the wall, τ_w, average velocity, $\langle v_z \rangle$, and volumetric flow rate, Q. It then behooves the designer of equipment and processes to find short, less mathematical ways of characterizing the fluid behavior. For example, under

Fig. 39. Hysteresis loop behavior for thixotropic fluids.

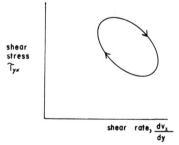

conditions of steady, fully developed flow in a horizontal pipe, we may wish to compare shear stress at the wall, τ_w, with shear rate at the wall, $(\partial v_z/\partial r)_w$. From equation (4.2-11) the shear stress at the wall is $\tau_w = D\Delta P/4L$. The rate of shear at the wall may be calculated from the Rabinowitsch–Mooney equation,[18] which is derived from the volumetric flow rate

$$Q = \int_0^R v_z 2\pi r \, dr = \pi \int_0^{R^2} v_z d(r^2) \tag{4.7-2}$$

Integrating by parts, we get equation

$$Q = \pi \left[v_z r^2 - \int r^2 dv_z \right] \Big|_0^{R^2} \tag{4.7-3}$$

Using the boundary condition $v_z = 0$ ar $r = R$, we can simplify equation (4.7-3), and with some rearrangement we get

$$\frac{Q}{\pi R^3} = \frac{8Q}{\pi D^3} = \frac{1}{\tau_w^3} \int_0^{\tau_w} \tau^2 f(\tau) d\tau \tag{4.7-4}$$

where $f(\tau) = -dv_z/dr$ is the shear rate. If equation (4.7-4) is multiplied by τ_w^3 and then differentiated with respect to τ_w (using the Leibnitz rule for differentiating an integral), the result after some simplification is

$$3\left(\frac{8Q}{\pi D^3}\right) + \frac{D\Delta P}{4L} \frac{d(8Q/\pi D^3)}{d(D\Delta P/4L)} = f(\tau_w) = \left(-\frac{dv_z}{dr}\right)_w \tag{4.7-5}$$

Equation (4.7-5) is the Rabinowitsch–Mooney equation giving the shear rate at the wall, $(-dv_z/dr)_w$, in terms of easily measured quantities. This equation holds for steady laminar flow in a circular pipe, where the fluid is time independent. Equation (4.7-5) can be further rearranged to yield

$$\left(-\frac{dv_z}{dr}\right)_w = \frac{8\langle v \rangle}{D} \frac{3n' + 1}{4n'} \tag{4.7-6}$$

where

$$n' = \frac{d\ln(D\Delta P/4L)}{d\ln(8\langle v\rangle/D)} \tag{4.7-7}$$

Equation (4.7-7) can be compared to the Newtonian flow situation. The Hagen–Poiseville equation (4.2-17), which is the governing equation for

[18] A. H. P. Skelland, *Non-Newtonian Flow and Heat Transfer*, John Wiley, New York (1967), p. 50.

the Newtonian case, can be rearranged to yield

$$\ln(D\Delta P/4L) = \ln\left(\frac{8\langle v\rangle}{D}\right) + \ln(\mu/g_c) \qquad (4.7\text{-}8)$$

and

$$\frac{d\ln(D\Delta P/4L)}{d\ln(8\langle v\rangle/D)} = 1.0 \qquad \left(\begin{array}{l}\text{Newtonian fluid:}\\ \mu = \text{viscosity} = \text{constant}\end{array}\right) \qquad (4.7\text{-}9)$$

Comparing equations (4.7-7) and (4.7-9), we note that for a Newtonian fluid $n' = 1$, and if this value is inserted into equation (4.7-6), we confirm that the shear rate at the wall for a Newtonian fluid is given by

$$\left(-\frac{dv_z}{dr}\right)_w = \frac{8\langle v\rangle}{D} \qquad (4.7\text{-}10)$$

Generally, for non-Newtonian fluids n' is not equal to 1.0; indeed n' is not necessarily constant, though it may be constant over a limited flow range. Thus, in seeking to obtain a correlation of shear stress at the wall, τ_w, and shear rate at the wall, $(-dv_z/dr)_w$, it is necessary to evaluate n' for use in equation (4.7-6). From equation (4.7-7), a logarithmic plot of $D\Delta P/4L$ (which is τ_w) versus $8\langle v\rangle/D$ (pseudo shear rate) determines n'. Another form for equation (4.7-7) is given by

$$\tau_w = \frac{D\Delta P}{4L} = K'\left(\frac{8\langle v\rangle}{D}\right)^{n'} \qquad (4.7\text{-}11)$$

From equation (4.7-4) it is possible to derive specific flow relationships simply by substituting appropriately for the shear rate $f(\tau) = -dv_z/dr$. For example, for Newtonian fluids, $\tau = -\mu\, dv_z/dr = -\mu f(\tau)$. Thus $f(\tau) = -\tau/\mu$, which can be substituted into equation (4.7-4) to yield, for Newtonian fluids,

$$\frac{Q}{\pi R^3} = \frac{R\Delta P}{8\mu L} \qquad (4.7\text{-}12)$$

For power-law fluids in laminar flow in circular pipes, we get from equation (3.4-1) the shear stress–shear rate relationship

$$\tau_{rz} = m\left(-\frac{dv_z}{dr}\right)^n \qquad (4.7\text{-}13)$$

or

$$-\frac{dv_z}{dr} = f(\tau_{rz}) = f(\tau) = \left(\frac{\tau}{m}\right)^{1/n}$$

so that from equation (4.7-4) with some manipulations we get for a power-law fluid

$$\frac{Q}{\pi R^3} = \frac{n}{3n+1}\left(\frac{\tau_w}{m}\right)^{1/n} \tag{4.7-14}$$

In a similar manner it is possible to derive relationships for a Bingham plastic. Table XIII summarizes some non-Newtonian relationships.

In discussing the flow of fluids it has been tacitly assumed that entrance effects were negligible and that there were no expansions, contractions, and fittings to be considered. Entrance effects arise when a fluid enters a pipe

TABLE XIII[a]

Flow Relationships for Power-Law and Bingham Fluids[b]

$\dfrac{Q}{\pi R^3} = \dfrac{\tau_w}{4\eta}\left[1 - \dfrac{4}{3}\left(\dfrac{\tau_y}{\tau_w}\right) + \dfrac{1}{3}\left(\dfrac{\tau_y}{\tau_w}\right)^4\right]$	Bingham fluid—Circular pipe
$\dfrac{Q}{WH^2} = \dfrac{\tau_w}{6\eta}\left[1 - \dfrac{3}{2}\left(\dfrac{\tau_y}{\tau_w}\right) + \dfrac{1}{2}\left(\dfrac{\tau_y}{\tau_w}\right)^3\right]$	Bingham fluid—Flat Plates
$\dfrac{Q}{\pi R^3} = \dfrac{n}{(3n+1)}\left(\dfrac{\tau_w}{m}\right)^{1/n}$	Power-law fluid—Circular pipes
$\dfrac{Q}{WH^2} = \left(\dfrac{1}{2}\right)\dfrac{n}{(2n+1)}\left(\dfrac{\tau_w}{m}\right)^{1/n}$	Power-law fluid—Flat plates
$v_z = \left(\dfrac{\Delta P}{2mL}\right)^{1/n}\dfrac{n}{n+1}[R^{n+1/n} - r^{n+1/n}]$	Power-law fluid—Circular pipes
$v_z = \left(\dfrac{\Delta P}{mL}\right)^{1/n}\dfrac{n}{n+1}\left[\left(\dfrac{H}{2}\right)^{n+1/n} - h^{n+1/n}\right]$	Power-law fluid—Flat plates

[a] A. H. P. Skelland, *Non-Newtonian Flow and Heat Transfer*, John Wiley, New York (1967), p. 110.
[b] See sketch for flat plate dimensions:

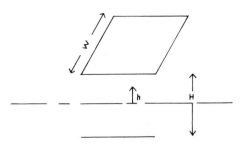

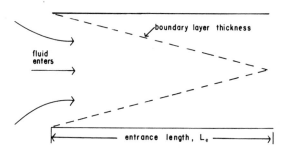

Fig. 40. Entry region in pipe flow.

as shown in Figure 40. The thickness of the boundary layer is zero at the entrance and increases along the tube. Flow is said to be fully developed beyond the point where the boundary layers converge on the centerline. The entrance length, L_e, is usually defined as the distance from the inlet to the point where the fluid velocity is 99% of the fully developed flow velocity (at the centerline.) The pressure loss per unit length along the entrance region is greater than where the flow is fully developed. This is the result of two entrance effects: (a) the conversion of pressure into kinetic energy, and (b) the increase in fluid friction caused by the high velocity gradients in the vicinity of the wall.

In working with power-law fluids, m and n were parameters introduced in equation (3.4-1). The parameters K' and n' introduced in equation (4.7-11) described a generalized macroscopic approach. The relationship between n and n' and m and K' can be shown by beginning with the power-law equation for flow in a circular pipe:

$$\tau = m\left(-\frac{dv_z}{dr}\right)^n \tag{4.7-15}$$

If we take logarithms of equation (4.7-15), the result is

$$\log \tau = \log m + n \log\left(-\frac{dv_z}{dr}\right) \tag{4.7-16}$$

which can be differentiated with respect to $(-dv_z/dr)$ to yield

$$\frac{d \log \tau}{d(-dv_z/dr)} = n$$

where m and n are assumed constant. It is possible to operate similarly on equation (4.7-6), and if logarithms are taken we obtain

$$\log\left(-\frac{dv_z}{dr}\right)_w = \log\left(\frac{8\langle v \rangle}{D}\right) + \log\left(\frac{3n' + 1}{4n'}\right) \tag{4.7-17}$$

Differentiating as before, with respect to $\log \tau_w$, equation (4.7-17) becomes

$$\frac{d \log(-dv_z/dr)_w}{d \log \tau_w} = \frac{d \log(8\langle v \rangle/D)}{d\tau_w} + \frac{d \log[(3n'+1)/4n']}{d \log \tau_w} \quad (4.7\text{-}18)$$

From $d \log \tau/d(-dv_z/dr) = n$, we can write

$$\frac{d \log \tau_w}{d(-dv_z/dr)_w} = n \quad (4.7\text{-}19)$$

which can be combined with equations (4.7-7) and (4.7-18) to yield

$$\frac{1}{n} = \frac{1}{n'} + \frac{d \log[(3n'+1)/4n']}{d \log \tau_w} \quad (4.7\text{-}20)$$

If n' is constant, then equation (4.7-20) yields $n = n'$. For a power-law fluid, it can be shown[19] that $n = n'$ identically and $K' = m[(3n'+1)/4n']^{n'}$. Thus n and m data from viscometric or tube flow data can be converted to n' and K' data. It is also possible to redefine a modified Reynolds number, $N_{Re'}$, to yield a generalized non-Newtonian friction factor–Reynolds number laminar flow relationship:

$$f = \frac{16}{N_{Re'}} \quad (4.7\text{-}21)$$

In arriving at this result, we combine equation (4.7-11) and the friction factor definition:

$$f = \frac{g_c(D\Delta P/4L)}{\rho v_z/2} = \frac{g_c \tau_w}{\rho v_z^2/2}$$

$$= \frac{16 g_c K' 8^{n'-1}}{D^{n'} v_z^{2-n'} \rho} \quad (4.7\text{-}22)$$

If the modified Reynolds number is defined by

$$N_{Re'} = \frac{D^{n'} \langle v_z \rangle^{2-n'} \rho}{g_c K' 8^{n'-1}} \quad (4.7\text{-}23)$$

then equation (4.7-21) is obtained.

For the Newtonian case, $n' = 1$ and $K' = \mu/g_c$. With equations (4.7-21) and (4.7-23) we find experimentally that the Newtonian concepts

---

[19] *Ibid.*, p. 172.

carry over as a special case and generally laminar flow prevails below $N_{Re'} = 2100$. For power-law fluids, it is often convenient to work in terms of $N_{Re'}$ when seeking friction factors and trying to establish whether flow is laminar or not. For turbulent flow, the power-law empirical relationship is found to be[20]

$$\left(\sqrt{\frac{1}{f}}\right)_{\substack{\text{power}\\\text{law}}} = 4 \log\left[(N_{Re'})_{\substack{\text{power}\\\text{law}}}(\sqrt{f})_{\substack{\text{power}\\\text{law}}}\right] - 0.40 \qquad (4.7\text{-}24)$$

4.8. Viscoelastic Non-Newtonian Properties

Some non-Newtonian fluids with normal stresses (τ_{xx}, τ_{yy}, τ_{zz}) and elastic as well as viscous characteristics are called viscoelastic fluids. For a viscoelastic fluid issuing as a jet, we observe that instead of the usual contraction, there is an expansion of the jet. It is observed that stirred viscoelastic fluids may actually climb up the rotating stirrer (Weissenberg effect). Flour dough, jellies, polymers, and polymer melts are examples of viscoelastic fluids. The number and importance of these fluids is increasing as fluid properties are measured more accurately.

When a purely viscous fluid is stressed and then unstressed, it recovers quickly to its zero-stress condition. Viscoelastic fluids may recover only part of their deformation upon removal of the stress. When nylon is extruded through fine perforations to make thread, the resulting thread cross section may be considerably larger than that of the hole through which it was extruded. This is the result of elastic recovery. The behavior of viscoelastic fluids depends upon their history, requiring the derivatives of shear stress and shear rate in the rheological equation. These fluids give stresses normal to the planes of shear, and the magnitude of these normal stresses varies with shear rate (τ_{xx}, τ_{yy}, τ_{zz} are not necessarily equal to zero).

In mechanical terms, two empirical viscoelastic models commonly proposed are the Maxwell body and the Voight body, defined by equations (4.8-1) and (4.8-2) respectively:

$$\bar{\bar{\tau}} = \mu \frac{d\bar{\bar{\gamma}}}{dt} - \frac{\mu}{k} \frac{d\bar{\bar{\tau}}}{dt} \qquad (4.8\text{-}1)$$

$$\bar{\bar{\tau}} = \mu \frac{d\bar{\bar{\gamma}}}{dt} + \bar{\bar{\gamma}} \qquad (4.8\text{-}2)$$

[20] *Ibid.*, p. 199.

where

$$\bar{\bar{\tau}} = \text{shear stress tensor}$$
$$\bar{\bar{\gamma}} = \text{shear rate tensor}$$
$$d\bar{\bar{\gamma}}/dt = \text{rate of shear tensor}$$
$$k = \text{rigidity modulus}$$
$$\mu = \text{viscosity}$$

There are other empirical models applicable specifically for low shear rates and others for high shear rates. These models and others generally require experimental determination of the parameters introduced. Where there is steady laminar flow, such as in a circular pipe of constant cross section, the elastic properties will exert an influence only as end effects at the inlet and outlet. In most of these cases it is usually sufficient for design purposes to impute purely viscous behavior to the fluid. The elastic effects become increasingly important in turbulent flow or in flow through fittings and through channels of varying cross section. Since the various rheological models require some empiricism, it becomes important to measure the non-vanishing normal stresses such as τ_{xx}, τ_{yy}, τ_{zz} (rectangular coordinates). The normal stresses are generally expressed as[21]

$$\tau_{xx} = -P + P_{xx} \qquad (4.8\text{-}3)$$

$$\tau_{yy} = -P + P_{yy} \qquad (4.8\text{-}4)$$

$$\tau_{zz} = -P + P_{zz} \qquad (4.\text{-}85)$$

where

$$P = \text{hydrostatic pressure} = -\tfrac{1}{3}(\tau_{xx} + \tau_{yy} + \tau_{zz})$$

$P_{xx}, P_{yy}, P_{zz} = $ "deviatoral stress" (representing deviations from isotropic behavior)

It follows from equations (4.8-3), (4.8-4), and (4.8-5) that

$$P_{xx} + P_{yy} + P_{zz} = 0 \qquad (4.8\text{-}6)$$

There is a "Weissenberg hypothesis," given by[22]

$$P_{yy} = P_{zz} \qquad (4.8\text{-}7)$$

which, if accepted, allows the rheological properties to be specified as

---

[21] *Ibid.*, p. 50.
[22] *Ibid.*, p. 51.

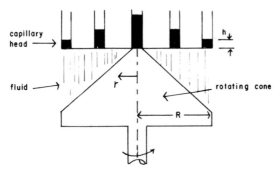

Fig. 41. A sketch of a Rheogoniometer.

functions of shear rate. These measurements can be made in an instrument called a rheogoniometer, sketched in Figure 41. It is a sophisticated plate-and-cone viscometer. Normal stresses in the fluid are obtained from the height of rise of the fluid in the capillary tubes. From a force balance it is possible to arrive at[23]

$$\varrho \frac{g}{g_c} \left(- \frac{dh}{d \ln r} \right) = P_{xx} + P_{yy} - 2P_{zz} \qquad (4.8\text{-}8)$$

where h is the height of the rise of the fluid in the capillary. From equation (4.8-8) we note that a plot of h versus $\ln r$ should yield slopes of $-(1/\varrho)$ $\times (P_{xx} + P_{yy} - 2P_{zz})g_c/g$, which can be set equal to $3(P_{zz}/\varrho)g_c/g$ by use of equation (4.8-6). Thus we can determine P_{zz} experimentally. From equation (4.8-7) we determine P_{yy} once P_{zz} is known. Then equation (4.8-6) yields P_{xx}.

If a viscoelastic fluid is sent through a long cylindrical pipe, it emerges as a horizontal jet. If surface tension and gravitational effects are negligible, one can measure the expanded diameter of the jet and deduce the normal stresses.[24] The elastic properties of viscoelastic fluids generally are neglected for flow in long tubes. In turbulent flow the elastic behavior may come into play.

4.9. Macroscopic Two-Phase Flow (Solid–Liquid)

The flow equations can be extended to multiphase systems if the assumption is made that the mass, momentum, and energy balances apply in each phase. The problem then resolves itself into an attempt to match

[23] *Ibid.*
[24] *Ibid.*, p. 53.

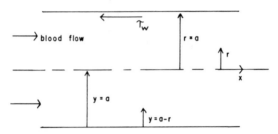

Fig. 42. Blood flow analysis-two phase flow.

the solutions at the interfaces. There are essentially four two-phase systems: liquid–gas, liquid–liquid, solid–gas, and solid–liquid. Liquid–gas two-phase transport takes place in equipment such as evaporators, coolers, condensers and some pipe flow situations. Liquid–liquid extraction equipment is quite commonly used. Solid–gas transport occurs in fluidized beds and pipe transport. Solid–liquid slurries travel in pipes, settle by gravity, occur in fluidized beds, etc. The solid–liquid systems are usually treated as homogeneous fluids, albeit non-Newtonian ones. Sometimes there is a relative velocity between fluid and solid, with the liquid phase usually moving more rapidly.

Suspensions can be considered homogeneous fluids if the settling velocity of the solids is less than 0.002 to 0.005 ft/sec.[25] In the flow of dilute suspensions of spheres in pipes at low Reynolds numbers, it has been found that radial forces cause the particles to move away from the pipe wall, producing a clear zone (a solids-depleted zone) next to the wall. This phenomenon is observed in aqueous suspensions of paper pulp, and even blood has been described as behaving in this manner.[26] This solids-depleted zone near the wall induces an apparent slip at the wall since it appears that the bulk material is sliding over a marginal layer of fluid. Many observers have noticed that when blood vessels of a living animal are viewed under a microscope, the red cells appear to be confined to an axial stream which is surrounded by a sheath of cell-poor plasma.

This behavior can be analyzed in a manner similar to that presented in Chapter 3. Figure 42 represents a fluid such as blood in laminar flow through a tube. From Tables I and III the governing equation is given by

$$\tau_{rz} = \frac{r}{2} \frac{dP}{dx} \qquad (4.9\text{-}1)$$

[25] R. S. Brodkey, *The Phenomena of Fluid Motion*, Addison-Wesley, Reading, Mass. (1967), p. 622.

[26] D. Hershey and S. J. Cho, *J. Appl. Physiol.* **21**, 27–32 (1966).

or, with $y = a - r$,

$$\tau_{rz} = \frac{a - y}{2} \frac{dP}{dx} \tag{4.9-2}$$

Next we substitute in the rheological behavior

$$\text{shear rate, } \dot{\gamma} = \frac{dv_x}{dy} = f(\tau_{rz})$$

where $f(\tau_{rz}) = (1/\mu)\tau_{rz}$ for a Newtonian fluid and $f(\tau_{rz}) = (\tau_{rz}/m)^{1/n}$ for a power-law fluid. For steady, laminar flow of blood in rigid tubes, if the red cells tended to migrate away from the wall there would be produced a region near the wall that had less red cells per unit volume than the bulk region. In this region close to the wall, the shear rate $\dot{\gamma}$ will be different from the homogeneous representation $f(\tau_{rz})$ by a factor $g(\tau_{rz}, y)$:

$$\dot{\gamma} = \frac{dv_x}{dy} = f(\tau_{rz}) + g(\tau_{rz}, y) \tag{4.9-3}$$

with $g(\tau_{rz}, y) = 0$ for $y > \varepsilon$, where ε is the apparent thickness of the cell-diminished zone adjacent to the wall. The velocity profile v_x can be obtained from equation (4.9-3) by separating the variables and integrating between $y = 0$ and $y = y$. The result is

$$v_x(y) = \int_0^y f(\tau_{rz})dy + \int_0^\varepsilon g(\tau_{rz}, y)dy$$

This equation has used the property of the g function indicated above, and also $v_x = 0$ at $y = 0$. The integral $\int_0^y f(\tau_{rz})dy$ is the velocity obtained by assuming a homogeneous condition. It is assumed that the $f(\tau_{rz})$ function is essentially unchanged by the small number of red cells that have migrated into the main flow area. Since we are dealing with velocities, it is helpful and consistent to define a so-called slip velocity of a region of thickness ε next to the wall. This slip velocity may be expressed by the equation

$$S = \int_0^\varepsilon g(\tau_{rz}, y)dy \tag{4.9-4}$$

where for a Newtonian fluid $g(\tau_{rz}, y) = \tau_{rz}/\mu$, and μ is the viscosity of the fluid next to the wall. Since the plasma fluid near the wall is Newtonian, the S function can be written as

$$S = \int_0^\varepsilon \frac{\tau_{rz}}{\mu} dy = \int_0^\varepsilon \frac{\tau_w}{a\mu} (a - y)dy \tag{4.9-5}$$

where from equation (4.9-2) we have used the relationship

$$\frac{\tau_{rz}}{\tau_w} = \frac{a - y}{a} \tag{4.9-6}$$

Equation (4.9-5) can be rearranged to yield

$$\xi = \frac{S}{\tau_w} = \int_0^\varepsilon \frac{a - y}{a\mu} \, dy \tag{4.9-7}$$

where ξ is called the slip coefficient. If the integration is performed in equation (4.9-7), the result is

$$\varepsilon^2 - 2a\varepsilon + 2a\mu\xi = 0 \tag{4.9-8}$$

From equation (4.9-8) it is possible to calculate the thickness ε of the cell-diminished zone near the wall if the slip coefficient is known. The slip coefficient ξ can be evaluated starting with an expression for the volumetric flow rate Q:

$$Q = \int_0^\varepsilon 2\pi(a - y)v_x dy \tag{4.9-9}$$

which can be integrated by parts and, after some rearrangement, yields[27]

$$\frac{Q}{\pi a^3 \tau_w} = \frac{\xi}{a} + \phi \tag{4.9-10}$$

where

$$\phi = \frac{1}{\tau_w^4} \int_0^{\tau_w} \tau_{rz}^2 f(\tau_{rz}) d\tau_{rz} \tag{4.9-11}$$

In arriving at equation (4.9-10), we have used $v_x = 0$ at $y = 0$ and $(a-y) \simeq a$ near the wall. Thus the slip coefficient can be evaluated from steady-flow-tube experiments and equation (4.9-10). For a range of tubes of different radii a, first a plot of $Q/\pi a^3 \tau_w$ against τ_w is made. If the flow regime behaves as if it were homogeneous, $Q/\pi a^3 \tau_w$ will be independent of radius a since $\xi = 0$ and hence the term ξ/a will not appear in equation (4.9-10). If the inhomogeneous two-phase model is reasonable, then $Q/\pi a^3 \tau_w$ should be distinct for each value of a since $\xi/a \neq 0$. Figure 43 shows this to be true for blood.[28] We next plot $Q/\pi a^3 \tau_w$ versus $1/a$ with τ_w as a parameter. This

[27] *Ibid.*
[28] *Ibid.*

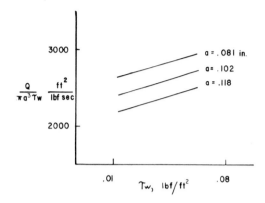

Fig. 43. Flow data for blood showing inhomogeneous condition.

is shown in Figure 44. The slope of the curves in Figure 44 are ξ values. Thus at selected values of τ_w corresponding ξ values are obtained, and hence the thickness ε of the marginal zone can be calculated from equation (4.9-8). Figure 45 shows the change in thickness of the cell-diminished zone near the wall.

Fluidized beds have been mentioned as examples of solid–liquid and gas–liquid two-phase contactors. In fluidized beds, it is of course extremely important to characterize the mixing process—in terms of residence time distributions for the particles. It is possible to use radioisotope methods, colored particles, impregnated particles, and ion-exchange resins to follow the position of the particles. Visual studies at the wall suggest a general downward motion at the wall, and it has further been observed that a large amount of solid material moves upward with slow velocities and smaller

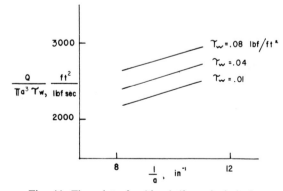

Fig. 44. Flow data for blood (for calculating).

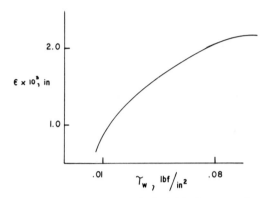

Fig. 45. Thickness of the cell-diminished zone at the
wall for blood flow.

amounts of solids move upward at higher velocities. The solids motion is
quite complex and apparently determined by gas or liquid flow.[29]

In discussing the interaction of particles with the containing walls,
it is necessary to note the particle shape, orientation, and position as well
as the geometry of the containing walls. Some recent work analyzed a
spinning sphere moving in a stationary fluid and determined that the particle experiences a lift force F_L, orthogonal to its direction, given by[30]

$$F_L = \pi a^3 \varrho \bar{\Omega} \times \bar{U}[1 + 0(N_{Re})] \qquad (4.9\text{-}12)$$

where

$$a = \text{radius of the sphere}$$
$$\bar{\Omega} = \text{angular velocity}$$
$$\bar{U} = \text{velocity of the sphere}$$
$$\varrho = \text{fluid density}$$
$$N_{Re} = \text{Reynolds number} = \varrho\bar{U}a/\mu$$
$$\mu = \text{fluid viscosity}$$
$$O(N_{Re}) = \text{terms that are functions of the Reynolds number}$$

For small values of N_{Re} the transverse force F_L is independent of the viscosity. For a sphere falling in a quiescent fluid it can be predicted that a
sphere located exactly on the center line axis will not rotate, but one located
off center will rotate.

[29] R. S. Brodkey, *op. cit.*, p. 620.
[30] J. Happel and H. Brenner, *op. cit.*, p. 316.

In studying the behavior of a multiple assemblage of spheres, the vessel wall effect is important. The simplest type of assemblage has the particles randomly distributed throughout the cross section, falling at the same velocity and maintaining fixed positions relative to each other. For this situation it can be shown that in pumping this fluid the following equation is valid[31]:

$$\Delta P_s = N_m L 8\pi \mu a \left(U_{mf} - \frac{3U}{4} \right) \qquad (4.9\text{-}13)$$

where

ΔP_s = excess (increase) in pressure due to the presence of the particles

N_m = mean number of particles per unit volume (a constant)

μ = viscosity of the fluid

a = radius of a particle

U_{mf} = average or superficial velocity of the fluid

U = particle velocity

For the case of sedimentation, $U_{mf} = 0$. Equation (4.9-13) may be expressed in terms of the fractional void, ε, starting with

$$\varepsilon = 1 - \frac{q(\frac{4}{3})\pi a^3}{\pi R_0^2 L} \qquad (4.9\text{-}14)$$

where q is the number of particles contained in a cylindrical section of length L, radius R_0, and fractional void volume ε. Thus the number of particles per unit volume, N_m, is given by

$$N_m = \frac{q}{\pi R_0^2 L} = \frac{3(1 - \varepsilon)}{4\pi a^3} \qquad (4.9\text{-}15)$$

Upon combining equation (4.9-15) with (4.9-13), the result is

$$\Delta P_s = 6(1 - \varepsilon) \frac{\mu L}{a^2} \left(U_{mf} - \frac{3U}{4} \right) \qquad (4.9\text{-}16)$$

For a dilute assemblage of particles equation (4.9-16) is simplified to[32]

$$\Delta P_s = 4.5(1 - \varepsilon)\mu U_{mf} \frac{L}{a^2} \qquad (4.9\text{-}17)$$

[31] *Ibid.*, p. 363.
[32] *Ibid.*, p. 364.

If the spheres are free to move relative to each other in the axial direction, with the possibility of nonuniform radial distribution of particles, it can be shown[33] that

$$\Delta P_s = 12\pi\mu a N_m L \frac{U_{\mathrm{TS}}}{U_{\mathrm{OF}}} \tag{4.9-18}$$

where U_{TS} is the terminal settling velocity of the sphere in an unbounded, quiescent fluid, and U_{OF}, is the fluid velocity at the center line. In terms of a fractional void volume ε, equation (4.9-18) becomes

$$\Delta P_s = \frac{9\mu L U_{\mathrm{TS}}(1 - \varepsilon)}{2 U_{mf} a^2} \tag{4.9-19}$$

The simplest of the models is that which assumes that the radial particle distribution is uniform. In pneumatic conveying and pumping of slurries this situation is approximated.

4.10. Boundary Layer Flow Analysis

Prandtl in 1904 introduced the concept of the boundary layer in an approximation to the Navier–Stokes equations for laminar flow at high Reynolds numbers. There have been many solutions of flow problems using the boundary layer approach, such as flow in entrance regions in pipes and flow over variously shaped wings and surfaces. When the Reynolds number is large the viscous forces can be considered to be confined to a small region near the boundary surface. Thus the flow is divided into a boundary region and an outer region. In the outer region (the main stream) we assume ideal, nonviscous flow under inertial forces only. In the boundary layer it is assumed that viscous forces are usually controlling.

For flow over a flat plate as sketched in Figure 46 the boundary conditions based on boundary layer concepts are given by

$$v_x = v_y = 0 \qquad \text{at } y = 0 \quad \text{for all } x \tag{4.10-1}$$

$$v_x = v_\infty \qquad \text{at } y = \delta \tag{4.10-2}$$

$$v_x = v_\infty \qquad \text{as } y \to \infty \tag{4.10-3}$$

The simplification of the Navier–Stokes equations applicable to boundary

[33] *Ibid.*, p. 365.

Fig. 46. Boundary layer flow over a flat
plate.

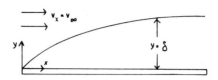

layer analysis begins by writing the governing equations from Table VIII.
These are given by

$$\frac{\partial v_x}{\partial t} + v_x \frac{\partial v_x}{\partial x} + v_y \frac{\partial v_x}{\partial y} = -\frac{1}{\varrho}\frac{dP}{dx} + \nu\left(\frac{\partial^2 v_x}{\partial x^2} + \frac{\partial^2 v_x}{\partial y^2}\right) \quad (4.10\text{-}4)$$

$$\frac{\partial v_y}{\partial t} + v_x \frac{\partial v_y}{\partial x} + v_y \frac{\partial v_y}{\partial y} = -\frac{1}{\varrho}\frac{dP}{dy} + \nu\left(\frac{\partial^2 v_y}{\partial x^2} + \frac{\partial^2 v_y}{\partial y^2}\right) \quad (4.10\text{-}5)$$

Next we apply an order-of-magnitude analysis, comparing all terms to a
unit length, x, unit time, t, and velocity v_∞. Thus v_x, which varies from
zero to v_∞, is of order 1. The distance y in the boundary layer varies between
zero and δ (the boundary layer thickness) and is of order δ. The distance x
is of order 1. Extending these ideas to the derivative $\partial v_x/\partial y$, we get the result
$1/\delta$. Similarly, for second derivatives,

$$\frac{\partial^2 v_x}{\partial y^2} = \frac{\partial}{\partial y}\left(\frac{\partial v_x}{\partial y}\right) = \frac{1}{\delta}\left(\frac{1}{\delta}\right) = \frac{1}{\delta^2}$$

The term $\partial v_x/\partial x$ is of order $1/1$ and

$$\frac{\partial^2 v_x}{\partial x^2} = \frac{\partial}{\partial x}\left(\frac{\partial v_x}{\partial x}\right) = \frac{1}{1}\left(\frac{1}{1}\right) = \frac{1}{1}$$

From the equation of continuity, Table I, we get

$$\frac{\partial v_x}{\partial x} + \frac{\partial v_y}{\partial y} = 0 \quad (4.10\text{-}6)$$

We have already noted that $\partial v_x/\partial x$ is of order $1/1$, so that $\partial v_y/\partial y$ is of order
$1/1$. Since y is of order δ, it becomes necessary for v_y to also be of order δ.
Thus $\partial^2 v_y/\partial y^2$ becomes of order

$$\frac{\partial}{\partial y}\left(\frac{\partial v_y}{\partial y}\right) \quad \text{or} \quad \frac{1}{\delta}\left(\frac{1}{1}\right) = \frac{1}{\delta}$$

Finally $\partial v_y/\partial x$ is of order $\delta/1 = \delta$ and

$$\frac{\partial^2 v_y}{\partial x^2} = \frac{\partial}{\partial x}\left(\frac{\partial v_y}{\partial x}\right) = \frac{1}{1}\left(\frac{\delta}{1}\right) = \delta$$

Applying this order-of-magnitude analysis to equations (4.10-4) and (4.10-5), we conclude that $\partial^2 v_x/\partial x^2$ is negligible in equation (4.10-4) and $\partial^2 v_y/\partial x^2$ can be ignored in equation (4.10-5). Further, equation (4.10-5) is of smaller order than (4.10-4) and may be neglected in some analyses. Another way of applying an order-of-magnitude analysis to equations (4.10-4) and (4.10-5) is to redefine the variables explicitly so that they are dimensionless. For example, equations (4.10-7) and (4.10-8) show how this might be done:

$$x^* = \frac{x}{L} \qquad y^* = \frac{y}{L} \tag{4.10-7}$$

$$v_x^* = \frac{v_x}{v_\infty} \qquad v_y^* = \frac{v_y}{v_\infty} \qquad P^* = \frac{P}{\rho v_\infty^2} \tag{4.10-8}$$

At steady state with equations (4.10-7) and (4.10-8) applied to (4.10-4), (4.10-5), and (4.10-6), we get

$$v_x^* \frac{\partial v_x^*}{\partial x^*} + v_y^* \frac{\partial v_x^*}{\partial y^*} = -\frac{\partial P^*}{\partial x^*} + \frac{\mu}{\rho L v_\infty} \left(\frac{\partial^2 v_x^*}{\partial x^{*2}} + \frac{\partial^2 v_x^*}{\partial y^{*2}} \right) \tag{4.10-9}$$

$$v_x^* \frac{\partial v_y^*}{\partial x^*} + v_y^* \frac{\partial v_y^*}{\partial y^*} = -\frac{\partial P^*}{\partial y^*} + \frac{\mu}{\rho L v_\infty} \left(\frac{\partial^2 v_y^*}{\partial x^{*2}} + \frac{\partial^2 v_y^*}{\partial y^{*2}} \right) + \frac{g_y L}{v_\infty^2} \tag{4.10-10}$$

$$\frac{\partial v_x^*}{\partial x^*} + \frac{\partial v_y^*}{\partial y^*} = 0 \tag{4.10-11}$$

An order-of-magnitude analysis based on the assumption that the boundary layer thickness is very small, $\delta \ll 1$, yields the result that v_x^* is of order 1, v_y^* is of order δ, x^* is of order 1, and y^* is of order δ. This analysis applied to equation (4.10-11) yields

$$\frac{1}{1} + \frac{\delta}{\delta} = 0 \tag{4.10-12}$$

and we conclude that neither term in equation (4.10-11) can be neglected. In equation (4.10-9) a similar order-of-magnitude analysis leads to

$$(1)\left(\frac{1}{1}\right) + (\delta)\left(\frac{1}{\delta}\right) = \frac{\partial P^*}{\partial x^*} + \left(\frac{\mu}{\rho L v_\infty}\right)\left[\left(\frac{1^2}{1^2}\right) + \left(\frac{1^2}{\delta^2}\right)\right] \tag{4.10-13}$$

We can conclude from equation (4.10-13) that $\partial^2 v_x^*/\partial x^{*2}$ can be neglected in (4.10-9). In some applications $\partial P^*/\partial x^*$ is approximately equal to zero (P is independent of x and y).

For steady flow in circular pipes, the boundary layer approach and boundary layer equations allow for the analysis of the entrance region. The boundary layer relations obtained from Tables I and VIII (using rectangular coordinates) are given by

$$v_x \frac{\partial v_x}{\partial x} + v_y \frac{\partial v_x}{\partial y} = -\frac{1}{\varrho} \frac{dP}{dx} + \nu \left(\frac{\partial^2 v_x}{\partial x^2} + \frac{\partial^2 v_x}{\partial y^2} \right) \qquad (4.10\text{-}14)$$

$$\frac{\partial v_x}{\partial x} + \frac{\partial v_y}{\partial y} = 0 \qquad (4.10\text{-}15)$$

If we assume

$$v_x \frac{\partial v_x}{\partial x} + v_y \frac{\partial v_x}{\partial y} = \nu \beta^2 v_x$$

and let $(1/\varrho\nu)dP/dx = \alpha$, equation (4.10-14) becomes

$$\frac{\partial^2 v_x}{\partial x^2} + \frac{\partial^2 v_x}{\partial y^2} - \beta^2 v_x = \alpha \qquad (4.10\text{-}16)$$

Equation (4.10-16) is a form of the Bessel equation with a solution given by[34]

$$v_x = AI_0(\beta r) - \frac{\alpha}{\beta^2} \qquad (4.10\text{-}17)$$

which can be used with the boundary conditions to solve for the velocity profile and thickness of the boundary layer. In some flow configurations, the pressure is found to be increasing in the direction of flow (contrary to the usual situation). This "adverse" pressure gradient causes a rapid deceleration of the fluid until the fluid undergoes a flow reversal. These drastic changes in flow patterns cause considerable alterations in the predicted flow and pressure drop.

There are of course many approaches to solving the boundary layer equations. One of the most common is similarity analysis. For example, suppose we were working with a simplified version of the boundary layer and continuity equations:

$$v_x \frac{\partial v_x}{\partial x} + v_y \frac{\partial v_x}{\partial y} = \nu \frac{\partial^2 v_x}{\partial y^2} \qquad (4.10\text{-}18)$$

$$\frac{\partial v_x}{\partial x} + \frac{\partial v_y}{\partial y} = 0 \qquad (4.10\text{-}19)$$

---

[34] R. S. Brodkey, *op. cit.*, p. 132.

with boundary conditions

$$v_x = 0 \qquad v_y = 0 \qquad \text{for } y = 0 \qquad (4.10\text{-}20)$$

$$v_x = v_\infty \qquad\qquad \text{for } y \to \infty \qquad (4.10\text{-}21)$$

For this situation we can use a similarity transformation as shown[35]:

$$\frac{v_x}{v_\infty} = \frac{dF(\eta)}{d\eta} = F'(\eta) \qquad (4.10\text{-}22)$$

$$\eta = y \sqrt{\frac{v_\infty}{\nu x}} \qquad (4.10\text{-}23)$$

$$F(\eta) = F'(\eta) = 0 \qquad \text{at } \eta = 0 \qquad (4.10\text{-}24)$$

Equations (4.10-18) through (4.10-24) can be combined to yield[36]

$$\frac{FF''}{2} + F''' = 0 \qquad (4.10\text{-}25)$$

with boundary conditions

$$F'(\eta) = 0 \qquad \text{at } \eta = 0 \qquad (4.10\text{-}26)$$

$$F(\eta) = 0 \qquad \text{at } \eta = 0 \qquad (4.10\text{-}27)$$

$$F'(\eta) = 1 \qquad \text{as } \eta \to \infty \qquad (4.10\text{-}28)$$

The solution of these equations is best attempted by numerical techniques. A basic weakness of this similarity transformation lies in its requirement that one "guess" at the proper form of the transformation, and the need to combine two boundary conditions into one new one.

There are some systematic procedures for applying the similarity transformation techniques to problems such as the boundary layer equations. One of these is called the free parameter method,[37] applied here to flow past a flat plate. Equations (4.10-4) and (4.10-5) are applicable, with boundary conditions shown in equations (4.10-1), (4.10-2), and (4.10-3). The term containing dP/dx is obtained from the Bernoulli macroscopic flow equation, applied to the outer, bulk flow region. From equation (4.6-13) (Bernoulli equation) we obtain dP/dx by differentiating the equation with respect to x

[35] A. G. Hansen, *Similarity Analysis of Boundary Value Problems in Engineering*, Prentice-Hall, Englewood Cliffs, N. J. (1964), p. 7.

[36] *Ibid.*, p. 8.

[37] *Ibid.*, p. 9.

or utilize the Bernoulli equation applied to a differential length along the plate. The result is given by

$$\langle v_x \rangle \frac{d\langle v_x \rangle}{dx} = \frac{1}{\varrho} \frac{dP}{dx} \tag{4.10-29}$$

where $\langle v_x \rangle$ is the average or turbulent bulk velocity in the region beyond the boundary layer. The pressure found at a value of x in the outer region (main stream) using equation (4.10-29) is the same as that in the boundary layer at the same value of x along the plate. If equation (4.10-29) is substituted into (4.10-4), if $\partial^2 v_x/\partial x^2$ is neglected as described in the previous discussion, and if we work only with equation (4.10-4) and neglect the effect of (4.10-5), we get the governing equations

$$v_x \frac{\partial v_x}{\partial x} + v_y \frac{\partial v_x}{\partial y} = \langle v_x \rangle \frac{d\langle v_x \rangle}{dx} + v \frac{d^2 v_x}{dy^2} \tag{4.10-30}$$

$$\frac{\partial v_x}{\partial x} + \frac{\partial v_y}{\partial y} = 0 \tag{4.10-31}$$

The boundary conditions are

$$v_x = v_y = 0 \qquad \text{at } y = 0 \tag{4.10-32}$$

$$v_x = \langle v_x \rangle \qquad \text{at } y \to \infty \tag{4.10-33}$$

In addition we specify the velocity at a point, i.e.,

$$v_x = v_x(x_0, y) = f_0(y) \tag{4.10-34}$$

From equation (4.10-31) we can rearrange the terms to yield

$$v_y = - \int_0^y \frac{\partial v_x}{\partial x} \, dy = - \frac{\partial}{\partial x} \int_0^y v_x dy \tag{4.10-35}$$

which has utilized $v_y = 0$ at $y = 0$. If equation (4.10-35) is substituted into (4.10-30), the result is

$$v_x \frac{\partial v_x}{\partial x} - \left(\frac{\partial}{\partial x} \int_0^y v_x dy \right) \frac{\partial v_x}{\partial y} = \langle v_x \rangle \frac{\partial \langle v_x \rangle}{\partial x} + v \frac{\partial^2 v_x}{\partial y^2} \tag{4.10-36}$$

The first step in the free-parameter similarity analysis is to express the dependent variable as a product of functions, one of which involves the similarity parameter η. In this case the velocity v_x is expressed *a priori* as

$$v_x = \langle v_x \rangle F'(\eta) \tag{4.10-37}$$

and

$$\eta = \eta(x, y) \tag{4.10-38}$$

The aim is to reduce equation (4.10-36) to an ordinary differential equation and achieve constant boundary conditions for $F'(\eta)$:

$$\frac{v_x}{\langle v_x \rangle} = F'(\eta) = 0 \qquad \text{at } y = 0 \tag{4.10-39}$$

and

$$\frac{v_x}{\langle v_x \rangle} = F'(\eta) = 1 \qquad \text{as } y \to \infty \tag{4.10-40}$$

From equation (4.10-38) we assume the following form for η:

$$\eta = yg(x) \tag{4.10-41}$$

where $g(x)$ is an arbitrary function. Applying equation (4.10-41) to (4.10-39) and (4.10-40) yields $\eta = 0$ for $y = 0$, and $\eta \to \infty$ for $y \to \infty$, so that the requirements on $F'(\eta)$ remain unchanged. To illustrate the general free-parameter method [without applying a "serendipic" function such as equation (4.40-41)] we substitute equation (4.10-37) into (4.10-36). The result, after dividing by $\langle v_x \rangle d\langle v_x \rangle / dx$, is[38]

$$F'^2 + \frac{\langle v_x \rangle}{d\langle v_x \rangle/dx} \frac{\partial \eta}{\partial x} F'F'' - \underbrace{\frac{1}{d\langle v_x \rangle/dx} \frac{\partial \eta}{\partial y} F'' \left\{ \frac{\partial}{\partial x} \int_0^y \langle v_x \rangle F' dy \right\}}_{(A)}$$

$$= 1 + \frac{v}{d\langle v_x \rangle/dx} \left[F''' \left(\frac{\partial \eta}{\partial y} \right)^2 + F'' \frac{\partial^2 \eta}{\partial y^2} \right] \tag{4.10-42}$$

We now seek to reduce equation (4.10-42) by postulating conditions on the coefficients of F, F', F'', etc. This will allow us to specify conditions on η and arrive at an F equation which must be solved. We begin with the highest derivative, in this case F''', and require the coefficient of F''' to be nonzero (a function of η). The requirement is expressed by

$$\frac{v}{d\langle v_x \rangle/dx} \left(\frac{\partial \eta}{\partial y} \right)^2 = f_1(\eta) \tag{4.10-43}$$

or

$$\frac{\partial \eta}{\partial y} = \sqrt{\frac{\partial \langle v_x \rangle}{\partial x} \Big/ v} \, (f_1)^{1/2} \tag{4.10-44}$$

---

[38] *Ibid.*, p. 15.

Separating the variables and integrating equation (4.10-44) results in

$$\int (f_1)^{-1/2} d\eta = \sqrt{\frac{d\langle v_x \rangle}{dx}\bigg/ v}\, y - C_1 \qquad (4.10\text{-}45)$$

where C_1 is an integration constant. Equation (4.10-45) can be rearranged to yield

$$y = \sqrt{\frac{v}{d\langle v_x \rangle / dx}} \left[\int \frac{1}{\sqrt{f_1}}\, d\eta + C_1 \right] \qquad (4.10\text{-}46)$$

By combining equation (4.10-46) with part (A) of equation (4.10-42) and by performing the differentiation indicated (with the Leibnitz rule), we can transform equation (4.10-42) to[39]

$$(F')^2 - \left[1 - \frac{1}{2} \frac{\langle v_x \rangle \langle v_x \rangle''}{\langle v_x \rangle' \langle v_x \rangle'} \right] \int_0^\eta \frac{F'}{\sqrt{f_1}}\, d\eta\, \sqrt{f_1}F''$$

$$= 1 + f_1 F''' + \tfrac{1}{2} f_1' F'' \qquad (4.10\text{-}47)$$

where $\langle v_x \rangle' = d\langle v_x \rangle / dx$ and $\langle v_x \rangle'' = d^2 \langle v_x \rangle / dx^2$.

In order to convert equation (4.10-47) to an ordinary differential equation in η, let

$$\left[1 - \frac{1}{2} \frac{\langle v_x \rangle \langle v_x \rangle''}{\langle v_x \rangle' \langle v_x \rangle'} \right] = \text{constant} = C_2 \qquad (4.10\text{-}48)$$

or[40]

$$\langle v_x \rangle = (C_5 x + C_6)^n \qquad (4.10\text{-}49)$$

where $n = 2/(2C_2 - 1)$ or $C_2 = (n + 1)/2n$. If initially $C_2 = \tfrac{1}{2}$ in equation (4.10-48), we get instead of (4.10-49),

$$\langle v_x \rangle = \exp(C_3 x + C_4) \qquad (4.10\text{-}50)$$

Either result is valid in producing a linear ordinary differential equation from (4.10-47). We still have not elicited the relationship between y and η. In equation (4.10-46) $f_1(\eta)$ was arbitrary, so that we can adjust C_1 in (4.10-46) for our convenience. Let us choose $C_1 = 0$ and $f_1(\eta) = C_2$ in equation (4.10-46), which yields

$$y = \eta \sqrt{\frac{v}{\langle v_x \rangle' C_2}} \qquad (4.10\text{-}51)$$

[39] *Ibid.*, p. 17.
[40] *Ibid.*, p. 17.

From equation (4.10-51) we can obtain

$$\eta = y \sqrt{\frac{\langle v_x \rangle' C_2}{\nu}} = y \sqrt{\left[\frac{d \ln\langle v_x \rangle}{dx}\right]\left(\frac{\langle v_x \rangle}{\nu}\right) C_2} \qquad (4.10\text{-}52)$$

If $\langle v_x \rangle$ is as given by equation (4.10-49), we can rearrange (4.10-52) to yield

$$\eta = \frac{y}{C_5 x + C_6} \sqrt{C_5\left(\frac{n+1}{2}\right)} (N_{\text{Re}_x})^{1/2} \qquad (4.10\text{-}53)$$

where

$$N_{\text{Re}_x} = \text{Reynolds number} = \frac{\langle v_x \rangle (C_5 x + C_6)}{\nu} \qquad (4.10\text{-}54)$$

With equations (4.10-53) and (4.10-54) substituted into (4.10-47), we get

$$\beta(1 - F'^2) + FF'' + F''' = 0 \qquad (4.10\text{-}55)$$

where

$$\beta = 2n/(n+1) \qquad (4.10\text{-}56)$$

and the boundary conditions are now

$$F'(\eta) = F(\eta) = 0 \qquad \text{at } \eta = 0 \qquad (4.10\text{-}57)$$

$$F'(\eta) = 1 \qquad \text{as } \eta \to \infty \qquad (4.10\text{-}58)$$

Equation (4.10-55) is called the Falkner–Skan equation and is usually solved numerically. Thus, if we know F, we can get v_x by equation (4.10-37).

Another solution for the boundary layer equations can be obtained from group theory[41] which is also classified as a similarity analysis. Starting with equations (4.10-30) and (4.30-31), the first step in the group theory technique is to write

$$\phi_1 = v_x \frac{\partial v_x}{\partial x} + v_y \frac{\partial v_x}{\partial y} - \langle v_x \rangle \frac{\partial \langle v_x \rangle}{\partial x} - \nu \frac{\partial^2 v_x}{\partial y^2}$$

$$= \phi_1\left(x, y, v_x, v_y, \ldots, \frac{\partial^2 v_x}{\partial y^2}\right) \qquad (4.10\text{-}59)$$

$$\phi_2 = \frac{\partial v_x}{\partial x} + \frac{\partial v_y}{\partial y} = \phi_2\left(x, y, v_x, v_y, \ldots, \frac{\partial v_x}{\partial x}\right) \qquad (4.10\text{-}60)$$

and attempt to find transformations such that ϕ_1 and ϕ_2 are conformally

---

[41] *Ibid.*, p. 59.

invariant, i.e., do not change with coordinate transformation. This is illustrated by the procedure which follows.

We set up one-parameter transformations:

$$\bar{x} = a^n x \tag{4.10-61}$$

$$\bar{y} = a^m y \tag{4.10-62}$$

$$\bar{v}_x = a^p v_x \tag{4.10-63}$$

$$\bar{v}_y = a^q v_y \tag{4.10-64}$$

$$\langle \bar{v}_x \rangle = a^r \langle v_x \rangle \tag{4.10-65}$$

Substituting transformation equations (4.10-61) through (4.10-65) into $\phi_1(\bar{x}, \bar{y}, \bar{v}_x, \ldots, \partial^2 \bar{v}_x / \partial \bar{y}^2)$, we get

$$
\phi_1 = \bar{v}_x \frac{\partial \bar{v}_x}{\partial \bar{x}} + \bar{v}_y \frac{\partial \bar{v}_x}{\partial \bar{y}} - \langle \bar{v}_x \rangle \frac{d \langle \bar{v}_x \rangle}{d \bar{x}} - \nu \frac{\partial^2 \bar{v}_x}{\partial \bar{y}^2}
$$

$$
= a^{2p-n} v_x \frac{\partial v_x}{\partial x} + a^{p+q-m} v_y \frac{\partial v_x}{\partial y} - a^{-n+2r} \langle v_x \rangle \frac{d \langle v_x \rangle}{dx} - \nu a^{p-2m} \frac{\partial^2 v_x}{\partial y^2}
$$

$$\tag{4.10-66}$$

If we require that the exponents of a in all the terms in equation (4.10-66) be equal, we produce a transformation which is said to be conformally invariant. Symbolically this transformation property is given by

$$
\phi_j \left(\bar{x}, \bar{y}, \bar{v}_x, \ldots, \frac{\partial^2 \bar{v}_x}{\partial \bar{y}^2} \right) = H_j \left(x, y, v_x, \ldots, \frac{\partial^2 v_x}{\partial y^2} \right) \phi_j \left(x, y, v_x, \ldots, \frac{\partial^2 v_x}{\partial y^2} \right)
$$

$$\tag{4.10-67}$$

For the present case, the result of requiring that the exponents of a in equation (4.10-66) be equal is the series of equations

$$p = r \tag{4.10-68}$$

$$\frac{p}{n} = 1 - 2 \frac{m}{n} \tag{4.10-69}$$

$$\frac{q}{n} = -\frac{m}{n} \tag{4.10-70}$$

It follows from equation (4.10-67) that

$$
\phi_1 \left(\bar{x}, \bar{y}, \bar{v}_x, \bar{v}_y, \ldots, \frac{\partial^2 \bar{v}_x}{\partial \bar{y}^2} \right) = a^{2p-n} \phi_1 \left(x, y, v_x, v_y, \ldots, \frac{\partial^2 v_x}{\partial y^2} \right) \tag{4.10-71}
$$

and

$$\bar{x} = a^n x = Ax \tag{4.10-72}$$

$$\bar{y} = a^m y = A^{m/n} y \tag{4.10-73}$$

$$\bar{v}_x = a^p v_x = A^{p/n} v_x = A^{1-2m/n} v_x \tag{4.10-74}$$

$$\bar{v}_y = a^q v_y = A^{q/n} v_y = A^{-m/n} v_y \tag{4.10-75}$$

$$\langle \bar{v}_x \rangle = a^r \langle v_x \rangle = A^{r/n} \langle v_x \rangle = A^{1-2m/n} \langle v_x \rangle \tag{4.10-76}$$

If we similarly operate upon ϕ_2, the result is

$$\phi_2\left(\bar{x}, \bar{y}, \ldots, \frac{\partial \bar{v}_y}{\partial \bar{y}}\right) = A^{-2m/n} \phi_2\left(x, y, \frac{\partial v_y}{\partial y}\right) \tag{4.10-77}$$

and thus ϕ_2 is also conformally invariant. We now also seek a variable η such that it too is invariant under the transformation:

$$\eta(\bar{x}, \bar{y}) = \eta(x, y) \tag{4.10-78}$$

A consistent guess for the η function might be

$$\eta = y x^s \tag{4.10-79}$$

Substituting equation (4.10-79) into (4.10-78), we obtain

$$\bar{y} \bar{x}^s = y x^s \tag{4.10-80}$$

With equations (4.10-72) and (4.10-73) substituted into (4.10-80), we have

$$\bar{y} \bar{x}^s = (A^{m/n} y)(Ax)^s = A^{m/n+s} y x^s \tag{4.10-81}$$

which requires

$$\frac{m}{n} + s = 0 \tag{4.10-82}$$

or

$$s = -\frac{m}{n} \tag{4.10-83}$$

so that equation (4.10-79) is now

$$\eta = y x^{-m/n} \tag{4.10-84}$$

Next we postulate a solution of our original problem as shown by

$$v_x = y^b F_1(\eta) \tag{4.10-85}$$

$$v_y = x^c F_2(\eta) \tag{4.10-86}$$

$$\langle v_x \rangle = x^d F_3(\eta) \tag{4.10-87}$$

or

$$F_1(\eta) = v_x y^{-b} \tag{4.10-88}$$

$$F_2(\eta) = v_y x^{-c} \tag{4.10-89}$$

$$F_3(\eta) = \langle v_x \rangle x^{-d} \tag{4.10-90}$$

and

$$F_1(\eta) = (A^{2m/n-1}\bar{v}_x)(A^{-m/n}\bar{y})^{-b} = A^{(m/n)(2+b)-1}\bar{v}_x\bar{y}^{-b} \tag{4.10-91}$$

$$F_2(\eta) = (A^{m/n}\bar{v}_y)(A^{-1}\bar{x})^{-c} = A^{m/n+c}\bar{v}_y\bar{x}^{-c} \tag{4.10-92}$$

$$F_3(\eta) = (A^{2m/n-1}\langle \bar{v}_x \rangle)(A^{-1}\bar{x})^{-d} = A^{2m/n-1+d}\langle \bar{v}_x \rangle\bar{x}^{-d} \tag{4.10-93}$$

For an invariance condition on F_1, F_2, and F_3, the exponents on A must be zero, yielding the following values for b, c, and d:

$$b = -2 + \frac{n}{m} \tag{4.10-94}$$

$$c = -\frac{m}{n} \tag{4.10-95}$$

$$d = -2\frac{m}{n} + 1 \tag{4.10-96}$$

All that remains is to substitute equations (4.10-94), (4.10-95), and (4.10-96) into (4.10-85), (4.10-86), and (4.10-87) and then solve for F_1, F_2, and F_3 by combining these equations with the boundary layer equations (4.10-30) and (4.10-31). Since we originally postulated that $\langle v_x \rangle$ is a function of only x, we assume from equation (4.10-87) that $F_3(\eta) = C_0 = $ constant, and (4.10-30) and (4.10-31) become[42]

$$F_1 F_1' \left(-\frac{m}{n} \right) \eta^{n/m+1} + \eta^2 F_2 F_1' + F_2 F_1 \left(\frac{n}{m} - 2 \right) \eta$$

$$- F_3^2 \left[1 - 2 \left(\frac{m}{n} \right) \right] \eta^{4-m/n} - F_3 F_3' \left(-\frac{m}{n} \right) \eta^{5-n/m}$$

$$- \nu \left\{ \left(\frac{n}{m} - 2 \right) \left(\frac{n}{m} - 3 \right) F_1 + 2 \left(\frac{n}{m} - 2 \right) \eta F_1' + F_1'' \eta^2 \right\} = 0 \tag{4.10-97}$$

[42] *Ibid.*, p. 63.

and

$$F_2'\eta^2 - \left(\frac{m}{n}\right)\eta^{n/m+1}F_1' = 0 \qquad (4.10\text{-}98)$$

Equation (4.10-98) can be integrated by parts to yield

$$F_2 = \frac{m}{n}\left[nF' - \left(\frac{n}{m} - 1\right)F\right] \qquad (4.10\text{-}99)$$

where a new function, $F(\eta)$, is defined as

$$F'(\eta) = \eta^{n/m-2}F_1 \qquad (4.10\text{-}100)$$

Substituting equations (4.10-99) and (4.10-100) into (4.10-97) yields

$$\left[1 - 2\frac{m}{n}\right]F'^2 - \left(1 - \frac{m}{n}\right)FF'' - C_0^2\left[1 - 2\frac{m}{n}\right] - \nu F''' = 0 \qquad (4.10\text{-}101)$$

From equations (4.10-84) and (4.10-90) we can write

$$\eta = y\sqrt{\frac{\langle v_x\rangle}{C_0 x}} \qquad (4.10\text{-}102)$$

or in dimensionless form,

$$\eta = y\sqrt{\frac{\langle v_x\rangle}{\nu x}} \qquad (4.10\text{-}103)$$

where $C_0 = \nu =$ kinematic viscosity.

If for simplification we also redefine F'

$$F^{*\prime} = \frac{F'}{C_0} = \frac{F'}{\nu} \qquad (4.10\text{-}104)$$

then equation (4.10-101) becomes

$$(F^{*\prime})^2 - 1 - \frac{[1 - m/n]}{[1 - 2(m/n)]}F^*F^{*\prime\prime} - F^{*\prime\prime\prime} = 0 \quad (4.10\text{-}105)$$

Thus, we can solve numerically for F^* in equation (4.10-105), which yields F in equation (4.10-104), which will allow us to determine F_2 in equation (4.10-99) and F_1 in equation (4.10-100). With F_1 and F_2 established, we can determine v_x and v_y from equations (4.10-85) and (4.10-86), subject to the boundary conditions imposed upon the problem, which will allow a determination of the m/n ratio.

There are some apparent advantages to the group theory method of solution. It seems to be easily applied. However, the boundary conditions are not taken into account in any way until the entire analysis is completed. A more important disadvantage rests in the uncertainty in choosing proper transformation groups and the seemingly indirect approach.

In the solution of some boundary layer problems, approximate methods are sometimes used to solve these nonlinear governing equations. One of these approximate methods is the "integral" (Von Karman) method. To illustrate this procedure, let us return to the two-dimensional incompressible boundary layer equations

$$\frac{\partial v_x}{\partial t} + v_x \frac{\partial v_x}{\partial x} + v_y \frac{\partial v_x}{\partial y} = -\frac{1}{\varrho}\frac{dP}{dx} + \frac{\mu}{\varrho}\frac{\partial^2 v_x}{\partial y^2} \qquad (4.10\text{-}106)$$

$$\frac{\partial P}{\partial y} = 0 \qquad (4.10\text{-}107)$$

$$\frac{\partial v_x}{\partial x} + \frac{\partial v_y}{\partial y} = 0 \qquad (4.10\text{-}108)$$

In equation (4.10-106) we have neglected the $\partial^2 v_x/\partial x^2$ as usual. The v_y boundary layer equation has been simplified drastically in arriving at (4.10-107) by assuming the order of magnitude of most of the terms to be negligibly small. Equation (4.10-108) is the familiar continuity relationship. The boundary conditions are the usual ones, i.e., $v_x = v_\infty$ at $y = \delta$ and $v_x = v_\infty$ at $y \to \infty$. Integrating equation (4.10-106) from $y = 0$ to $y = \delta(x)$ gives

$$\int_0^\delta \frac{\partial v_x}{\partial t}\, dy + \int_0^\delta \left(v_x \frac{\partial v_x}{\partial x} + v_y \frac{\partial v_x}{\partial y}\right) dy$$

$$= \left(-\frac{1}{\varrho}\frac{dP}{dx}\right)\delta - \frac{\mu}{\varrho}\int_0^\delta \frac{\partial^2 v_x}{\partial y^2}\, dy \qquad (4.10\text{-}109)$$

The last term in equation (4.10-109) may be simplified

$$\int_0^\delta \frac{\partial^2 v_x}{\partial y^2}\, dy = \frac{\partial v_x}{\partial y}\Big|_0^\delta = \frac{\partial v_x}{\partial y}\Big|_{y=\delta} - \frac{\partial v_x}{\partial y}\Big|_{y=0} = -\frac{\partial v_x}{\partial y}\Big|_{y=0} \qquad (4.10\text{-}110)$$

since

$$\frac{\partial v_x}{\partial y}\Big|_{y=\delta} = 0$$

Similarly, from equation (4.10-109) we can, with integration by parts, arrive at

$$\int_0^\delta v_y \frac{\partial v_x}{\partial y} dy = v_x v_y \Big|_0^\delta - \int_0^\delta v_x \frac{\partial v_y}{\partial y} dy \qquad (4.10\text{-}111)$$

More information is needed before the evaluation of (4.10-111) can be completed. From equation (4.10-108) $\partial v_y/\partial y = -\partial v_x/\partial x$, and if we separate the variables and integrate, the result is

$$\int_0^{v_y} dv_y = -\int_0^\delta \frac{\partial v_x}{\partial x} dy \qquad (4.10\text{-}112)$$

or

$$v_y \Big|_{y=\delta} = -\int_0^\delta \frac{\partial v_x}{\partial x} dy \qquad (4.10\text{-}113)$$

Note that in equation (4.10-112) we have used $v_y = 0$ at $y = 0$. Substituting equations (4.10-108) and (4.10-113) into (4.10-111), we get

$$\int_0^\delta v_y \frac{\partial v_x}{\partial y} dy = -v_\infty \int_0^\delta \frac{\partial v_x}{\partial x} dy + \int_0^\delta v_x \frac{\partial v_x}{\partial x} dy \qquad (4.10\text{-}114)$$

In the region outside of the boundary layer we assume a uniform velocity v_∞, and as discussed previously we determine $\partial P/\partial x$ from v_∞ and the Bernoulli equation. (Another way of arriving at the Bernoulli equation is to rewrite equation (4.10-106) for the region beyond the boundary layer, with $v_x = v_\infty$ and $\partial v_\infty/\partial y = 0$.) The following result is a form of the Bernoulli equation:

$$-\frac{1}{\varrho} \frac{\partial P}{\partial x} = \frac{\partial v_\infty}{\partial t} + v_\infty \frac{\partial v_\infty}{\partial x} \qquad (4.10\text{-}115)$$

which can be integrated,

$$\int_0^\delta -\frac{1}{\varrho} \frac{\partial P}{\partial x} dy = -\frac{1}{\varrho} \frac{\partial P}{\partial x} \delta = \int_0^\delta \frac{\partial v_\infty}{\partial t} dy + v_\infty \int_0^\delta \frac{\partial v_\infty}{\partial x} dy \qquad (4.10\text{-}116)$$

Substitution of equations (4.10-110), (4.10-114), and (4.10-116) into the boundary layer equation (4.10-109) yields

$$\frac{\mu}{\varrho} \left(\frac{\partial v_x}{\partial y} \right) \Big|_{y=0} = \int_0^\delta (v_x - v_\infty) dy + v_\infty \frac{\partial}{\partial x} \int_0^\delta v_x dy - \frac{\partial}{\partial x} \int_0^\delta v_x^2 dy$$

$$+ \frac{\partial v_\infty}{\partial x} \int_0^\delta v_\infty dy \qquad (4.10\text{-}117)$$

If we define "displacement thickness" by

$$\delta^* = \int_0^\delta \left(1 - \frac{v_x}{v_\infty}\right) dy \qquad (4.10\text{-}118)$$

and a "momentum thickness," λ, by

$$\lambda = \int_0^\delta \frac{v_x}{v_\infty} \left(1 - \frac{v_x}{v_\infty}\right) dy \qquad (4.10\text{-}119)$$

we can convert equation (4.10-117) to[43]

$$\frac{\tau_w}{\varrho v_\infty^2} = \frac{1}{v_\infty^2} \frac{\partial}{\partial t}(v_\infty \delta^*) + \frac{\partial \lambda}{\partial x} + \frac{1}{v_\infty}[2\lambda + \delta^*]\frac{dv_\infty}{dx} \qquad (4.10\text{-}120)$$

where

$$\tau_w = \text{shear stress at the wall} = \mu \left.\frac{\partial v_x}{\partial y}\right|_{y=0}$$

This approach now needs a velocity profile in the boundary layer so that equation (4.10-120) will give values of the boundary layer thickness, δ, as a function of x. This method abandons the attempt to satisfy the boundary layer equation everywhere, but chooses a reasonable velocity which satisfies the boundary conditions, usually a linear form.

Assignments in Chapter 4

4.1. Verify equations (4.1-7), (4.1-8), (4.1-10), (4.2-8), (4.2-10), (4.2-13), (4.2-14), (4.2-17), (4.2-19), (4.2-20), (4.2-22), (4.2-23), (4.2-28), (4.3-6), (4.3-25), (4.4-11), (4.5-11), (4.5-19), (4.6-10), (4.7-4), (4.7-5), (4.7-6), (4.7-12), (4.7-14), (4.9-8), (4.9-14), (4.10-9), (4.10-10), (4.10-11), (4.10-68), (4.10-69), (4.10-70), (4.10-102), (4.10-111).

4.2. Solve equation (4.4-10) by a Laplace transform technique.

[43] W. F. Ames, *Non-linear Partial Differential Equations in Engineering*, Academic Press, New York (1965), p. 273.

4.3. For a power-law fluid flowing in a concentric-cylinder viscometer as shown in the diagram,

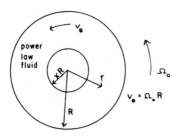

(a) Evaluate Δ_{ij}. Show that only $\Delta_{r\theta}$ is nonzero.

(b) Show that the governing $\tau_{r\theta}$ equation is

$$0 = \frac{1}{r^2} \frac{d}{dr} (r^2 \tau_{r\theta})$$

(c) Find $\tau_{r\theta}$ for a power-law fluid.

(d) With the appropriate boundary conditions on v_θ, and torque $T = \tau_{r\theta}\big|_{r=R} \times (2\pi RL)(R)$, get v_θ/r.

(e) Find the Ω_0 equation as a function of T, n, m, R, and $\varkappa$.

(f) From (e) describe experiments that will allow the evaluation of n and m.

4.4. For a wetted-wall column operating as shown in the diagram,

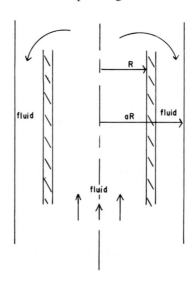

(a) Obtain the velocity profile equation for a Newtonian fluid

$$v_z = \frac{\varrho(g/g_c)R^2}{4\mu}\left[1 - \left(\frac{r}{R}\right)^2 + 2a^2 \ln\left(\frac{r}{R}\right)\right]$$

(b) From $Q = \int_R^{aR} v_z 2\pi r\, dr$, find the volumetric flow rate, Q.

4.5. Calculate the pressure drop required to cause a fluid to flow in a horizontal circular pipe.

Data: $D = 3$ cm

$Q = 1.1$ liters/sec

$\varrho = 0.935$ g/cm³

$\mu = 1.95$ cp (centipoise)

$T = 20°C$

4.6. Calculate the pressure drop required to pump water through a circular pipe with the following specifications:

$T = 20°C$

$D = 25$ cm

$L = 1234$ m (with four $90°$ elbows and two $45°$ elbows)

$Q = 1.97$ m³/sec

4.7. Calculate the pressure drop required to pump water up a circular pipe with the following specifications:

$Q = 18$ gal/min

$T = 68°F$

$\mu = 1.002$ cp

$D = 3.068$ in.

$\varrho = 0.998$ g/ml

$L = 95$ ft (with two $45°$ elbows)

Height pumped $= 35.3$ ft

4.8. Calculate the horsepower needed to pump water from one reservoir to another (see diagram).

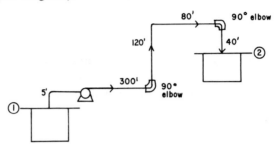

Data:

$$Q = 12 \text{ ft}^3/\text{min}$$
$$D = 4 \text{ in.}$$
$$\varrho = 62.4 \text{ lb/ft}^3$$
$$\mu = 1.0 \text{ cp}$$

4.9. Water is flowing in turbulent flow in a U-shaped pipe bend as shown in the diagram. Calculate the force exerted by the flowing fluid on the pipe.

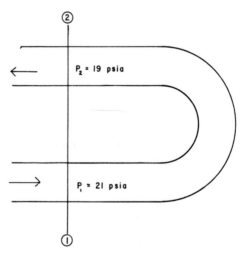

Data:

$$T = 68°\text{F}$$
$$\varrho = 62.4 \text{ lbm/ft}^3$$
$$\mu = 1.0 \text{ cp}$$
$$Q = 3 \text{ ft}^3/\text{sec}$$

4.10.[44] For the data collected and shown below, calculate the volumetric flow rate to be expected through a circular pipe.

τ_{xy} (lbf/ft^2)	dv_x/dy (sec^{-1})	
0.3	0.000	$D = 2$ in.
0.4	0.644	$\Delta P/L = 20$ lbf/ft^3
0.6	5.78	
1.0	31.50	
1.5	92.5	
2.0	185.8	
2.5	311.5	
3.0	469.0	

4.11. Given the following fluid properties (Bingham plastic), find P_2 (see diagram):

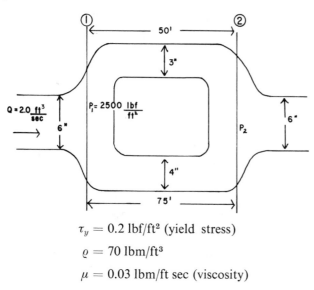

$$\tau_y = 0.2 \text{ lbf/ft}^2 \text{ (yield stress)}$$

$$\varrho = 70 \text{ lbm/ft}^3$$

$$\mu = 0.03 \text{ lbm/ft sec (viscosity)}$$

4.12.[45] A Bingham plastic paint is flowing in a laminar film down a plane vertical wall. Determine the velocity profile.

[44] A. H. P. Skelland, *Non-Newtonian Flow and Heat Transfer*, John Wiley, New York (1967), p. 105.

[45] *Ibid.*, p. 100.

4.13.[46] Given the pipe flow data shown below, determine $\Delta P/L$.

τ_{rz} (lbf/ft^2)	dv_z/dr (sec^{-1})
2.0	5
2.79	20
3.48	50
4.10	100
6.05	500
8.43	1000

$$D = 2 \text{ in.}$$
$$\langle v_z \rangle = 1.5 \text{ ft/sec}$$

Is the flow laminar?

4.14. The following data are given:

Plant data	mass flow rate $= 940$ lbm/sec
	$D = 12$ in.
	$\varrho = 60$ lbm/ft^3

Laboratory pipe flow data (viscometry)

Mass flow rate (lbm/hr)	ΔP (lbf/ft^2)
40	3.95
100	5.21
200	6.41
400	7.88
600	8.85

$$D = 0.25 \text{ in.}$$
$$L = 125 \text{ in.}$$

Is the flow laminar? Is the fluid non-Newtonian? Determine $\Delta P/L$.

[46] *Ibid.*, p. 152.

4.15.[47] Given the pipe flow data shown below, determine the τ_{rz} versus dv_z/dr curve.

Q (ft³/sec)	ΔP (lbf/in²)	
		$\varrho = 50 \text{ lbm/ft}^3$
		$L = 10 \text{ ft}$
0.0	3.9	$D = 0.255 \text{ in.}$
0.000187	5.2	
0.000374	6.11	
0.000935	7.15	
0.002240	9.75	
0.00375	12.75	
0.0043	14.3	
0.00486	18.2	
0.00655	33.5	
0.00935	59.2	

4.16. Solve equation (4.4-17) by a Laplace transform technique.

For Further Reading

Similarity Analysis of Boundary Value Problems in Engineering, by A. G. Hansen, Prentice-Hall, Englewood Cliffs, N. J., 1964.

Transport Phenomena, by R. B. Bird, W. E. Stewart, and E. N. Lightfoot, John Wiley, New York, 1960.

The Phenomena of Fluid Motion, by R. S. Brodkey, Addison-Wesley, Reading, Massachusetts, 1967.

Low Reynolds Number Hydrodynamics, by J. Happel and H. Brenner, Prentice-Hall, Englewood Cliffs, N. J., 1965.

Non-Newtonian Flow and Heat Transfer, by A. H. P. Skelland, John Wiley, New York, 1967.

Process Analysis and Simulation, by D. M. Himmelblau and K. B. Bischoff, John Wiley, New York, 1968.

Non-linear Partial Differential Equations in Engineering, by W. F. Ames, Academic Press, New York, 1965.

Non-Newtonian Fluids, by W. L. Wilkinson, Pergamon Press, New York, 1960.

[47] *Ibid.*, p. 64.

Chapter 5

Derivation of the Mass Transport Equations

There are times when the description of an occurrence in abstract terms is required. Usually the abstraction is written in mathematical symbolism, giving what is sometimes called the "governing equation." Depending upon one's level of education and sophistication, the governing equation may be an algebraic equation or perhaps a vector or tensor equation in the three-dimensional space of our environment. If the dimensionality is more than three, we may write an n-dimensional matrix equation expressing n equations in compact form.

Once a phenomenon has been described in a mathematical equation, the next step in the analysis is to describe the behavior of this phenomenon at some time in its temporal history along with its condition at the boundaries. The mathematical abstraction (a model of the actual phenomenon) is thus tied concretely to physical reality and we may predict the behavior under a myriad of possible conditions. We may extrapolate, interpolate, and formulate.

In this chapter we begin with a derivation of an unsteady-state mass transport equation, which includes contributions due to molecular diffusion, convection, and chemical reaction. Next some transport theories such as penetration and surface renewal are derived. The chapter concludes with some multicomponent considerations, turbulent transport, and a macro-scopic approach to the analysis of mass transport processes.

5.1. Governing Equation for Unsteady-State Diffusional Mass Transport with Chemical Reaction and Convective Flow

Diffusion in the liquid state is generally attributed to hydrodynamic or activated-state mechanisms. According to hydrodynamic theories, the diffusion of solute A through medium B can be written as[1]

$$D_{AB} = \varkappa T\left(\frac{v_A}{F_A}\right) \qquad (5.1\text{-}1)$$

where

D_{AB} = diffusion coefficient of A in B

v_A/F_A = mobility of A

T = temperature

$\varkappa$ = Boltzmann's constant

From Stokes' law, we can get[2]

$$F_A = 6\pi\mu_B v_A R_A \qquad (5.1\text{-}2)$$

where

F_A = drag forces between solute and solvent molecules

μ_B = viscosity of the pure solvent B

v_A = velocity (mobility) of component A

R_A = radius of the diffusing particle

If equation (5.1-2) is substituted into (5.1-1), we get the Stokes–Einstein equation

$$D_{AB} = \frac{\varkappa T}{6\pi R_A \mu_B} \qquad (5.1\text{-}3)$$

which is usually acceptable as a predictor of the diffusion coefficient for systems of large spherical solute particles. If the solute and solvent particles are assumed all alike (self-diffusion), and if they are assumed to be arranged in cubic lattices, then we get the following equation[3] for predicting the self-diffusion coefficient D_{AA}:

$$D_{AA} = \frac{\varkappa T}{2\pi\mu_A}\left(\frac{N}{V}\right)^{1/3} \qquad (5.1\text{-}4)$$

where N is the number of molecules contained in volume V, and μ_A is the viscosity of the system.

[1] T. W. Drew and J. W. Hoopes, ed., *Advances in Chemical Engineering*, Vol. I, Academic Press, New York (1956), p. 195.

[2] *Ibid.*, p. 195.

[3] *Ibid.*, p. 196.

Another theory of diffusion is the activated-state hypothesis, which attempts to explain the transport processes in liquids. From these concepts we get the diffusion coefficient equation[4]

$$D = \frac{\lambda_1 \varkappa T}{\lambda_2 \lambda_3} \tag{5.1-5}$$

where $\lambda_1, \lambda_2, \lambda_3$ are distances characterizing the spacing between layers of molecules in the hypothesized quasicrystalline liquid lattice. If $\lambda_1 = \lambda_2 = \lambda_3 = V/N$, then equation (5.1-5) becomes

$$D = \frac{\varkappa T}{\mu} \left(\frac{N}{V} \right)^{1/3} \tag{5.1-6}$$

The activated-state theories assume transport is by a unimolecular process. Others have suggested that the basic mechanism of liquid diffusion may be by a bimolecular collision mechanism.

With this brief introduction, we now present a general mass transport equation which incorporates not only diffusion but also convective mass transport and chemical reaction. Convective mass transport implies the movement of mass by virtue of fluid flow, carrying the component of interest in and out of the region of interest. Chemical reactions also alter the mass present. In Chapter 6 we will show mass transport applications, but in this chapter we will only derive the governing equations.

For a liquid of constant density ϱ containing component A, the concentration C_A of this component A in the liquid is described by the mass transport equation

$$\frac{\partial C_A}{\partial t} + v \cdot \vec{V} C_A = D_A \vec{V}^2 C_A + R_A \tag{5.1-7}$$

where

$$v = \text{fluid velocity vector}$$
$$\bar{v} \cdot \vec{V} C_A = \text{convective mass transport contribution}$$
$$D_A \vec{V}^2 C_A = \text{molecular diffusion contribution}$$
$$R_A = \text{chemical reaction contribution}$$
$$D_A = \text{diffusion coefficient of A in the liquid}$$

Equation (5.1-7) is a vector equation, which can be broken down by standard

[4] *Ibid.*, p. 197.

TABLE XIV

Governing Mass Transport Relationships [Equation (5.1-7)]

Rectangular coordinates $=(x, y, z)$

$$\frac{\partial C_A}{\partial t} + \left(v_x \frac{\partial C_A}{\partial x} + v_y \frac{\partial C_A}{\partial y} + v_z \frac{\partial C_A}{\partial z} \right) = D_A \left(\frac{\partial^2 C_A}{\partial x^2} + \frac{\partial^2 C_A}{\partial y^2} + \frac{\partial^2 C_A}{\partial z^2} \right) + R_A \quad \text{(A)}$$

Cylindrical coordinates (r, θ, z)

$$\frac{\partial C_A}{\partial t} + \left(v_r \frac{\partial C_A}{\partial r} + v_\theta \frac{1}{r} \frac{\partial C_A}{\partial \theta} + v_z \frac{\partial C_A}{\partial z} \right)$$

$$= D_A \left(\frac{1}{r} \frac{\partial}{\partial r} \left(r \frac{\partial C_A}{\partial r} \right) + \frac{1}{r^2} \frac{\partial^2 C_A}{\partial \theta^2} + \frac{\partial^2 C_A}{\partial z^2} \right) + R_A \qquad \text{(B)}$$

Spherical coordinates (r, θ, ϕ)

$$\frac{\partial C_A}{\partial t} + \left(v_r \frac{\partial C_A}{\partial r} + v_\theta \frac{1}{r} \frac{\partial C_A}{\partial \theta} + v_\phi \frac{1}{r \sin \theta} \frac{\partial C_A}{\partial \phi} \right)$$

$$= D_A \left(\frac{1}{r^2} \frac{\partial}{\partial r} \left(r^2 \frac{\partial C_A}{\partial r} \right) + \frac{1}{r^2 \sin \theta} \frac{\partial}{\partial \theta} \left(\sin \theta \frac{\partial C_A}{\partial \theta} \right) + \frac{1}{r^2 \sin^2 \theta} \frac{\partial^2 C_A}{\partial \phi^2} \right) + R_A$$

$$\text{(C)}$$

mathematical techniques into rectangular, cylindrical, or spherical coordinates as given in Table XIV.

The derivation of the general governing equation (5.1-7) can be accomplished by working strictly with vector quantities. However, it is perhaps more useful in the early stages of transport analysis to derive equation (5.1-7) by the seemingly more tedious route of starting with rectangular coordinates, deriving equation (5.1-7) in this coordinate system, and then generalizing to vector notation.

Given a volume element of dimensions $\Delta x \Delta y \Delta z$ as shown in Figure 47 a general statement of mass balance takes the form

$$\text{summation} \vert_{\text{in}} - \text{summation} \vert_{\text{out}} + \begin{Bmatrix} + \text{ generation} \\ - \text{ depletion} \end{Bmatrix} = \text{rate of accumulation}$$

$$\text{(5.1-8)}$$

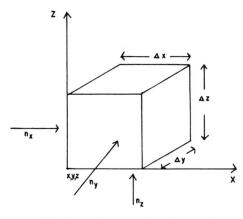

Fig. 47. Differential element showing mass fluxes.

In terms of mass fluxes n_x, n_y, and n_z, equation (5.1-8) translates into

$$\underbrace{[(n_x \Delta y \Delta z)\,|_x + (n_y \Delta x \Delta z)\,|_y + (n_z \Delta x \Delta y)\,|_z]}_{\text{summation}\,|_{\text{in}}}$$

$$- \underbrace{[(n_x \Delta y \Delta z)\,|_{x+\Delta x} + (n_y \Delta x \Delta z)\,|_{y+\Delta y} + (n_z \Delta x \Delta y)\,|_{z+\Delta z}]}_{\text{summation}\,|_{\text{out}}} + \underbrace{r_A \Delta x \Delta y \Delta z}_{\substack{\text{generation by} \\ \text{chemical reaction}}}$$

$$= \underbrace{\frac{\partial}{\partial t}\,(\varrho_A \Delta x \Delta y \Delta z)}_{\text{rate of accumulation}} \qquad\qquad (5.1\text{-}9)$$

In equation (5.1-9), n_x, n_y, and n_z are mass flux terms, mass/(area · time), so that multiplying the flux by the area through which it travels produces mass/time. The term r_A is the rate of *generation* of component A by chemical reaction (per unit volume), and ϱ_A is the concentration of component A (mass/volume) within the differential element. If equation (5.1-9) is divided by $\Delta x \Delta y \Delta z$ and the volume dimensions are allowed to approach zero in the usual mathematical sense, the result, written for component A, is

$$\frac{\partial \varrho_A}{\partial t} + \left(\frac{\partial n_{Ax}}{\partial x} + \frac{\partial n_{Ay}}{\partial y} + \frac{\partial n_{Az}}{\partial z} \right) = r_A \qquad\qquad (5.1\text{-}10)$$

Here use has been made of the definition of the derivative:

$$\lim_{\Delta x \to 0} \frac{n_{Ax}\,|_x - n_{Ax}\,|_{x+\Delta x}}{\Delta x} = \frac{d n_{Ax}}{dx} \qquad\qquad (5.1\text{-}11)$$

In vector notation we can generalize equation (5.1-10) to

$$\frac{\partial \varrho_A}{\partial t} + \nabla \cdot n_A = r_A \qquad (5.1\text{-}12)$$

Equation (5.1-12) can be expressed in other coordinate systems, depending upon the geometry desired for the description of the transport phenomenon. Table XV presents equation (5.1-12) in rectangular, cylindrical, and spherical coordinates.

For two-component (or multicomponent) systems, equation (5.1-12) can be written for each component. When these equations are added together for all components and generalized, the result is

$$\frac{\partial \varrho}{\partial t} + (\nabla \cdot \varrho \bar{v}) = 0 \qquad (5.1\text{-}13)$$

where

$$\varrho \bar{v} = \text{mass flux of the mixture} = n_A + n_B + \cdots$$

$$\bar{v} = \text{mass flow rate of the mixture}$$

$$\varrho = \text{mass density of the mixture} = \varrho_A + \varrho_B + \cdots$$

$$\bar{v}_A = \text{mass flow rate of component A}$$

$$r_A + r_B + \cdots = 0 \text{ (conservation of mass)}$$

TABLE XV

Transport Equations in Terms of Mass Fluxes [Equation (5.1-12)]

Rectangular coordinates (x, y, z)

$$\frac{\partial \varrho_A}{\partial t} + \left(\frac{\partial n_{Ax}}{\partial x} + \frac{\partial n_{Ay}}{\partial y} + \frac{\partial n_{Az}}{\partial z} \right) = r_A \qquad (A)$$

Cylindrical coordinates (r, θ, z)

$$\frac{\partial \varrho_A}{\partial t} + \left(\frac{1}{r} \frac{\partial}{\partial r} (rn_{Ar}) + \frac{1}{r} \frac{\partial n_{A\theta}}{\partial \theta} + \frac{\partial n_{Az}}{\partial z} \right) = r_A \qquad (B)$$

Spherical coordinates (r, θ, ϕ)

$$\frac{\partial \varrho_A}{\partial t} + \left(\frac{1}{r^2} \frac{\partial}{\partial r} (r^2 n_{Ar}) + \frac{1}{r \sin \theta} \frac{\partial}{\partial \theta} (n_{A\theta} \sin \theta) + \frac{1}{r \sin \theta} \frac{\partial n_{A\phi}}{\partial \phi} \right) = r_A \qquad (C)$$

For a fluid of constant mass density, $\varrho = $ constant and equation (5.1-13) becomes

$$\nabla \cdot \bar{v} = 0 \qquad \varrho = \text{constant} \qquad (5.1\text{-}14)$$

In molar units, the derivation leading from equation (5.1-9) to (5.1-12) can be analogously performed, yielding

$$\frac{\partial C_A}{\partial t} + \nabla \cdot N_A = R_A \qquad (5.1\text{-}15)$$

where

$C_A = $ molar concentration of A, moles/volume

$N_A = $ molar flux of A, moles/area $\cdot$ time

$R_A = $ chemical reaction *producing* component A (per unit volume)

For multicomponent systems equation (5.1-15) can be generalized by adding the contributions of all the components:

$$\frac{\partial C}{\partial t} + (\nabla \cdot Cv) = R_A + R_B + \cdots \qquad (5.1\text{-}16)$$

where

$Cv = $ molar flux of the mixture $= N_A + N_B + \cdots$

$v = $ molar flow rate of the mixture

$C = $ total molar concentration $= C_A + C_B + \cdots$

$N_A = $ molar flux of component A $= C_A v_A$

$C_A = $ molar concentration of component A

$v_A = $ molar flow rate of component A

$R_A + R_B + \cdots = $ total molar change due to chemical reaction (not necessarily equal to zero)

For a system with constant molar density ($C = $ constant), from equation (5.1-16) we get

$$\nabla \cdot v = \frac{1}{C}(R_A + R_B + \cdots) \qquad (5.1\text{-}17)$$

In order to express equations (5.1-12) and (5.1-15) in alternative and perhaps more useful forms, the fluxes n_A(mass) and N_A(moles) can be expressed as molecular transport (diffusion) and convective flow:

$$n_A = \underbrace{-D_A \nabla \varrho_A}_{\substack{\text{molecular flux} \\ \text{(diffusion)}}} + \underbrace{\varrho_A \bar{v}}_{\substack{\text{convective} \\ \text{flux}}} \tag{5.1-18}$$

and

$$N_A = \underbrace{-D_A \nabla C_A}_{\substack{\text{molecular flux} \\ \text{(diffusion)}}} + \underbrace{C_A v}_{\substack{\text{convective} \\ \text{flux}}} \tag{5.1-19}$$

where

D_A = diffusion coefficient of component A in the system

ϱ_A = mass density (concentration) of component A

C_A = molar density (concentration) of component A

$\bar{v}$ = total mass velocity

v = total molar velocity

The equation for the molecular (diffusion) flux is sometimes referred to as Fick's first law of diffusion.

From the relationships

$$\varrho \bar{v} = \varrho_A \bar{v}_A + \varrho_B \bar{v}_B + \cdots = n_A + n_B + \cdots \tag{5.1-20}$$

and

$$C v = C_A v_A + C_B v_B + \cdots = N_A + N_B + \cdots \tag{5.1-21}$$

we can convert equations (5.1-18) and (5.1-19) to the forms

$$n_A = -D_A \nabla \varrho_A + w_A(n_A + n_B + \cdots) \tag{5.1-22}$$

$$N_A = -D_A \nabla C_A + x_A(N_A + N_B + \cdots) \tag{5.1-23}$$

where

$$w_A = \text{mass fraction} = \varrho_A/\varrho$$

$$x_A = \text{mole fraction} = C_A/C$$

If the flux relationships shown by equations (5.1-20), (5.1-21), (5.1-22), and (5.1-23) are substituted into (5.1-12) and (5.1-15), after some rearrangement we obtain

$$\frac{\partial \varrho_A}{\partial t} + (\nabla \cdot \varrho_A \bar{v}) = (\nabla \cdot \varrho D_A \nabla w_A) + r_A \tag{5.1-24}$$

$$\frac{\partial C_A}{\partial t} + (\nabla \cdot C_A v) = (\nabla \cdot C D_A \nabla x_A) + R_A \tag{5.1-25}$$

For dilute liquid solutions, where ϱ and D_A can be considered constant, equation 5.1-24 becomes

$$\frac{\partial \varrho_A}{\partial t} + \varrho_A (\nabla \cdot \bar{v}) + \bar{v} \cdot \nabla \varrho_A = D_A \nabla^2 \varrho_A + r_A \qquad (5.1\text{-}26)$$

With constant density, $\nabla \cdot \bar{v} = 0$ from equation (5.1-14), and hence (5.1-26) is simplified to

$$\frac{\partial \varrho_A}{\partial t} + \bar{v} \cdot \nabla \varrho_A = D_A \nabla^2 \varrho_A + r_A \qquad (5.1\text{-}27)$$

If the mass concentration of A, ϱ_A, is divided by the molecular weight of A, $(MW)_A$, then the molar concentration results:

$$C_A = \varrho_A / (MW)_A \qquad (5.1\text{-}28)$$

Applying this operation to equation (5.1-27), we get the general mass transport relationship, equation (5.1-7), which was introduced earlier and presented in Table XIV:

$$\frac{\partial C_A}{\partial t} + v \cdot \nabla C_A = D_A \nabla^2 C_A + R_A \qquad (5.1\text{-}7)$$

Equation (5.1-7) is the starting point for most liquid-phase mass transport analyses. It results from a mass balance and the assumption of constant D_A and ϱ, with the mass velocity profile $\bar{v}$ needed before (5.1-7) can be fully utilized.

If there is a constant molar density C and if D_A is constant, then from equation (5.1-25) we can get

$$\frac{\partial C_A}{\partial t} + C_A (\nabla \cdot v) + (v \cdot \nabla C_A) = D_A \nabla^2 C_A + R_A \qquad (5.1\text{-}29)$$

The assumptions of constant C and D_A are more common for gases with constant temperature and pressure (approximating ideal gas behavior). With equation (5.1-17) substituted into (5.1-29), we can produce a general equation useful for gas phase transport:

$$\frac{\partial C_A}{\partial t} + (v \cdot \nabla C_A) = D_A \nabla^2 C_A + R_A - \frac{C_A}{C} (R_A + R_B + \cdots) \qquad (5.1\text{-}30)$$

Equation (5.1-30) implies constant molar density C and constant diffusion coefficient D_A, with v, the molar velocity profile, needed.

The general procedure to be followed in the analysis of a mass transport problem involves a sketch of the physical situation, the writing of the appropriate governing equation [(5.1-7) or (5.1-29)], the stating of the boundary and initial conditions, the elimination of the negligible or inappropriate terms, and finally the solution of the transport equations. The result is a mathematical model of the physical reality.

There are many complications that can be superimposed upon the straight-forward mass (or molar) analysis presented here. If there is energy (heat) transport concurrently with mass transport, then an energy equation must be written and developed in a similar manner to the equations shown here. The mass and energy transport equations must be solved simultaneously if they are coupled or solved separately if they are independent. It may also be uncertain in multicomponent mixtures whether the diffusion coefficient is actually a constant. It may be that $D_{A\text{-mixture}}$ (diffusion coefficient for component A in the mixture) is not known but must be estimated from binary data such as D_{A-B}, D_{A-C}, D_{A-D}, etc., where D_{A-B} represents the diffusion coefficient of component A in solvent B, with solvents C and D indicated similarly.

In the derivation presented here, only ordinary (molecular) diffusion and convective transport were considered. It is quite possible that other gradients may be driving mass across the boundaries. For example, there can be temperature gradients (thermal diffusion effects), pressure gradients, electrical gradients, etc. All of these effects, if they are significant, must be accounted for in some phenomenological equation, usually expressed as a linear combination of effects.

In arriving at the various equations presented thus far in the chapter, we have distinguished between $\bar{v}$, the mass velocity, and v the molar velocity. It is useful before leaving this section to present the definitions of these velocities. From a weighted-residuals concept of elementary statistics, the averaged mass velocity $\bar{v}$ (velocity profile) is

$$\bar{v} = \frac{\sum_{i=1}^{n} \varrho_i \bar{v}_i}{\sum_{i=1}^{n} \varrho_i} = \frac{\varrho_A \bar{v}_A + \varrho_B \bar{v}_B + \cdots}{\varrho_A + \varrho_B + \cdots} = \frac{\varrho_A \bar{v}_A + \varrho_B \bar{v}_B + \cdots}{\varrho} \qquad (5.1\text{-}31)$$

where $\varrho = \sum_{i=1}^{n} \varrho_i =$ total mass concentration, mass/volume, and $\varrho\bar{v}$ is the total mass flux. The mass velocity, $\bar{v}$, is usually used in liquid flow situations.

Similarly, for molar flow, the averaged molar flow (velocity profile) is

$$v = \frac{\sum\limits_{i=1}^{n} C_i v_i}{\sum\limits_{i=1}^{n} C_i} = \frac{C_A v_A + C_B v_B + \cdots}{C_A + C_B + \cdots} = \frac{C_A v_A + C_B v_B + \cdots}{C} \qquad (5.1\text{-}32)$$

where $C = \sum_{i=1}^{n} C_i =$ total molar concentration, moles/volume, and Cv is the total molar flux. Fick's first law of molecular diffusion (relative to the total system) can be expressed as

$$J_A = C_A(v_A - v) = -CD_A \nabla x_A \qquad (5.1\text{-}33)$$

where J_A is the molar flux of component A relative to the total system and equation (5.1-33) can be regarded as the defining equation for D_A.

For completeness we conclude this section by presenting the energy equation which is analogous to the general mass transport relationship, equation (5.1-7). This is[5]

$$\underbrace{\frac{\partial}{\partial t}\left(\varrho U + \frac{\varrho v^2}{2}\right)}_{\text{energy accumulation}} = \underbrace{(\nabla \cdot \bar{e})}_{\substack{\text{energy transport} \\ \text{through the surface}}} + \underbrace{\sum_{i=1}^{n}(\bar{n}_i \cdot \bar{g}_i)}_{\text{energy generation}} \qquad (5.1\text{-}34)$$

where

$$\bar{e} = -k\nabla T + \sum_{i=1}^{n} \frac{n_i}{M_i}\,\bar{h}_i$$

$U =$ internal energy per unit mass

$k =$ thermal conductivity

$T =$ temperature

$n_i =$ mass flux of component i

$M_i =$ molecular weight of component i

$\bar{h}_i =$ partial molar enthalpy of component i

$\varrho =$ mass density (or concentration), mass/volume

$v =$ mass velocity

$\bar{g}_i =$ energy generation of component i.

[5] D. M. Himmelblau and K. B. Bischoff, *Process Analysis and Simulation*, John Wiley, New York (1968), p. 13.

If a mechanical energy relationship is also needed to describe the phenomenon completely, this is given by[6]

$$\frac{\partial}{\partial t}\,(\varrho v^2) = \underbrace{-(\nabla \cdot \bar{T})}_{\substack{\text{transport} \\ \text{of mechanical} \\ \text{energy through} \\ \text{the surface}}} + \underbrace{P(\nabla \cdot v) + \bar{\bar{\tau}} : \nabla \bar{v} + \varrho(\nabla \cdot \bar{g})}_{\text{mechanical energy generation}} \qquad (5.1\text{-}35)$$

$$\underbrace{\phantom{\frac{\partial}{\partial t}\,(\varrho v^2)}}_{\substack{\text{accumulation} \\ \text{of mechanical} \\ \text{energy}}}$$

where

$$\bar{T} = \varrho v^2 \bar{v}/2 + P\bar{v} + \bar{\bar{\tau}} \cdot \bar{v}$$

$$P = \text{pressure}$$

$$\bar{\bar{\tau}} = \text{shear stress tensor}$$

$$\bar{v} = \text{mass velocity vector}$$

$$\bar{g} = \text{gravity force vector}$$

5.2. Penetration Theories of Mass Transport

Mass transport and chemical kinetics are frequently controlled by diffusion of a reactant through a film or within a catalyst pellet. In electrochemistry, diffusion may be the limiting process in corrosion processes. In accounting for mass transport across gas–liquid interfaces, some analyses utilize "penetration" theories, usually attributed to Higbie or Dankwerts.[7] It is hypothesized that the gas–liquid interface contains small liquid elements, which are continuously brought up to the surface from the bulk liquid. Each element of fluid at the surface is considered to have a concentration equal to that within the bulk system. Absorption from the gas to the liquid is by unsteady-state diffusion. The time of stay at the surface for a liquid element is given by

$$t^* = \frac{d}{v_b} \qquad (5.2\text{-}1)$$

where d is the length of contacting surface and v_b is the velocity of the fluid element going to the surface. For mass transport in gas bubbles, the governing transport equation for each fluid element from Table XIV is given by

$$\frac{\partial C}{\partial t} = D\,\frac{\partial^2 C}{\partial x^2} \qquad (5.2\text{-}2)$$

---

[6] *Ibid.*, p. 13.

[7] G. Astarita, *Mass Transfer with Chemical Reaction*, Elsevier, Amsterdam (1967), p. 4.

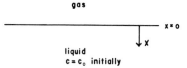

Fig. 48. Gas–liquid interface for penetration
model.

where we have used rectangular geometry to simulate a flat contact surface
and diffusion is assumed in only one direction. The situation is sketched
in Figure 48. The time t in equation (5.2-2) is the elapsed time after the
liquid element has reached the surface. The boundary conditions associated
with equation (5.2-2) are

$$C = C_0 \qquad \text{at } t = 0 \tag{5.2-3}$$

$$C = C^0 \qquad \text{at } x = 0 \text{ (gas–liquid interface)} \tag{5.2-4}$$

$$C = \text{finite} \qquad \text{as } x \to \infty \tag{5.2-5}$$

where C_0 is the initial concentration and C^0 is the interfacial concentration.
Equations (5.2-2) through (5.2-5) can be solved by Laplace transform
techniques, $\mathscr{L}_{t \to p}$, to yield initially

$$D \frac{d^2\bar{C}}{dx^2} = p\bar{C} - C_0 \tag{5.2-6}$$

where $\mathscr{L}_{t \to p}C = \bar{C}$. Equation (5.2-6) can be solved by a number of techniques, including another application of the Laplace transform. Then the
inverse Laplace transform operation is performed. The final result is

$$C = (C^0 - C_0)\,\text{erfc}\!\left(\frac{x}{2\sqrt{Dt}}\right) + C_0 \tag{5.2-7}$$

From equations (5.2-7) and (5.1-19) we can obtain the flux through the
interface at $x = 0$:

$$N\big|_{x=0} = -D\frac{\partial C}{\partial x}\bigg|_{x=0} = (C^0 - C_0)\sqrt{\frac{D}{\pi t}} \tag{5.2-8}$$

The amount of material absorbed into the fluid element (per unit area) is

$$w = \int_0^{t*} N\big|_{x=0}\,dt \tag{5.2-9}$$

and the average flux $\langle N\big|_{x=0}\rangle$ is given by the mean value theorem. The result is

$$\langle N\,|_{x=0}\rangle = \frac{\displaystyle\int_0^{t*} N\,|_{x=0}\,dt}{\displaystyle\int_0^{t*} dt} = (C^0 - C_0)2\sqrt{\frac{D}{\pi t^*}} \qquad (5.2\text{-}10)$$

From a definition of an absorption (mass transfer) coefficient given by

$$\langle N\,|_{x=0}\rangle = k_L(C^0 - C_0) \qquad (5.2\text{-}11)$$

we can compare equations (5.2-10) and (5.2-11) and evaluate k_L:

$$k_L = 2\sqrt{\frac{D}{\pi t^*}} \qquad (5.2\text{-}12)$$

If t^*, the time of stay on the surface, can be estimated, then this approach is useful. Note that t^* for this model is the same for all liquid elements. Another model replaces equation (5.2-10) with

$$\langle N\,|_{x=0}\rangle = \int_0^\infty N\,|_{x=0}\,\psi(t)dt \qquad (5.2\text{-}13)$$

where $\psi(t)dt$ is the fraction of the total surface which is made up of surface elements whose age is between t and $t + dt$. The function $\psi(t)$ is called the age distribution function and satisfies the condition

$$\int_0^\infty \psi(t)dt = 1 \qquad (5.2\text{-}14)$$

If the $\psi(t)$ function can be inferred, then this approach can become useful. One reasonable assumption for the form of the $\psi(t)$ function is

$$\frac{d\psi}{dt} = -s\psi \qquad (5.2\text{-}15)$$

where the rate of disappearance of surface elements is proportional to the number of elements of that age that are present (s is a proportionality factor). From equation (5.2-15) we get

$$\psi = \alpha_1 e^{-st} \qquad (5.2\text{-}16)$$

where α_1 is an integration constant. From equations (5.2-14) and (5.2-16) we get $\alpha_1 = s$. With equation (5.2-16) and (5.2-8) substituted into (5.12-13), and performing the integration required, we get

$$\langle N\,|_{x=0}\rangle = \sqrt{Ds}(C^0 - C_0) \qquad (5.2\text{-}17)$$

where s relates to the rate of surface renewal and $1/s$ is a measure of the average life of surface elements. For this model, comparing equations (5.2-11) and (5.2-17), we get

$$k_L = \sqrt{Ds} \qquad (5.2\text{-}18)$$

5.3. Multicomponent and Turbulent Convection Mass Transport

For a multicomponent fluid, an equation of continuity is required for each chemical species [equation (3.1-4)] and an equation of motion for the mixture [equation (3.2-12)]. The equation of motion must account for individual species effects such as ions that interact with electric fields. We should also include an energy equation that accounts for the work done by diffusion. The equations of motion, mass transport, and energy transport for a mixture with v components are[8]:

Equation of motion

$$\varrho\left(\frac{\partial \bar{v}}{\partial t} + \bar{v} \cdot \nabla \bar{v}\right) = -\nabla P - (\nabla \cdot \bar{\bar{\tau}}) + \sum_{i=1}^{v} \varrho_i F_i \qquad (5.3\text{-}1)$$

Equation of mass transport

$$\frac{\partial \varrho_i}{\partial t} + \bar{v} \cdot \nabla \varrho_i = -\varrho_i(\nabla \cdot \bar{v}) - (\nabla \cdot \bar{j}_i) + k_i \qquad (5.3\text{-}2)$$

Equation of energy transport

$$\frac{\partial \bar{U}}{\partial t} + \bar{v} \cdot \nabla \bar{U} = -(\nabla \cdot \bar{q}) + P(\nabla \cdot \bar{v}) - (\bar{\bar{\tau}} : \nabla \bar{v}) + \sum_{i=1}^{v} (\bar{j}_i \cdot F_i) \quad (5.3\text{-}3)$$

where

$\quad \varrho_i = $ mass density of species i

$\quad k_i = $ rate of production of species i by chemical reaction

$\quad F_i = $ external body force on the ith component

$\quad \bar{v} = $ mass average velocity

$\quad \varrho = $ overall density of the fluid

$\quad \bar{U} = $ internal energy of the mixture

$\quad \bar{j}_i = $ mass flux of species i with respect to the average velocity of the system [from equation (5.1-33)]

[8] T. W. Drew and J. W. Hoopes, *op. cit.*, p. 165.

The energy balance, equation (5.3-3), may also be written in terms of temperature[9]:

$$\varrho C_v \left(\frac{\partial T}{\partial t} + \bar{v} \cdot \nabla T \right)$$

$$= -(\nabla \cdot \bar{q}) - P(\nabla \cdot \bar{v}) - (\bar{\bar{\tau}} : \nabla \bar{v}) + \sum_{i=1}^{\nu} (\bar{j}_i \cdot \bar{F}_i)$$

$$+ \left(P - T \frac{\partial P}{\partial T} \right)(\nabla \cdot \bar{v})$$

$$+ \sum_{i=1}^{\nu} [(\nabla \cdot \bar{j}_i) + k_i] \left[\frac{U_i}{M_i} + \left(P - T \frac{\partial P}{\partial T} \right) \frac{V_i}{M_i} \right] \quad (5.3\text{-}4)$$

where $(\partial P/\partial T)$ implies constant volume and composition and U_i and V_i are partial molal internal energy and partial molal volume respectively. The term

$$\sum_{i=1}^{\nu} (\bar{j}_i \cdot \bar{F}_i)$$

is identically zero if the species are subjected to the same external force such as gravity. Both $(P - T\,\partial P/\partial T)(\nabla \cdot \bar{v})$ and

$$\sum_{i=1}^{\nu} [(\nabla \cdot \bar{j}_i) + k_i] \left[\frac{U_i}{M_i} + \left(P - T \frac{\partial P}{\partial T} \right)\left(\frac{V_i}{M_i} \right) \right]$$

are identically zero if the fluid is an ideal gas and $P(\nabla \cdot \bar{v})$ and

$$\sum_{i=1}^{\nu} [(\nabla \cdot \bar{j}_i) + k_i] \left[\frac{U_i}{M_i} + \left(P - T \frac{\partial P}{\partial T} \right) \frac{V_i}{M_i} \right]$$

are identically zero for incompressible fluids.

The mass flux $\bar{j}_i$ may be the sum of perhaps four components: $\bar{j}_i^{(x)}$, $\bar{j}_i^{(P)}$, $\bar{j}_i^{(F)}$, and $\bar{j}_i^{(T)}$:

$$\bar{j}_i = \bar{j}_i^{(x)} + \bar{j}_i^{(P)} + \bar{j}_i^{(F)} + \bar{j}_i^{(T)} \quad (5.3\text{-}5)$$

where

$\bar{j}_i^{(x)}$ = mass flux due to ordinary molecular diffusion

$\bar{j}_i^{(P)}$ = mass flux due to pressure diffusion

$\bar{j}_i^{(F)}$ = mass flux due to force diffusion (external forces)

$\bar{j}_i^{(T)}$ = mass flux due to thermal diffusion

[9] *Ibid.*, p. 166.

The expressions for the above fluxes are[10]

$$\bar{j}_i^{(x)} = \frac{C^2}{\varrho RT} \sum_{j=1}^{v} M_i M_j D_{ij} \left[x_j \sum_{\substack{k=1 \\ k \neq j}}^{v} \left(\frac{\partial G_j}{\partial x_k} \right) \nabla x_k \right] \qquad (5.3\text{-}6)$$

$$\bar{j}_i^{(P)} = \frac{C^2}{\varrho RT} \sum_{j=1}^{v} M_i M_j D_{ij} \left[x_j M_j \left(\frac{V_j}{M_j} - \frac{1}{\varrho} \right) \nabla P \right] \qquad (5.3\text{-}7)$$

$$\bar{j}_i^{(F)} = \frac{C^2}{\varrho RT} \sum_{j=1}^{v} M_i M_j D_{ij} \left[x_j M_j \left(F_j - \sum_{k=1}^{v} \frac{\varrho_k}{\varrho} F_k \right) \right] \qquad (5.3\text{-}8)$$

$$\bar{j}_i^{(T)} = -D_i^T \nabla \ln T \qquad (5.3\text{-}9)$$

where

G_i = partial molal free energy of component i

V_j = partial molal volume of component j

M_j = molecular weight of component j

C = total molar concentration

ϱ = total mass density

x_i = mole fraction of species i

D_{ij} = multicomponent diffusion coefficients

D_i^T = multicomponent thermal diffusion coefficient

The forced diffusion term $\bar{j}_i^{(F)}$ is important in ionic systems, and if gravity is the only external force, then $\bar{j}_i^{(F)} = 0$.

It should be pointed out that the equations of change are valid for describing turbulent as well as laminar conditions. For example, in the diffusion of component A in a nonreacting binary mixture we can write the mass transport equation from equation (5.1-24):

$$\frac{\partial \varrho_A}{\partial t} + (\nabla \cdot \varrho_A \bar{v}) = (\nabla \cdot \varrho D_A \nabla w_A) \qquad (5.1\text{-}24)$$

where $\bar{j}_A = -\varrho D_A \nabla w_A$ from equation (5.1-33). In turbulent flow ϱ_A and $\bar{v}$ are rapidly varying functions because of turbulent fluctuations. Equation (5.1-24) may be rewritten for turbulent flow by relations such as

$$\varrho_A = \bar{\varrho}_A + \varrho_A' \qquad (5.3\text{-}10)$$

$$\bar{v} = \bar{\bar{v}} + v' \qquad (5.3\text{-}11)$$

[10] *Ibid.*, p. 168.

where $\bar{\varrho}_A$ and $\bar{v}$ are time-averaged variables and ϱ_A' and v' represent turbulent fluctuations. Substitution of equations (5.3-10) and (5.3-11) into (5.1-24) yields

$$\frac{\partial \bar{\varrho}_A}{\partial t} + \nabla \cdot \bar{\varrho}_A \bar{v} + \nabla \cdot \{\bar{j}_A + \bar{j}_A^{(t)}\} + \left(\frac{\partial \varrho_A'}{\partial t} + \nabla \cdot \bar{\varrho}_A v' + \nabla \cdot \varrho_A' \bar{v} \right) = 0$$

(5.3-12)

where $\bar{j}_A^{(t)}$ = turbulent mass flux = $\varrho_A' v'$, and $\bar{j}_A = -\varrho D_A \nabla w_A$. If equation (5.3-12) is averaged by the mean value theorem over a time interval long enough to account for the full range of turbulent fluctuations, then equation (5.3-12) becomes

$$\frac{\partial \bar{\varrho}_A}{\partial t} + (\nabla \cdot \bar{\varrho}_A \bar{v}) + (\nabla \cdot \{\bar{j}_A + \bar{j}_A^{(t)}\}) = 0 \qquad (5.3\text{-}13)$$

where

$$\bar{j}_A^{(t)} = \overline{\varrho_A' v'} \quad \text{(averaged)}$$

For two-component systems we can write for turbulent flow conditions the following equations, which are definitions for D_{AB} and $D_{AB}^{(t)}$:

$$\bar{j}_A = \text{molecular flux} = -D_{AB} \nabla \bar{\varrho}_A \qquad \text{from (5.1-33)} \qquad (5.3\text{-}14)$$

$$\bar{j}_A^{(t)} = \text{turbulent flux} = -D_{AB}^{(t)} \nabla \bar{\varrho}_A \qquad\qquad\qquad\qquad (5.3\text{-}15)$$

In turbulent flow $D_{AB}^{(t)}$ is greater than D_{AB}.

5.4. Some Macroscopic Transport Approaches

In many mass transport situations we do not wish (or are not able) to write the governing equation in terms of microscopic (infinitesimal) points in the system. In these cases we adopt the macroscopic point of view, writing equations which are concerned with the outer boundaries such as the surrounding surface area. Such an analysis, for an unsteady-state macroscopic approach to mass transport,[11] leads to

$$\underbrace{\frac{dm_i}{dt}}_{\substack{\text{rate of} \\ \text{accumulation}}} = \underbrace{-\Delta(\varrho_i \langle v \rangle S)}_{\substack{\text{mass transport through} \\ \text{the surface}}} + w_i^{(m)} + \underbrace{r_i V_{\text{tot}}}_{\substack{\text{mass generation} \\ \text{by chemical} \\ \text{reaction}}} \qquad (5.4\text{-}1)$$

[11] D. M. Himmelblau and K. B. Bischoff, op. cit., p. 30.

where

m_i = total mass of component i

Δ = difference between locations 1 and 2

ϱ_i = mass density of component i

$\langle v \rangle$ = average velocity

S = surface area

$w_i^{(m)}$ = rate of interphase mass transfer of component i

r_i = rate of production of component i by chemical reaction

V_{tot} = total volume of the system

The macroscopic balances for momentum, energy, and mechanical energy are given by the following equations:[12]

Macroscopic momentum balance

$$\underbrace{\frac{dP_i}{dt}}_{\substack{\text{rate of} \\ \text{accumulation}}} = \underbrace{-\Delta(\varrho\langle v^2\rangle S_i + \langle p\rangle S_i) - F_i^{(m)}}_{\substack{\text{momentum transfer through} \\ \text{the surface}}} + \underbrace{m_{\text{tot}}g_i + F_i}_{\substack{\text{momentum} \\ \text{generation}}} \qquad (5.4\text{-}2)$$

where

P_i = total momentum of the system

p = pressure

$F_i^{(m)}$ = rate of interphase momentum transfer

m_{tot} = total mass of the system

g_i = gravitational acceleration

F_i = force exerted by the fluid on the system boundaries

Macroscopic energy balance

$$\underbrace{\frac{dE_{\text{tot}}}{dt}}_{\substack{\text{rate of} \\ \text{accumulation}}} = \underbrace{-\Delta\left[\left(\hat{H} + \frac{1}{2}\frac{\langle v^3\rangle}{\langle v\rangle} + \varphi\right)(\varrho\langle v\rangle S)\right] + Q^{(m)} - W + Q}_{\text{energy transport through the surface}} + \underbrace{S_R}_{\substack{\text{energy} \\ \text{generation}}}$$

$$(5.4\text{-}3)$$

[12] *Ibid.*, p. 30.

where

E_{tot} = total energy of the system = $U_{\text{tot}} + K_{\text{tot}} + \varphi_{\text{tot}}$

U_{tot} = total internal energy of the system

K_{tot} = total kinetic energy of the system

φ_{tot} = total potential energy of the system

$\hat{H}$ = partial molal enthalpy

φ = partial potential energy

Q = interphase heat transfer by conduction

W = work done by the system

$Q^{(m)}$ = interphase energy transfer accompanying interphase mass transfer

S_R = volumetric energy generation

ϱ = total mass density

$\langle v \rangle$ = average velocity

S = surface area through which fluid flows

Mechanical energy balance (constant mass density)

$$\underbrace{\frac{d}{dt}(K_{\text{tot}} + \varphi_{\text{tot}})}_{\text{rate of accumulation}} = \underbrace{\varDelta \left[\left(\frac{1}{2} \frac{\langle v^3 \rangle}{\langle v \rangle} + \varphi + \frac{P}{\varrho} \right)(\varrho \langle v \rangle S) \right] + B^{(m)} - W}_{\text{mechanical energy transport through the surface}} - \underbrace{E_v}_{\substack{\text{mechanical} \\ \text{energy} \\ \text{depletion}}}$$

$$(5.4\text{-}4)$$

where

$B^{(m)}$ = rate of interphase mechanical energy transport associated with mass transport

E_v = friction loss

Macroscopic energy balance (isothermal with no interfacial mass transfer)

$$\underbrace{\frac{d}{dt}(K_{\text{tot}} + \varphi_{\text{tot}} + A_{\text{tot}})}_{\text{rate of accumulation}} = \underbrace{-\varDelta \left[\left(\frac{1}{2} \frac{\langle v^3 \rangle}{\langle v \rangle} + \varphi + \hat{G} \right)(\varrho \langle v \rangle S) \right] - W}_{\text{energy transport through the surface}} - \underbrace{E_v}_{\substack{\text{energy} \\ \text{depletion}}}$$

$$(5.4\text{-}5)$$

where

K_{tot} = total kinetic energy of the system

φ_{tot} = total potential energy of the system

A_{tot} = total Helmholtz free energy of the system

$\langle v \rangle$ = average velocity

$\hat{G}$ = partial molal Gibbs free energy

ϱ = total mass density

S = surface area through which the fluid flows

W = work done by the system

E_v = frictional loss

In solving these macroscopic equations one generally needs a set of n boundary conditions for each nth-order derivative with respect to each space variable. The same requirement is set for the time variable. The total boundary and temporal conditions required are the sum of the space and time requirements. Some typical mass transport boundary conditions are

$C = C_0$ at $x = 0$ an equilibrium condition   (5.4-6)

$N_A \big|_{x=0} = k(C - C^*)$ mass flux, N_A, is related to a concentration driving force   (5.4-7)

$N_A \big|_{x=0} = R_A$ flux at the surface is caused by a chemical reaction   (5.4-8)

$v = 0$ at $r = R$ no-slip fluid condition at the wall   (5.4-9)

$\tau \big|_{x=0_-} = \tau \big|_{x=0_+}$ momentum flux is continuous across a boundary   (5.4-10)

$v \big|_{x=0_-} = v \big|_{x=0_+}$ velocity is continuous across a boundary   (5.4-11)

$T = T_0$ at $x = 0$ an equilibrium condition   (5.4-12)

$q \big|_{x=0_-} = q \big|_{x=0_+}$ heat flux is continuous across a boundary   (5.4-13)

$T \big|_{x=0_-} = T \big|_{x=0_+}$ continuity of temperature   (5.4-14)

$q \big|_{x=0} = h(T - T^*)$ heat flux is related to a temperature driving force   (5.4-15)

Assignments in Chapter 5

5.1. Verify equations (5.1-10), (5.1-24), (5.1-25), (5.1-26), (5.1-29), (5.2-6), (5.2-7), (5.2-8), (5.2-10), and (5.2-17).

5.2. A steady-state tubular chemical reactor is operating with first-order kinetics. Write a governing transport equation with some boundary conditions.

5.3. In an adsorption process, porous adsorbent spheres fall through a rising stream of liquid. Assume that the liquid is in plug flow. Assume that the concentration of the adsorbed material, A, in the sphere surface is proportional to the concentration of A in the bulk liquid.

(a) Set up the governing transport equations with boundary conditions for the solid and the liquid.
(b) Knowing the velocity of fall of the spheres, relate the mass transport to the location of the sphere.

5.4. For an isothermal plug flow reactor, with chemical reaction

$$A + B \underset{k_2}{\overset{k_1}{\rightleftharpoons}} R$$

set up the governing transport equations to find the length of the reactor.

5.5.[13] A tank contains $100 \, \text{ft}^3$ of hexane in which $2 \, \text{lb/ft}^3$ of pentane is dissolved. A pump adds $10 \, \text{ft}^3/\text{min}$ of hexane solution (containing $1 \, \text{lb/ft}^3$ of pentane) to the tank. The tank is depleted at a rate of $5 \, \text{ft}^3/\text{min}$. Set up the governing equations.

5.6. A is fed at rate F to a well mixed tank, where it decomposes to B. A product stream at rate P is drawn off continuously. Write the governing equations.

For Further Reading

Advances in Chemical Engineering, T. W. Drew and J. W. Hoopes, eds., Volume I, Academic Press, New York, 1956.

Transport Phenomena, by R. B. Bird, W. C. Stewart, and E. N. Lightfoot, John Wiley, New York, 1960.

Process Analysis and Simulation, by D. M. Himmelblau and K. B. Bischoff, John Wiley, New York, 1968.

Mass Transfer with Chemical Reaction, by G. Astarita, Elsevier, Amsterdam, 1967.

[13] D. M. Himmelblau and K. B. Bischoff, *Process Analysis and Simulation*, John Wiley, New York (1968), p. 40.

Chapter 6

Transport Analysis in Mass Transport Phenomena

In this chapter we utilize the equations and principles presented in Chapter 5 to analyze mass transport situations involving diffusion, diffusion plus chemical kinetics, and diffusion plus chemical kinetics and convection.

6.1. Diffusion Phenomena

The steady-state evaporation of a liquid through a stationary gas film offers an opportunity to analyze a representative phenomenon. The physical situation is sketched in Figure 49. In rectangular coordinates, from Table XV the governing equation for the gas phase is

$$\frac{dN_{Az}}{dz} = 0 \qquad (6.1\text{-}1)$$

where we are dealing with a one-dimensional diffusion phenomenon (in molar units) and there is no chemical reaction. We have also assumed constant temperature and pressure (total concentration, C = constant). From equation (5.1-23) with $N_B = 0$ (stagnant film) we get

$$N_A = \frac{CD_A}{1 - x_A} \frac{dx_A}{dz} \qquad (6.1\text{-}2)$$

where $Cx_A = C_A$ and $\nabla C_A = C\nabla x_A = C(dx_A/dz)$. Upon substituting equation (6.1-2) into (6.1-1), and after some simplification, we obtain

$$\frac{d}{dz} \left(\frac{1}{1 - x_A} \frac{dx_A}{dz} \right) = 0 \qquad (6.1\text{-}3)$$

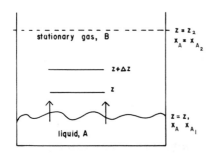

Fig. 49. Evaporation of a liquid into a stationary gas film.

From equation (6.1-3) the term in parentheses becomes

$$\frac{1}{1 - x_A} \frac{dx_A}{dz} = C_1 = \text{constant} \tag{6.1-4}$$

By separating the variables in equation (6.1-4) and integrating, we find

$$-\ln(1 - x_A) = C_1 z + C_2 \tag{6.1-5}$$

With boundary conditions such as

$$x_A = x_{A_1} \quad \text{at } z = z_1 \tag{6.1-6}$$

$$x_A = x_{A_2} \quad \text{at } z = z_2 \tag{6.1-7}$$

we can evaluate C_1 and C_2 in (6.1-5). The final result is

$$\left(\frac{1 - x_A}{1 - x_{A_1}}\right) = \left(\frac{1 - x_{A_2}}{1 - x_{A_1}}\right)^{(z-z_1)/(z_2-z_1)} \tag{6.1-8}$$

Many times it is not the mole fraction x_A that is desired but an average value, $\langle x_A \rangle$, averaged over some interval z. This requires the use of the mean value theorem as presented in Chapter 1. Thus the average value of x_A, over the interval z_1 to z_2, is given by

$$\langle x_A \rangle = \frac{\displaystyle\int_{z_1}^{z_2} x_A \, dz}{\displaystyle\int_{z_1}^{z_2} dz} \tag{6.1-9}$$

It is also often of interest to obtain from the mathematical model the flux N_A across a particular surface, $N_A |_z$. Thus the flux of A leaving the liquid surface at $z = z_1$ is obtained from equation (6.1-2):

$$N_A \big|_{z=z_1} = \left[\frac{-CD_A}{1 - x_A} \frac{dx_A}{dz} \right] \bigg|_{z=z_1}$$

$$= + \frac{CD_A}{x_{B_1}} \frac{dx_B}{dz} \bigg|_{z=z_1} = \frac{CD_A}{(z_2 - z_1)} \ln \frac{x_{B_2}}{x_{B_1}} \qquad (6.1\text{-}10)$$

From equation (6.1-1) it is apparent that N_A, the flux of A, is constant with respect to the distance z. This means that N_A is the same at $z = z_1$ and $z = z_2$, or more generally, N_A is the same for all values of z. Equation (6.1-1) could have also been derived by invoking equation (5.1-8) and the differential element shown in Figure 49. Since there is no generation or depletion term and there is steady state, equation (5.1-8) becomes

$$(N_A S) \big|_z - (N_A S) \big|_{z+\Delta z} = 0 \qquad (6.1\text{-}11)$$

where S is the cross-sectional area and is constant. If equation (6.1-11) is divided by Δz and we let Δz shrink to zero, equation (6.1-1) results.

6.2. Unsteady-State One-Dimensional Diffusion

The governing equation for unsteady-state one-dimensional phenomena (in rectangular coordinates) is

$$\frac{\partial C}{\partial t} = \alpha \frac{\partial^2 C}{\partial x^2} \qquad (6.2\text{-}1)$$

where we have used Table XIV to get (6.2-1). There is no convection or chemical reaction. A number of mathematical techniques are available for solving equation (6.2-1). One method involves a substitution of variables

$$q = xt^n \qquad (6.2\text{-}2)$$

where n is any constant. By the chain rule of calculus, we can generate the terms in equation (6.2-1) using (6.2-2). This is given by the following equations:

$$\frac{\partial C}{\partial x} = \frac{\partial C}{\partial q} \frac{\partial q}{\partial x} \qquad (6.2\text{-}3)$$

$$\frac{\partial^2 C}{\partial x^2} = \frac{\partial}{\partial x} \left(\frac{\partial C}{\partial x} \right) = \frac{\partial}{\partial q} \left[\frac{\partial C}{\partial x} \right] \frac{\partial q}{\partial x} \qquad (6.2\text{-}4)$$

$$\frac{\partial C}{\partial t} = \frac{\partial C}{\partial q} \frac{\partial q}{\partial t} \qquad (6.2\text{-}5)$$

Combining equation (6.2-1) through (6.2-5), we get

$$\alpha t^{2n} \frac{d^2C}{dq^2} = nqt^{-1} \frac{dC}{dq} \tag{6.2-6}$$

Choosing $n = -\frac{1}{2}$ simplifies equation (6.2-6) considerably, yielding

$$\alpha \frac{d^2C}{dq^2} = -\frac{1}{2}q \frac{dC}{dq} \tag{6.2-7}$$

The solution of equation (6.2-7) can readily be obtained by a substitution of variables:

$$P = \frac{dC}{dq} \tag{6.2-8}$$

and

$$\frac{dP}{dq} = \frac{d^2C}{dq^2} \tag{6.2-9}$$

Equation (6.2-7) with (6.2-8) and (6.2-9) can be converted to

$$\alpha \frac{dP}{dq} = -\frac{1}{2}qP \tag{6.2-10}$$

which can be integrated by the separation of variables technique to yield, ultimately,

$$\frac{dC}{dq} = Ae^{-q^2/4\alpha} \tag{6.2-11}$$

where A is an integration constant. Integrating (6.2-11) once more solves the original equation (6.2-1). The result is

$$C = B \operatorname{erf}\left(\frac{x}{2\sqrt{\alpha t}}\right) + K \tag{6.2-12}$$

where B and K are arbitrary constants that will be evaluated from boundary and initial conditions.

Equation (6.2-1) could also be solved by the Laplace transform technique discussed in Chapter 2. Suppose that as an initial condition we have

$$C = C_0 \qquad \text{at } t = 0 \tag{6.2-13}$$

By taking the Laplace transform of equation (6.2-1), $\mathcal{L}_{t \to s}$, we get

$$s\bar{C}(x, s) - C_0 = \alpha \frac{d^2\bar{C}(x, s)}{dx^2} \tag{6.2-14}$$

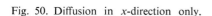

Fig. 50. Diffusion in x-direction only.

Equation (6.2-14) is a linear second-order ordinary differential equation having as a solution

$$\bar{C}(x, s) = Ae^{-qx} + Be^{qx} + \frac{C_0}{s} \qquad (6.2\text{-}15)$$

where $q^2 = s/\alpha$. For diffusion in the x-direction as shown in Figure 50, the boundary conditions are given by

$$C = C_1 \qquad \text{at } x = 0 \qquad (6.2\text{-}16)$$

$$C = \text{finite} \qquad \text{as } x \to \infty \qquad (6.2\text{-}17)$$

In order to use the boundary conditions, we must first transform the variable $C(x, t)$ into $\bar{C}(x, s)$ by the Laplace transform operation $\mathscr{L}_{t \to s}$. Thus equations (6.2-16) and (6.2-17) become

$$\bar{C} = \frac{C_1}{s} \qquad \text{at } x = 0 \qquad (6.2\text{-}18)$$

$$\bar{C} = \text{finite} \qquad \text{as } x \to \infty \qquad (6.2\text{-}19)$$

Application of (6.2-19) to (6.2-15) yields the conclusion that $B = 0$. Substituting equation (6.2-18) into (6.2-15) yields the value for the constant A, and thus we get

$$\bar{C}(x, s) = (C_1 - C_0)\left[\frac{e^{-\sqrt{s/\alpha}\, x}}{s}\right] + \frac{C_0}{s} \qquad (6.2\text{-}20)$$

By going into the Laplace transform tables[1] we find the inverse of equation (6.2-20), $\mathscr{L}_{s \to t}^{-1}$. The result is

$$C(x, t) = (C_1 - C_0)\, \text{erfc}\left(\frac{x}{2\sqrt{\alpha t}}\right) + C_0 \qquad (6.2\text{-}21)$$

[1] G. E. Roberts and H. Kaufman, *Table of Laplace Transforms*, W. B. Saunders, Philadelphia (1966).

The flux, N_A, at $x = 0$, is given by

$$N_A = -D \frac{\partial C}{\partial x} \qquad \text{at } x = 0 \qquad (6.2\text{-}22)$$

In order to use equation (6.2-22) in the Laplace transform method, we need to take the transform $\mathscr{L}_{t \to s}$. The result from equation (6.2-22), if N_A is handled as a constant, becomes

$$\frac{N_A}{s} = -D \frac{d\bar{C}(x, s)}{dx} \qquad \text{at } x = 0 \qquad (6.2\text{-}23)$$

Substituting equation (6.2-15) into (6.2-23) (with $B = 0$) we can again evaluate the constant $A = N_A \sqrt{\alpha}/Ds^{3/2}$, and equation (6.2-15) becomes

$$\bar{C}(x, s) = \frac{N_A \sqrt{\alpha}}{Ds^{3/2}} e^{-\sqrt{s/\alpha}\, x} \qquad (6.2\text{-}24)$$

The inverse Laplace transform $\mathscr{L}_{s \to t}^{-1}$ of equation (6.2-24) yields[2]

$$C(x, t) = \frac{N_A \sqrt{\alpha}}{D} \int_0^t \frac{1}{\sqrt{\pi \tau}} e^{-x^2/4\alpha\tau} d\tau \qquad (6.2\text{-}25)$$

Another set of initial and boundary conditions that may go along with equation (6.2-1) is

$$C = C_0 \qquad \text{at } t = 0 \qquad (6.2\text{-}26)$$
$$C = C_0 \qquad \text{at } x = 0 \qquad (6.2\text{-}27)$$
$$C = C_1 \qquad \text{at } x = L \qquad (6.2\text{-}28)$$

Again equation (6.2-15) is the transformed result of operating on (6.2-1). This time B is not equal to zero in equation (6.2-15) since x is finite. Thus we use equations (6.2-27) and (6.2-28) to evaluate A and B in (6.2-15), noting that (6.2-26) has already been used in arriving at (6.2-15). In order to use equations (6.2-27) and (6.2-28) we need to transform them, $\mathscr{L}_{t \to s}$. The result is given by

$$\bar{C}(x, s) = \frac{C_0}{s} \qquad \text{at } x = 0 \qquad (6.2\text{-}29)$$
$$\bar{C}(x, s) = \frac{C_1}{s} \qquad \text{at } x = L \qquad (6.2\text{-}30)$$

[2] *Ibid.*

Using equations (6.2-29) and (6.2-30) in (6.2-15), we evaluate A and B and the result is

$$\bar{C}(x, s) = \left(\frac{C_1 - C_0}{s}\right)\left(\frac{e^{qx} - e^{-qx}}{e^{qL} - e^{-qL}}\right) + \frac{C_0}{s} \tag{6.2-31}$$

The inverse Laplace transform can again be found in the transform tables, but it might also be helpful to show how the inverse is obtained by the method of residues presented in Chapter 2. There is a simple pole at $s = 0$ in equation (6.2-31) and also for the roots of

$$e^{qL} - e^{-qL} = 0 \tag{6.2-32}$$

where $q^2 = s/\alpha$. If $qL = (s/\alpha)^{1/2}L$ is set equal to $in\pi$, where $n = 1, 2, 3, \ldots$, we can find the infinite number of poles which corresponds to this condition. Thus we get the relationships

$$\left(\frac{s}{\alpha}\right)^{1/2}L = in\pi \tag{6.2-33}$$

or

$$s = s_n = -\frac{n^2\pi^2\alpha}{L^2} \quad \text{(infinite number of poles)} \tag{6.2-34}$$

In arriving at equation (6.2-34), use has been made of the relationship

$$e^{qL} - e^{-qL} = e^{in\pi} - e^{-in\pi}$$

$$= (\cos n\pi + i \sin n\pi)$$

$$-(\cos n\pi - i \sin n\pi) = 0 \quad n = 1, 2, 3, \ldots \tag{6.2-35}$$

The application of the residue theorem for $s = 0$ in equation (6.2-31) yields

$$C(x, t) = (C_1 - C_0) \lim_{s \to 0} \left[\frac{e^{qx} - e^{-qx}}{e^{qL} - e^{-qL}}\right] + C_0 \tag{6.2-36}$$

where L'Hôpital's rule was used to perform the limiting process in equation (6.2-36). The result for $s = 0$ is finally given by

$$C(x, t) = \frac{(C_1 - C_0)}{L}x + C_0 \tag{6.2-37}$$

The inversion of equation (6.2-31) for $s = s_n$ can also be found and then added to the results for $s = 0$, finally giving the complete inversion of $\bar{C}(x, t)$ to $C(x, t)$. The simplest method of evaluating the s_n residues is to first express equation (6.2-31) in terms of hyperbolic functions. The result is

$$\bar{C}(x, s) = \frac{C_1 - C_0}{s_n} \frac{\sinh \sqrt{(s_n/\alpha)}\, x}{\sinh \sqrt{(s_n/\alpha)}\, L} + \frac{C_0}{s_n} \qquad (6.2\text{-}38)$$

with $s_n = n^2\pi^2\alpha/L^2$. By the methods shown in Chapter 2 the inversion can be performed. The final inversion result, for $s = 0$ and $s = s_n$, is[3]

$$\frac{C - C_0}{C_1 - C_0} = (C_1 - C_0) + \sum_{n=0}^{\infty} \frac{2}{n\pi} (-1)^n \sin \frac{n\pi x}{L} e^{-n^2\pi^2\alpha t/L^2} \qquad (6.2\text{-}39)$$

In studying mass and heat transport in spheres the governing equation is given by (5.1-7) and in spherical coordinates the result from Table XV is

$$D\left(\frac{\partial^2 C}{\partial r^2} + \frac{2}{r} \frac{\partial C}{\partial r}\right) = \frac{\partial C}{\partial t} \qquad (6.2\text{-}40)$$

with some typical boundary conditions such as

$$C = C_0 \qquad \text{at } t = 0 \quad r > 0 \qquad (6.2\text{-}41)$$

$$C = C_1 \qquad \text{as } t \to \infty \quad r > 0 \qquad (6.2\text{-}42)$$

or

$$\left.\begin{array}{ll} C = \text{finite} & \text{at } r = 0 \quad t > 0 \\ \dfrac{dC}{dr} = 0 & \text{at } r = 0 \quad t > 0 \end{array}\right\} \qquad (6.2\text{-}43)$$

Figure 51 illustrates this transport situation.

Another boundary condition which can be invoked is not as common as those given above. We "invent" a transfer coefficient, k, such that the mass transported is proportional to the concentration driving force:

$$W_A = 4\pi r_0^2 [k(C_s - C_1)] \qquad \text{at } r = r_0 \qquad (6.2\text{-}44)$$

where

$4\pi r_0^2$ = surface area through which the mass is transported

C_s = concentration of the surrounding fluid

C_1 = concentration on the solid surface

k = mass transport coefficient defined by equation (6.2-44)

[3] V. G. Jenson and G. V. Jeffreys, *Mathematical Methods in Chemical Engineering*, Academic Press, New York (1963), p. 290.

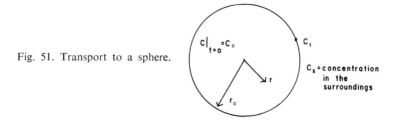

Fig. 51. Transport to a sphere.

C_s = concentration in the surroundings

Thus W_A in equation (6.2-44) goes across the surface into the sphere and then into the interior of the solid by diffusion. The flux, $N_A|_{r=r_0}$ at the boundary multiplied by the area through which it passes, $4\pi r_0^2$, is W_A, so that a composite boundary condition at $r = r_0$ is given by

$$\underbrace{4\pi r_0^2 [k(C_s - C_1)]|_{r=r_0}}_{\text{from fluid to solid surface}} = \underbrace{-4\pi r_0^2 D \overbrace{\frac{\partial C}{\partial r}}^{N_A}\Big|_{r=r_0}}_{\substack{\text{from solid surface} \\ \text{into solid interior}}} \qquad (6.2\text{-}45)$$

Returning to the original statement, equation (6.2-40) with the boundary conditions given, we can begin the mathematical solution. Thus far we have shown solution techniques called similarity transformations and also Laplace transform techniques. Now we show another method, called the separation of variables technique, which is expressed by

$$C(r, t) = R(r)T(t) \qquad (6.2\text{-}46)$$

By substituting equation (6.2-46) into (6.2-40) we can get

$$\frac{R''}{R} + \frac{2R'}{rR} = \frac{1}{D}\frac{T'}{T} \qquad (6.2\text{-}47)$$

where R' denotes $dR(r)/dr$ and T' denotes $dT(t)/dt$. From equation (6.2-47) we see that the left-hand side is a function of only r and the right-hand side is a function of t. Since r and t are independent variables, the equality must hold in equation (6.2-47) even though r and t can be varied separately and independently of each other. Therefore each side of equation (6.2-47) must be a constant, such as $-a^2$, chosen so that the resulting equations have reasonable forms. Thus equation (6.2-47) becomes

$$\frac{R''}{R} + \frac{2R'}{rR} = \frac{1}{D}\frac{T'}{T} = -a^2 \qquad (6.2\text{-}48)$$

and we must solve the equations

$$\frac{R''}{R} + \frac{2R'}{rR} = -a^2 \tag{6.2-49}$$

and

$$\frac{1}{D}\frac{T'}{T} = -a^2 \tag{6.2-50}$$

6.3. Some More Diffusion Problems Solved by Laplace Transform Techniques

Some diffusional transport equations can be written:

$$\frac{\partial U(x, t)}{\partial t} = \alpha \frac{\partial^2 U(x, t)}{\partial x^2} \tag{6.3-1}$$

$$U(x, t)\,|_{x=0} = U(0, t) = \begin{cases} 0 & t < 0 \tag{6.3-2} \\ U_0 & t > 0 \tag{6.3-3} \end{cases}$$

$$U(x, t)\,|_{x=\infty} = U(\infty, t) = 0 \tag{6.3-4}$$

The physical situation described by these equations may be as sketched in Figure 52, where at $t = 0$ we contact the surface with diffusant at a concentration U_0. The solution of equation (6.3-1) is accomplished in a straightforward manner by Laplace transform. Taking the Laplace transform of equation (6.3-1), $\mathscr{L}_{t \to s}$, we obtain

$$s\bar{U}(x, s) - U(x, 0) = \alpha \frac{d^2\bar{U}(x, s)}{dx^2} \tag{6.3-5}$$

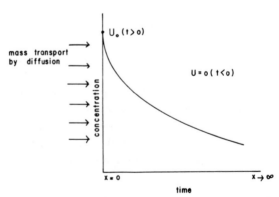

Fig. 52. Sketch of diffusion with an infinitely long path.

In Chapter 2 we noted that t_{0^-} $(t < 0)$ is the initial condition we use. Thus equation (6.3-5) becomes

$$s\bar{U}(x, s) = \alpha \frac{d^2\bar{U}(x, s)}{dx^2} \tag{6.3-6}$$

The solution of equation (6.3-6) is

$$\bar{U}(x, s) = C_1 e^{\sqrt{s/\alpha}\, x} + C_2 e^{-\sqrt{s/\alpha}\, x} \tag{6.3-7}$$

Application of boundary equations (6.3-3) and (6.3-4) to (6.3-7) evaluates C_1 and C_2, yielding $C_1 = 0$ and $C_2 = U_0/s$. If these values of C_1 and C_2 are substituted into equation (6.3-7) and the inverse Laplace transform performed, the final result is

$$U(x, t) = U_0\left[1 - \text{erf}\,\frac{x/\sqrt{\alpha}}{2\sqrt{t}}\right] \tag{6.3-8}$$

Another common situation with slightly different boundary conditions is given by

$$\frac{\partial U(x, t)}{\partial t} = \alpha \frac{\partial^2 U(x, t)}{\partial x^2} \tag{6.3-9}$$

$$U(0, t) = \begin{cases} 0 & t < 0 \tag{6.3-10} \\ U_0 & t > 0 \tag{6.3-11} \end{cases}$$

$$U(L, t) = 0 \tag{6.3-12}$$

Going through a similar procedure as shown above, we can get

$$\bar{U}(x, s) = \frac{(U_0/s)[e^{(L-x)\sqrt{s/\alpha}} - e^{-(L-x)\sqrt{s/\alpha}}]}{[e^{L\sqrt{s/\alpha}} - e^{-L\sqrt{s/\alpha}}]} \tag{6.3-13}$$

or, in another form,

$$\bar{U}(x, s) = \frac{U_0}{s}\,\frac{\sinh(L - x)\sqrt{s/\alpha}}{\sinh L\sqrt{s/\alpha}} \tag{6.3-14}$$

and finally,

$$U(x, t) = U_0\left[\frac{L - x}{L} + \frac{2}{\pi}\sum_{n=1}^{\infty}\frac{e^{-\alpha(\pi n/L)^2 t}}{(-1)^n n}\sin\frac{\pi n(L - x)}{L}\right] \tag{6.3-15}$$

If the unsteady-state diffusion is a bit more complicated in that the diffusion takes place in two directions, then the governing relation might be

$$\frac{\partial U(x, y, t)}{\partial t} = \alpha\left[\frac{\partial^2 U(x, y, t)}{\partial x^2} + \frac{\partial^2 U(x, y, t)}{\partial y^2}\right] \qquad (6.3\text{-}16)$$

with the boundary and initial conditions

$$U(0, y, t) = 0 \qquad (6.3\text{-}17)$$

$$U(L, y, t) = 0 \qquad (6.3\text{-}18)$$

$$U(x, 0, t) = 0 \qquad (6.3\text{-}19)$$

$$U(x, H, t) = \begin{cases} 0 & t < 0 \qquad (6.3\text{-}20) \\ U_0 & t > 0 \qquad (6.3\text{-}21) \end{cases}$$

$$U(x, y, 0) = 0 \qquad (6.3\text{-}22)$$

There are a number of ways of solving this set of equations, and the one that we show here is a two-step transform technique. We will first take the Laplace transform $\mathscr{L}_{t\to s}$ of equation (6.3-16) since we have an initial condition on t and then use a sine transform.[4] Operating on equation (6.3-16) with the Laplace transform, we get

$$s\bar{U}(x, y, s) - U(x, y, 0) = \alpha\left[\frac{\partial^2 \bar{U}(x, y, s)}{\partial x^2} + \frac{\partial^2 \bar{U}(x, y, s)}{\partial y^2}\right] \qquad (6.3\text{-}23)$$

By equation (6.3-22) we can simplify (6.3-23). Now we take the finite sine transform of equation (6.3-23), $\mathscr{S}_{x\to n}$. This yields

$$\frac{s}{\alpha}\,\bar{U}(n, y, s) = \frac{-\pi n}{L}\left[\frac{\pi n}{L}\,\bar{U}(n, y, s) + (-1)^n\bar{U}(L, y, s)\right.$$
$$\left. -\bar{U}(0, y, s) + \frac{d^2\bar{U}(n, y, s)}{dy^2}\right] \qquad (6.3\text{-}24)$$

where

$$\mathscr{L}_{t\to s}U(x, y, t) = \bar{U}(x, y, s) \qquad (6.3\text{-}25)$$

and

$$\mathscr{S}_{x\to n}\bar{U}(x, y, s) = \bar{\bar{U}}(n, y, s) \qquad (6.3\text{-}26)$$

---

[4] F. H. Raven, *Mathematics of Engineering Systems*, McGraw-Hill, New York (1966), p. 301.

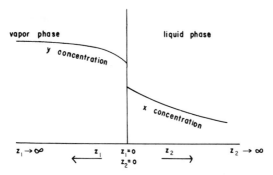

Fig. 53. Transport between liquid and vapor phases.

From equation (6.3-24) we can get $\bar{\bar{U}}(n, y, s)$ in the form

$$\bar{\bar{U}}(n, y, s) = C_1 e^{ay} + C_2 e^{-ay} \qquad (6.3\text{-}27)$$

which, after evaluating C_1 and C_2 and inverting twice,[5] yields

$$U(x, y, t) = \frac{2U_0}{\pi} \sum_{n=1}^{\infty} \left[\frac{\sinh(\pi n y/L)}{\sinh(\pi n H/L)} \right.$$
$$\left. - \sum_{m=1}^{\infty} \frac{[2\pi\alpha m \sin(\pi m y/H)]e^{K_{mn}t}}{(-1)^m H^2 K_{mn}} \right]\left[\frac{1 - (-1)^n}{n} \sin\frac{\pi n x}{L} \right]$$
$$(6.3\text{-}28)$$

where

$$K_{mn} = \alpha\pi^2\left(\frac{m^2}{H^2} + \frac{n^2}{L^2} \right)$$

Suppose a liquid and vapor are contacted so that there is mass transport by diffusion between phases, as sketched in Figure 53. Governing equations (6.3-29) through (6.3-35) are obtained from Table XIV, where x and y are concentrations (mole fractions).

$$D_V \frac{\partial^2 y}{\partial z_1^2} = \frac{\partial y}{\partial t} \qquad \text{(vapor phase)} \qquad (6.3\text{-}29)$$

$$D_L \frac{\partial^2 x}{\partial z_2^2} = \frac{\partial x}{\partial t} \qquad \text{(liquid phase)} \qquad (6.3\text{-}30)$$

$$y = mx \text{ at } z_1 = z_2 = 0 \qquad \text{(equilibrium at the interface)} \qquad (6.3\text{-}31)$$

[5] *Ibid.*, p. 306.

$$\varrho_V D_V \frac{\partial y}{\partial z_1} = -\varrho_L D_L \frac{\partial x}{\partial z_2} \quad \text{at } z_1 = z_2 = 0 \quad \text{(flux equation at the interface)}$$

$$(6.3\text{-}32)$$

$$\frac{\partial y}{\partial z_1} = \frac{\partial x}{\partial z_2} = 0 \quad \text{at } z_1 = z_2 = \infty \qquad (6.3\text{-}33)$$

$$y = y_0 \quad \text{at } t = 0 \qquad (6.3\text{-}34)$$

$$x = x_0 \quad \text{at } t = 0 \qquad (6.3\text{-}35)$$

The Laplace transform, $\mathscr{L}_{t \to p}$, of equations (6.3-29) and (6.3-30) yields

$$\frac{d^2 \bar{y}}{dz_1^2} - \lambda_V^2 \bar{y} = -\frac{1}{D_V} y_0 \qquad (6.3\text{-}36)$$

$$\frac{d^2 \bar{x}}{dz_2^2} - \lambda_L^2 \bar{x} = -\frac{1}{D_L} x_0 \qquad (6.3\text{-}37)$$

where

$$\lambda_V^2 = p/D_V \quad \text{and} \quad \lambda_L^2 = p/D_L \qquad (6.3\text{-}38)$$

Equations (6.3-36) and (6.3-37) can be solved to yield

$$\bar{y} = A_1 e^{\lambda_V z_1} + B_1 e^{-\lambda_V z_1} + \frac{y_0}{p} \qquad (6.3\text{-}39)$$

$$\bar{x} = A_2 e^{\lambda_L z_2} + B_2 e^{-\lambda_L z_2} + \frac{x_0}{p} \qquad (6.3\text{-}40)$$

which gives with the boundary conditions, $A_1 = A_2 = 0$. Furthermore, from the boundary conditions we can get

$$B_1 + \frac{y_0}{p} = mB_2 + m\frac{x_0}{p} \qquad (6.3\text{-}41)$$

and

$$-\varrho_V D_V B_1 \lambda_V = \varrho_L D_L B_2 \lambda_L \qquad (6.3\text{-}42)$$

which yields

$$B_1 = -\frac{(1/p)(y_0 - mx_0)}{1 + mD_V \lambda_V \varrho_V / D_L \lambda_L \varrho_L} \qquad (6.3\text{-}43)$$

and

$$B_2 = -\frac{(1/p)(y_0 - mx_0)}{m + D_L \lambda_L \varrho_L / D_V \lambda_V \varrho_V} \qquad (6.3\text{-}44)$$

so that equation (6.3-39) and (6.3-40) become

$$\bar{y} = \frac{y_0}{p} - \frac{(1/p)(y_0 - mx_0)}{1 + mD_V\varrho_V\lambda_V/D_L\lambda_L\varrho_L} e^{-\lambda_V z_1} \qquad (6.3\text{-}45)$$

$$\bar{x} = \frac{x_0}{p} - \frac{(1/p)(y_0/m - x_0)}{1 + D_L\lambda_L\varrho_L/mD_V\lambda_V\varrho_V} e^{-\lambda_L z_2} \qquad (6.3\text{-}46)$$

The inverse Laplace transform can be accomplished by looking up the appropriate form in the tables.

The rate of transport (flux) across the interface, for the vapor phase, is[6]

$$N_A = \frac{D_V P}{RT} \frac{\partial y}{\partial z_1}\bigg|_{z_1=0} = \frac{P}{RT} \frac{y_0 - mx_0}{(1/D_V)^{1/2} + m(\varrho_V/\varrho_L)(1/D_L)^{1/2}} \cdot \frac{1}{(\pi t)^{1/2}}$$

$$(6.3\text{-}47)$$

where P is the total pressure. The total amount of material transferred in time τ is given by[7]

$$W = \int_0^\tau N_A dt = \frac{P}{RT} \frac{y_0 - mx_0}{(1/D_V)^{1/2} + m(\varrho_V/\varrho_L)(1/D_L)^{1/2}} \left(\frac{4\tau}{\pi}\right)^{1/2} \qquad (6.3\text{-}48)$$

6.4. Some Diffusion Problems Solved by the Separation of Variables Technique

The diffusion equation with the convective and chemical reaction contributions absent is

$$\frac{\partial U(x, t)}{\partial t} = \alpha \frac{\partial^2 U(x, t)}{\partial x^2} \qquad (6.4\text{-}1)$$

By the separation of variables technique we assume that equation (6.4-1) can be written as

$$U(x, t) = X(x)T(t) \qquad (6.4\text{-}2)$$

so that

$$\frac{\partial U}{\partial t} = X \frac{\partial T}{\partial t} \qquad (6.4\text{-}3)$$

and

$$\frac{\partial^2 U}{\partial x^2} = T \frac{\partial^2 X}{\partial x^2} \qquad (6.4\text{-}4)$$

---

[6] W. R. Marshall and R. L. Pigford, *Applications of Differential Equations to Chemical Engineering Problems*, Univ. of Delaware, Dover (1947), p. 134.

[7] *Ibid.*

Substitution of equations (6.4-3) and (6.4-4) into (6.4-2), with some re-arrangement, gives

$$\frac{T'(t)}{T(t)} = \alpha \frac{X''(x)}{X(x)} \tag{6.4-5}$$

where by definition $dT/dt = T'$ and $d^2X/dx^2 = X''$. Note that T'/T is a function of t only, and X''/X is a function of x only. The two variables, t and x, are independent, so that if equation (6.4-5) is to be true always, it must be rewritten as

$$\frac{T'}{T} = \frac{X''}{X} = -k \tag{6.4-6}$$

where k is a constant. We write $-k$ since we ultimately wish to find a solution that is finite as $t \to \infty$. From equation (6.4-6) we get

$$T' + kT = 0 \tag{6.4-7}$$

and

$$X'' + \frac{k}{\alpha} X = 0 \tag{6.4-8}$$

Solving equations (6.4-7) and (6.4-8) with (6.4-2), we obtain

$$U(x, t) = Ce^{-\alpha\lambda^2 t}(A \cos \lambda x + B \sin \lambda x) \tag{6.4-9}$$

where C, A, and B are arbitrary constants and $\lambda^2 = k/\alpha$. If the complete statement of the problem comprises equation (6.4-1) and

$$U(0, t) = 0 \tag{6.4-10}$$

$$U(L, t) = 0 \tag{6.4-11}$$

$$U(x, 0) = f(x) \tag{6.4-12}$$

then we get $A = 0$ and $\sin \lambda L = 0$ or $\lambda L = n\pi$, where $n = 1, 2, 3, \ldots$. Thus we have evaluated λ as $\lambda_n = n\pi/L$. Equation (6.4-9) can now be written as

$$U_n(x, t) = C_n e^{-\alpha\lambda_n^2 t} \sin \lambda_n x \tag{6.4-13}$$

where $C_n = CB$. For linear relationships,

$$U(x, t) = U_1 + U_2 + \cdots + U_n = \sum_{i=1}^{n} U_i$$

where U_i are solutions of the original equation.

In light of this, we can write equation (6.4-13) as

$$U(x, t) = \sum_{n=1}^{\infty} C_n e^{-\alpha(\pi n/L)^2 t} \sin \frac{n\pi x}{L} \tag{6.4-14}$$

From equation (6.4-12) applied to (6.4-14) we get

$$U(x, 0) = f(x) = \sum_{n=1}^{\infty} C_n \sin \frac{n\pi x}{L} \tag{6.4-15}$$

Equation (6.4-15) can be written as a Fourier series expansion[8], where

$$C_n = \frac{2}{L} \int_0^L f(x) \sin \frac{n\pi x}{L} \, dx \tag{6.4-16}$$

so that the complete solution of equation (6.4-1) is given by (6.4-14) and (6.4-16).

The separation of variables method is sometimes applied to two-dimensional unsteady-state diffusion problems such as

$$\frac{1}{\alpha} \frac{\partial C}{\partial t} = \frac{\partial^2 C}{\partial x^2} + \frac{\partial^2 C}{\partial y^2} \tag{6.4-17}$$

with boundary and initial conditions

$$C = 1 \qquad t = 0 \qquad x > 0 \qquad y > 0 \tag{6.4-18}$$

$$C = 0 \qquad t > 0 \qquad x = 0 \qquad y > 0 \tag{6.4-19}$$

$$C = 0 \qquad t > 0 \qquad x > 0 \qquad y = 0 \tag{6.4-20}$$

In attempting to apply the separation of variables technique, our first inclination is to write

$$C = T(t)X(x)Y(y) \tag{6.4-21}$$

With equation (6.4-21) substituted into (6.4-17), we get

$$\frac{1}{\alpha} \frac{T'}{T} = \frac{X''}{X} + \frac{Y''}{Y} \tag{6.4-22}$$

where $T' = \partial T(t)/\partial t$, $X'' = \partial^2 X(x)/\partial x^2$, and $Y'' = \partial^2 Y(y)/\partial y^2$. By arguments similar to that used in arriving at equation (6.4-6), we get

$$\frac{1}{\alpha} \frac{T'}{T} = \frac{X''}{X} + \frac{Y''}{Y} = \lambda \tag{6.4-23}$$

[8] F. H. Raven, *op. cit.*, p. 317.

and hence

$$T = C_1 e^{\alpha \lambda t} \tag{6.4-24}$$

From equations (6.4-18), (6.4-21), and (6.4-24) we can produce

$$1 = C_1 X(x) Y(y) \tag{6.4-25}$$

Since x and y are independent variables, and noting the requirements of equations (6.4-19) and (6.4-20), we conclude that the separation of variables method has led to an inconsistency.

As a modification of this separation of variables procedure we first try a similarity transformation in order to reduce the (t, x, y) variables to (ξ, η). As a trial we choose the transformations

$$\xi = \frac{y}{t^n} \qquad n > 0 \tag{6.4-26}$$

$$\eta = \frac{x}{t^m} \qquad m > 0 \tag{6.4-27}$$

The constraints given previously as equations (6.4-18), (6.4-19), and (6.4-20) now become

$$C(\xi, \eta) = 1 \qquad \eta \to \infty \qquad \xi \to \infty \tag{6.4-28}$$

$$C(\xi, \eta) = 0 \qquad \eta = 0 \tag{6.4-29}$$

$$C(\xi, \eta) = 0 \qquad \xi = 0 \tag{6.4-30}$$

Equation (6.4-17) can now be transformed by the chain rule for differentiation, $C = C(\xi, \eta)$:

$$\frac{\partial C}{\partial t} = \frac{\partial C}{\partial \eta} \frac{\partial \eta}{\partial t} + \frac{\partial C}{\partial \xi} \frac{\partial \xi}{\partial t} = -\frac{m}{t} \eta \frac{\partial C}{\partial \eta} - \frac{n}{t} \xi \frac{\partial C}{\partial \xi} \tag{6.4-31}$$

$$\frac{\partial C}{\partial x} = \frac{\partial C}{\partial \eta} \frac{\partial \eta}{\partial x} + \frac{\partial C}{\partial \xi} \frac{\partial \xi}{\partial x} = \frac{\partial C}{\partial \eta} \frac{1}{t^m} \tag{6.4-32}$$

$$\frac{\partial^2 C}{\partial x^2} = \frac{1}{t^{2m}} \frac{\partial^2 C}{\partial \eta^2} \tag{6.4-33}$$

$$\frac{\partial^2 C}{\partial y^2} = \frac{1}{t^{2n}} \frac{\partial^2 C}{\partial \xi^2} \tag{6.4-34}$$

If equations (6.4-31) through (6.4-34) are substituted into (6.4-17), the result is

$$\frac{1}{\alpha}\left[-mt^{-1}\eta\,\frac{\partial C}{\partial \eta} - nt^{-1}\xi\,\frac{\partial C}{\partial \xi}\right] = t^{-2m}\frac{\partial^2 C}{\partial \eta^2} + t^{-2n}\frac{\partial^2 C}{\partial \xi^2} \quad (6.4\text{-}35)$$

If now we set $2m = 2n = 1$, or $m = n = \frac{1}{2}$, equation (6.4-35) simplifies to (6.4-36), which does not contain t explicitly. This new equation lends itself to a straightforward separation of variables approach:

$$-\frac{1}{2\alpha}\left(\eta\,\frac{\partial C}{\partial \eta} + \xi\,\frac{\partial C}{\partial \xi}\right) = \frac{\partial^2 C}{\partial \eta^2} + \frac{\partial^2 C}{\partial \xi^2} \quad (6.4\text{-}36)$$

Now let $C = H(\xi)F(\eta)$ and solve for $H(\xi)$ and $F(\eta)$.

6.5. Moving-Front Diffusion Models

There are some mass transport phenomena described by moving-front models, as illustrated in Figure 54. The approach is characterized by the hypothesis that an interface exists where the chemical reaction is occurring. The regions adjacent to the interface (front) are governed only by diffusion since the reaction is limited to the interface (which is moving). Thus if gas A diffuses into liquid S and reacts irreversibly and rapidly with component B dissolved in S, then a moving-front model is reasonable. As shown in Figure 54, the region to the left of the front contains only A diffusing through S (which contains reaction products), while in the region to the right of the front only B is present, diffusing in S. The governing differential equations, boundary, and initial conditions for this model are obtained

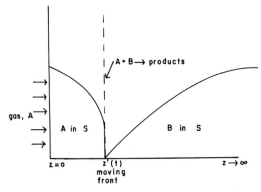

Fig. 54. Illustration of the moving-front model.

from Table XIV and are given by

$$\frac{\partial C_A}{\partial t} = D_{AS} \frac{\partial^2 C_A}{\partial z^2} \qquad 0 \le z \le z'(t) \tag{6.5-1}$$

$$\frac{\partial C_B}{\partial t} = D_{BS} \frac{\partial^2 C_B}{\partial z^2} \qquad z'(t) \le z \le z \le \infty \tag{6.5-2}$$

$$t = 0 \qquad C_B = C_{B0} \tag{6.5-3}$$

$$z = 0 \qquad C_A = C_{A0} \tag{6.5-4}$$

$$z = z'(t) \qquad C_A = 0 \tag{6.5-5}$$

$$z = z'(t) \qquad C_B = 0 \tag{6.5-6}$$

$$z = z'(t) \qquad -D_{AS} \frac{\partial C_A}{\partial z}\bigg|_{z=z'} = D_{BS} \frac{\partial C_B}{\partial z}\bigg|_{z=z'} \tag{6.5-7}$$

Equation (6.5-7) expresses the reaction stoichiometry A + B → products, implying that at the interface $N_A = -N_B$. The mathematical solution of this set of equations can be expressed as[9]

$$\frac{C_A}{C_{A0}} = a_1 + a_2 \, \mathrm{erf} \frac{z}{\sqrt{4 D_{AS} t}} \tag{6.5-8}$$

$$\frac{C_B}{C_{B0}} = b_1 + b_2 \, \mathrm{erf} \frac{z}{\sqrt{4 D_{BS} t}} \tag{6.5-9}$$

$$z'(t) = \sqrt{4 \alpha t} \tag{6.5-10}$$

with the constants a_1, a_2, b_1, b_2, and α given by

$$a_1 = 1 \tag{6.5-11}$$

$$a_2 = -\frac{1}{\mathrm{erf} \sqrt{\alpha/D_{AS}}} \tag{6.5-12}$$

$$b_1 = 1 - \frac{1}{1 - \mathrm{erf} \sqrt{\alpha/D_{BS}}} \tag{6.5-13}$$

$$b_2 = \frac{1}{1 - \mathrm{erf} \sqrt{\alpha/D_{BS}}} \tag{6.5-14}$$

$$1 - \mathrm{erf} \sqrt{\frac{\alpha}{D_{BS}}} = \frac{C_{B0}}{C_{A0}} \sqrt{\frac{D_{BS}}{D_{AS}}} \, \mathrm{erf} \sqrt{\frac{\alpha}{D_{AS}}} \, \exp\left\{ \frac{\alpha}{D_{AS}} - \frac{\alpha}{D_{AS}} \right\} \tag{6.5-15}$$

[9] R. B. Bird, W. E. Stewart, and E. N. Lightfoot, *Transport Phenomena*, John Wiley, New York (1960), p. 600.

In some moving-front examples equilibrium relationships may be controlling at the interface, as given by

$$C_B = QC_A + R \quad \text{at } z = z'(t) \quad (6.5\text{-}16)$$

where Q and R are constants. This equilibrium relationship can be invoked when, for example, the system obeys Henry's law. Typical applications of the moving-front concepts are in absorption phenomena where a single component may be absorbed from a gas by a liquid. It also has utility in studying the formation of a film of tarnish on a metal surface by reaction with a gas. The tarnishing film is assumed porous and allows the gas to diffuse through it to the fresh metal surface. The progressive freezing of a liquid as heat is removed from the surface is another example where moving-front models are useful in the analysis.

6.6. Diffusion and Phase Changes

Some important transport phenomena involve heat and mass transport with associated phase changes, such as freezing water, evaporation from a liquid surface, and tarnishing of metals. Diffusion in solids can involve particles diffusing into a medium which contains C_0 holes per unit volume. In each hole one particle can be bound and removed from the diffusion process. If the holes are immobile, the concentrations of unbound particles, C', and of unfilled holes, C, are given by

$$\frac{\partial C'}{\partial t} = \frac{\partial^2 C'}{\partial x^2} \quad (6.6\text{-}1)$$

$$\frac{\partial C}{\partial t} = KCC' \quad (6.6\text{-}2)$$

where K is the rate constant for an assumed second-order reaction between particles and holes. The KCC' term is usually small except in the very thin transition region (moving front), where the newly diffused particles are reacting with holes. The diffusion of thiosulfate ions into both ends of a capillary filled with a gel containing iodine (and starch to immobilize the iodine) has been studied with this model. In other experiments, the penetration of sodium sulfate into agar gel which contains lead acetate has been followed by watching the advancing free boundaries.

6.7. Diffusion with a Variable Diffusion Coefficient

If the diffusion coefficient is a function of concentration, then the governing transport equation is obtained from equation (5.1-24). This result, in rectangular coordinates,

$$\frac{\partial C}{\partial t} = \frac{\partial}{\partial x}\left(D(C)\frac{\partial C}{\partial x}\right) \tag{6.7-1}$$

We may alter equation (6.7-1) by a combination of variables given by

$$\eta = \frac{1}{2}\frac{x}{t^{1/2}} \tag{6.7-2}$$

so that by the chain rule the equations

$$\frac{\partial C}{\partial x} = \frac{1}{2t^{1/2}}\frac{\partial C}{\partial \eta} \tag{6.7-3}$$

$$\frac{\partial C}{\partial t} = \frac{-x}{4t^{3/2}}\frac{\partial C}{\partial \eta} \tag{6.7-4}$$

result. Combining the above equations gives

$$-2\eta\frac{dC}{d\eta} = \frac{d}{d\eta}\left(D(C)\frac{dC}{d\eta}\right) \tag{6.7-5}$$

We can redefine the variables in equation (6.7-1) by the relationships

$$\sigma = \frac{C}{C_0} \qquad \chi = \frac{x}{l} \qquad T = \frac{t}{l^2} \qquad F(C) = D(C) \tag{6.7-6}$$

and get

$$\frac{\partial \sigma}{\partial T} = \frac{\partial}{\partial \chi}\left\{F(C)\frac{\partial \sigma}{\partial \chi}\right\} \tag{6.7-7}$$

For the diffusion situation sketched in Figure 55, we can write the boundary and initial conditions

$$\sigma = 0 \qquad \text{at } T = 0 \quad 0 < \chi < 1 \tag{6.7-8}$$

$$\sigma = 1 \qquad \text{at } \chi = 1 \quad T > 0 \tag{6.7-9}$$

$$\frac{\partial \sigma}{\partial \chi} = 0 \qquad \text{at } \chi = 0 \quad T > 0 \tag{6.7-10}$$

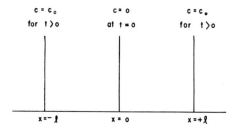

Fig. 55. Diffusion model for variable diffusion coefficient example.

Equations (6.7-8), (6.7-9), and (6.7-10) express the conditions that initially the concentration everywhere is $C = 0$ and thereafter the concentration on the boundaries is $C = C_0$. In addition, there is a symmetry of concentration at the center, $\partial C/\partial x = 0$, at $x = 0$. This set of equations can be solved approximately by the method of moments.[10] If we define a residual of equation (6.7-7) by

$$\text{residual} = \frac{\partial \sigma}{\partial T} - \frac{\partial}{\partial \chi}\left\{ F(C)\frac{\partial \sigma}{\partial \chi}\right\} \tag{6.7-11}$$

then the method of moments allows us to find $\sigma(x, T)$ subject to the constraints of

$$\int_0^1 \underbrace{\left[\frac{\partial \sigma}{\partial T} - \frac{\partial}{\partial \chi}\left\{ F(C)\frac{\partial \sigma}{\partial \chi}\right\}\right]}_{\text{residual}} d\chi = 0 \tag{6.7-12}$$

and

$$\int_0^1 \underbrace{\left[\frac{\partial \sigma}{\partial T} - \frac{\partial}{\partial \chi}\left\{ F(C)\frac{\partial \sigma}{\partial \chi}\right\}\right]}_{\text{residual}} \chi\, d\chi = 0 \tag{6.7-13}$$

Note that the limits on the integration are the range on χ. We define $x_0(t)$ as the point at which $C = 0$. Thus at $t = 0$, $x_0 = l$, and for $t = t_1$ we have $x_0 = 0$ (where t_1 is a short time). In the new variable system, the above statements are: at $T = 0$, $\chi_0 = l/l = 1$, and for $T = t_1/l = T_1$ we have $\chi_0 = 0$.

Next we assume a form for the solution of equation (6.7-7), given by

$$\sigma(\chi, T) = B(T)\{\chi - \chi_0(T)\}^2 + E(T)\{\chi - \chi_0(T)\}^3 \quad \text{for } \chi_0 \leq \chi \leq 1 \tag{6.7-14}$$

$$\sigma(\chi, T) = 0 \quad \text{for } 0 \leq \chi \leq \chi_0 \tag{6.7-15}$$

[10] J. Crank, *The Mathematics of Diffusion*, Oxford University Press, Oxford (1956), p. 178.

where $B(T)$, $E(T)$, and $\chi_0(T)$ are to be determined. From equations (6.7-14), (6.7-15), (6.7-8), and (6.7-9), we can get

$$\chi_0 = 1 \qquad T = 0 \tag{6.7-16}$$

and

$$U + V = 1 \tag{6.7-17}$$

where

$$B(1 - \chi_0)^2 = U \tag{6.7-18}$$

$$E(1 - \chi_0)^3 = V \tag{6.7-19}$$

From equation (6.7-12) with these results we can get[11]

$$\frac{d}{dT} \{(1 - \chi_0)(\tfrac{1}{3}U + \tfrac{1}{4}V)\} = \frac{(2U + 3V)}{1 - \chi_0} F(1) \tag{6.7-20}$$

If equation (6.7-20) is combined with (6.7-17) to eliminate V, we get

$$\frac{d}{dT} \{(1 - \chi_0)(1 + \tfrac{1}{3}U)\} = \frac{12}{1 - \chi_0}(1 - \tfrac{1}{3}U)F(1) \tag{6.7-21}$$

Proceeding analogously with equation (6.7-13) instead of (6.7-12), we obtain[12]

$$\frac{d}{dT} \{(1 - \chi_0)^2(1 + \tfrac{2}{3}U)\} = 20G(1) \tag{6.7-22}$$

where

$$G(\sigma) = \int_0^\sigma F(\sigma')d\sigma' \tag{6.7-23}$$

Note that $F(1)$ and $G(1)$ are values of F and G when $\sigma = 1$. Integration of equation (6.7-22) yields

$$(U + \tfrac{3}{2})\xi^2 = \alpha T \tag{6.7-24}$$

where

$$\alpha = 30G(1) \tag{6.7-25}$$

and

$$\xi = 1 - \chi_0 \tag{6.7-26}$$

Substitution of equations (6.7-24) and (6.7-26) into (6.7-21) yields

$$\frac{d}{dT} \left(\frac{1}{2}\xi + \frac{\alpha T}{3\xi} \right) = \frac{12}{\xi} \left(\frac{3}{2} - \frac{T}{3\xi^2} \right) F(1) \tag{6.7-27}$$

[11] Ibid., p. 179.
[12] Ibid., p. 180.

which has the solution[13]

$$\xi = \sqrt{T/\beta} \tag{6.7-28}$$

where β is determined from the quadratic formula

$$24\alpha\beta^2 F(1) + \{\alpha - 108F(1)\}\beta + \tfrac{3}{2} = 0 \tag{6.7-29}$$

Inserting equation (6.7-28) into (6.7-24) yields

$$U = \alpha\beta - \tfrac{3}{2} \tag{6.7-30}$$

and hence, from (6.7-17),

$$V = -\alpha\beta + \tfrac{5}{2} \tag{6.7-31}$$

With these results we get from (6.7-26)

$$\chi_0 = 1 - \xi = 1 - \sqrt{T/\beta} \tag{6.7-32}$$

Finally, with equation (6.7-32) substituted into (6.7-14), we get

$$\sigma(\chi, T) = (\alpha\beta - \tfrac{3}{2})\left\{\chi - 1 + \sqrt{\frac{T}{\beta}}\right\}^2 \left(\frac{\beta}{T}\right)$$

$$-(\alpha\beta - \tfrac{5}{2})\left\{\chi - 1 + \sqrt{\frac{T}{\beta}}\right\}^3 \left(\frac{\beta}{T}\right)^{3/2} \qquad 1 - \sqrt{\frac{T}{\beta}} \le \chi \le 1 \tag{6.7-33}$$

$$\sigma(\chi, T) = 0 \qquad 0 \le \chi \le 1 - \sqrt{T/\beta} \tag{6.7-34}$$

The transformation of equation (6.7-2) could have been applied in a more general form:

$$\eta = x^\alpha t^\beta \tag{6.7-35}$$

where α and β are constants to be determined so that the resulting equation obtained from equation (6.7-1) is free of x and t, containing only η as the independent variable. If this is done (with the chain rule for differentiation), the result is

$$\frac{\alpha\beta x}{t}\, \eta\, \frac{\partial C}{\partial \eta} = \alpha(\alpha - 1)\eta D\, \frac{\partial C}{\partial \eta} + \alpha^2 \eta^2 \frac{\partial}{\partial \eta}\left(D\, \frac{\partial C}{\partial \eta}\right) \tag{6.7-36}$$

In equation (6.7-36) the right-hand side is free of x and t, so that on the left

[13] *Ibid.*, p. 181.

side we only require that x/t be a function of η. We have a range of choices for this function, and a simple one is $x/t = \eta$, which implies $\alpha = 1$ and $\beta = -1$ in equation (6.7-35). An advantage of this combination is that the first term on the right-hand side of equation (6.7-36) vanishes. This transformation, equation (6.7-35), is called the Boltzmann transformation. With this transformation the original equation (6.7-1) now becomes a second-order nonlinear ordinary differential equation (6.7-5).

6.8. Chemical Kinetics

It might be helpful to review some simple concepts of kinetics. For example, for the chemical reaction

$$A + B \xrightarrow{k_1} \text{Products} \tag{6.8-1}$$

let us define the rate of consumption of component A by

$$r_A = \text{Rate of consumption of A (per unit volume)} = -\frac{dC_A}{dt} \tag{6.8-2}$$

The negative sign for r_A is introduced by convention in order to keep k_1 positive. For the kinetics of equation (6.8-1) we write

$$r_A = k_1 C_A C_B \tag{6.8-3}$$

where C_A and C_B are concentrations of A and B respectively. If the kinetics are a bit more complex,

$$A + B \underset{k_2}{\overset{k_1}{\rightleftharpoons}} D \tag{6.8-4}$$

then the rate of consumption of A is given by

$$r_A = k_1 C_A C_B - k_2 C_D \tag{6.8-5}$$

The r_A equations can become nonlinear, as indicated by the following examples:

$$2A \xrightarrow{k_1} \text{Products} \tag{6.8-6}$$

$$r_A = k_1 C_A^2 \tag{6.8-7}$$

and

$$4A + 2B \underset{k_2}{\overset{k_1}{\rightleftharpoons}} 5D \tag{6.8-8}$$

$$r_A = k_1 C_A{}^4 C_B{}^2 - k_2 C_D{}^5 \tag{6.8-9}$$

For more complicated situations, such as the consecutive reactions

$$A + 2B \underset{k_2}{\overset{k_1}{\rightleftharpoons}} 5D \tag{6.8-10}$$

$$D + A \underset{k_4}{\overset{k_3}{\rightleftharpoons}} 3E \tag{6.8-11}$$

we have for r_A, the rate of consumption of A,

$$r_A = k_1 C_A C_B{}^2 - k_2 C_D{}^5 + k_3 C_D C_A - k_4 C_E{}^3 \tag{6.8-12}$$

If we wish to work in terms of moles consumed (per unit volume), we let x be defined as the moles of A consumed in time t. We can then construct a time chart for the reactions and work in terms of x. For example, with equation (6.8-8) the chart is

Time	C_A	C_B	C_D
0	a	b	d
t	$a - x$	$b - x/2$	$d + \frac{5}{4}x$
$t \to \infty$	$a - x_\infty$	$b - x_\infty/2$	$d + \frac{5}{4}x_\infty$

where a, b, and d are the initial concentrations and $t \to \infty$ implies the approach to equilibrium. Equation (6.8-9) becomes

$$R_A = -\frac{dC_A}{dt} = k_1(a - x)^4\left(b - \frac{x}{2}\right)^2 - k_2\left(d + \frac{5}{4}x\right)^5 \tag{6.8-13}$$

where R_A has replaced r_A to indicate that the units are moles rather than mass. At equilibrium, $R_A = 0$, and if equation (6.8-13) is rearranged, the result is

$$\frac{(d + \frac{5}{4}x)^5}{(a - x)^4(b - x/2)^2} = \frac{k_2}{k_1} = K_{\text{equilibrium}} \tag{6.8-14}$$

where $K_{\text{equilibrium}}$ is the equilibrium constant.

6.9. Diffusion with Chemical Reaction

For diffusion plus chemical reaction phenomena a distinction needs to be made between homogeneous and heterogeneous reactions. If the reacting material is distributed (or dissolved) throughout the region of interest, one can speak of this as a homogeneous system and the reaction term appears explicitly in the governing equation. However, if the reaction occurs at a surface or boundary, then the governing transport equation does not include a chemical reaction term since there is no reaction occurring within the region of interest. In this heterogeneous case the chemical reaction is accounted for at the boundary as a boundary condition. For example, to analyze the diffusion and reaction process for catalyst particles, a number of models may be proposed:

(1) Reaction on the surface of the catalyst
   (a) with an instantaneous reaction
   (b) with a finite reaction rate

(2) Diffusion into the interior of the particles with the reaction on the interstitial surfaces.

Model (1) hypothesizes a transport model as illustrated in Figure 56. Within this gas film, component A moves toward the reactive surface, where $2A \rightarrow A_2$, and A_2, once formed, diffuses away from the catalyst. From Table XV in rectangular coordinates we can get the governing flux equation

$$\frac{dN_{Az}}{dz} = 0 \qquad (6.9\text{-}1)$$

Fig. 56. Diffusion of component A through a stagnant gas layer to the catalyst surface.

From the stoichiometry we know that

$$N_A = -2N_{A_2} \qquad (6.9\text{-}2)$$

Equation (6.9-2) states that for every two moles of A moving toward the reactive surface, there is one mole of A_2 moving away from the reactive surface (at steady state). Combining equation (6.9-2) with (5.1-23) (N_B is equivalent to N_{A_2}), we obtain

$$N_A = \frac{-CD}{1 - \frac{1}{2}x_A} \frac{dx_A}{dz} \qquad (6.9\text{-}3)$$

where $C_A = Cx_A$ and C is the total number of moles per unit volume and is constant. (This is ideal gas behavior at constant pressure and temperature.) Combining equations (6.9-1) and (6.9-3), we get

$$\frac{d}{dz}\left(\frac{1}{1 - \frac{1}{2}x_A} \frac{dx_A}{dz}\right) = 0 \qquad (6.9\text{-}4)$$

If we separate the variables and integrate twice, the result is

$$-2\ln(1 - \frac{1}{2}x_A) = C_1 z + C_2 \qquad (6.9\text{-}5)$$

where C_1 and C_2 are integration constants. For this model of diffusion with instantaneous chemical reaction at the surface, the boundary conditions are given by

$$x_A = x_{A0} \qquad \text{at } z = 0 \qquad \text{(equilibrium or saturation value)} \qquad (6.9\text{-}6)$$

and

$$x_A = 0 \qquad \text{at } z = \delta \qquad \text{(catalyst surface)} \qquad (6.9\text{-}7)$$

Solving equation (6.9-5) with (6.9-6) and (6.9-7) yields

$$(1 - \frac{1}{2}x_A) = (1 - \frac{1}{2}x_{A0})^{1-z/\delta} \qquad (6.9\text{-}8)$$

With equation (6.9-3) and (6.9-8) we can determine the flux at $z = 0$:

$$N_A = \frac{2CD}{\delta} \ln\left(\frac{1}{1 - \frac{1}{2}x_{A0}}\right) \qquad (6.9\text{-}9)$$

If the surface reaction in Figure 56 is not instantaneous, then at the surface $z = \delta$, x_A is not equal to zero. This requires the determination of the boundary condition at $z = \delta$. Since N_A is a constant [by equation

(6.9-1)] a reasonable assumption is that at $z = \delta$ there is a first-order reaction occurring which consumes A, giving $R_A = N_A$, or $N_A\big|_{z=\delta} = kC_A\big|_{z=\delta}$. Thus, as a replacement for equation (6.9-7) we have

$$x_A = \frac{N_A}{Ck} \quad \text{at } z = \delta \tag{6.9-10}$$

Solving equation (6.9-5) as before with the one new condition at $z = \delta$ yields

$$(1 - \tfrac{1}{2}x_A) = \left(1 - \tfrac{1}{2}\,\frac{N_A}{Ck}\right)^{z/\delta}(1 - \tfrac{1}{2}x_{A0})^{1-z/\delta} \tag{6.9-11}$$

and

$$N_A\big|_{z=0} = \frac{2CD}{\delta}\ln\!\left(\frac{1 - \tfrac{1}{2}N_A/Ck}{1 - \tfrac{1}{2}x_{A0}}\right) \tag{6.9-12}$$

Equation (6.9-12) is awkward to use since it involves a trial and error solution for the flux. An approximate result may be obtained by simplifying the logarithmic term with a Taylor series expansion:

$$f(x) = f(a) + \frac{\partial f}{\partial x}\bigg|_{x=a}\frac{(x-a)}{1!} + \frac{d^2 f}{\partial x^2}\bigg|_{x=a}\frac{(x-a)^2}{2!} + \cdots \tag{6.9-13}$$

If the logarithmic term in equation (6.9-12) is expanded in a Taylor series and only the first term retained (assume large values for k), then we get

$$N_A = \frac{2CD}{\delta(1 + D/k\delta)}\ln\frac{1}{1 - \tfrac{1}{2}x_{A0}} \tag{6.9-14}$$

It is interesting to compare the two models presented here (instantaneous *versus* finite reaction at the catalyst surface) for their predictions of the flux N_A. Dividing equation (6.9-9) by (6.9-14), we obtain

$$\frac{N_A\big|_{\text{instantaneous reaction}}}{N_A\big|_{\text{finite reaction}}} = \frac{1 + D/k\delta}{1} > 1 \tag{6.9-15}$$

which predicts, as expected, that there is more A transferred when A is consumed immediately and completely at the reactive surface.

If the catalyst is spherical and the reaction is homogeneous, Figure 57 is a sketch of the situation. The governing flux equation is obtained from Table XV, and in spherical coordinates at steady state the result is

$$\frac{1}{r^2}\frac{\partial}{\partial r}(r^2 N_{Ar}) = R_A \tag{6.9-16}$$

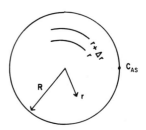

Fig. 57. Diffusion with a homogeneous reaction in a catalyst.

This equation could also have been derived by writing a material balance from equation (5.1-8):

$$(N_{Ar}4\pi r^2)\big|_r - (N_{Ar}4\pi r^2)\big|_{r+\Delta r} + R_A 4\pi r^2 \Delta r = 0 \qquad (6.9\text{-}17)$$

where $4\pi r^2 \Delta r$ is the differential volume with $4\pi r^2$ referring to the surface area. In the usual fashion, equation (6.9-17) is divided by $4\pi\Delta r$ and in the limit as $\Delta r \to 0$, the result is equation (6.9-16).

Since very little is known about the transport processes within the solid, the following equation is used as the defining equation for an apparent diffusion coefficient:

$$N_{Ar} = -D\frac{\partial C_A}{\partial r} \qquad (6.9\text{-}18)$$

Combining equations (6.9-16) and (6.9-18) and assuming $R_A = kC_A$ (first-order reaction), we have

$$D\frac{1}{r^2}\frac{d}{dr}\left(r^2\frac{dC_A}{dr}\right) = kC_A \qquad (6.9\text{-}19)$$

If the concentration C_A is defined by a similarity transformation,

$$\frac{C_A}{C_{AS}} = \frac{f(r)}{r} \qquad (6.9\text{-}20)$$

then equation (6.9-19) simplifies to

$$\frac{d^2f}{dr^2} = \left(\frac{k}{D}\right)f \qquad (6.9\text{-}21)$$

which yields a solution as shown:

$$\frac{C_A}{C_{AS}} = \frac{C_1}{r}\cosh\left(\sqrt{\frac{k}{D}}r\right) + \frac{C_2}{r}\sinh\left(\sqrt{\frac{k}{D}}r\right) \qquad (6.9\text{-}22)$$

where C_1 and C_2 are arbitrary constants. The boundary conditions are

$$C_A = C_{AS} \quad \text{at } r = R \quad \text{(saturated or equilibrium value)} \qquad (6.9\text{-}23)$$

$$\frac{dC_A}{dr} = 0 \quad \text{at } r = 0 \quad \text{(symmetry)} \qquad (6.9\text{-}24)$$

With equations (6.9-23) and (6.9-24), we can solve for C_1 and C_2 in (6.9-22) to yield

$$\frac{C_A}{C_{AS}} = \frac{R}{r} \, \frac{\sinh \sqrt{k/D}\,r}{\sinh \sqrt{k/D}\,R} \qquad (6.9\text{-}25)$$

and the flux at the surface is obtained from equation (6.9-18) and (6.9-25). The total mass transferred, w_A, is obtained from

$$w_A = (N_{Ar}4\pi r^2)\,|_{r=R} \qquad (6.9\text{-}26)$$

where $4\pi^2$ is the surface area. The final result of combining equations (6.9-18), (6.9-25), and (6.9-26) is

$$w_A = 4\pi RD C_{AS}\!\left(1 - \sqrt{\frac{k}{D}}\,R \coth \sqrt{\frac{k}{D}}\,R\right) \qquad (6.9\text{-}27)$$

In some liquid systems, diffusion and chemical reactions occur simultaneously. We may have a gas, A, sparingly soluble in a layer of liquid, B, dissolving and reacting by first-order kinetics. Figure 58 illustrates this, and the governing equation, obtained from Table XIV (rectangular co-

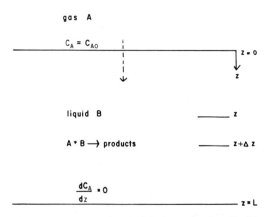

Fig. 58. Diffusion with chemical reaction in a liquid.

ordinates), is

$$D \frac{d^2 C_A}{dz^2} - k C_A = 0 \tag{6.9-28}$$

The reaction term is $R_A = k_1 C_A C_B \cong k C_A$ (pseudo-first-order).

Equation (6.9-28) could also have been derived by starting with a differential element and applying equation (5.1-8). The result is

$$(N_A S)\big|_z - (N_A S)\big|_{z+\Delta z} - k C_A S \Delta z = 0 \tag{6.9-29}$$

where S is the cross-sectional area and k is the rate constant. By the usual calculus limiting process (divide by $S\Delta z$ and take the limit as Δz goes to zero) we get

$$\frac{dN_A}{dz} + k C_A = 0 \tag{6.9-30}$$

For dilute solutions and A sparingly soluble in B, equation (5.1-23) is approximated by

$$N_A = -D \frac{dC_A}{dz} \tag{6.9-31}$$

Combining equations (6.9-30) and (6.9-31), we can get (6.9-28). With boundary conditions such as

$$C_A = C_{A0} \qquad \text{at } z = 0 \tag{6.9-32}$$

$$N_A = 0 \qquad \text{at } z = L \quad \text{(no flux through the wall)} \tag{6.9-33}$$

equation (6.9-28) can be readily solved. Note that equation (6.9-33) is the primary statement at $z = L$ but it implies the more useful form of the boundary condition, $N_A = -D dC_A/dz = 0$, or

$$\frac{dC_A}{dz} = 0 \qquad \text{at } z = L \tag{6.9-34}$$

The solution of equation (6.9-28) with (6.9-32) and (6.9-34) is

$$\frac{C_A}{C_{A0}} = \frac{\cosh b_1(1 - z/L)}{\cosh b_1} \tag{6.9-35}$$

where $b_1 = \sqrt{kL^2/D}$. In this case the flux N_A is not constant [see equation (6.9-30)], and it becomes of interest to determine the mass transfer rate into the liquid film, i.e., $N_A\big|_{z=0}$. From equations (6.9-31) and (6.9-35), the flux at the gas–liquid surface can be determined and is

$$N_A \mid_{z=0} = \left(\frac{DC_{A0}}{L} \right) b_1 \tanh b_1 \qquad (6.9\text{-}36)$$

In discussing simultaneous diffusion plus chemical reaction we have written the reaction as first-order so that the resulting governing equation is linear and solved readily. However, suppose a substance A is diffusing into a medium which contains another component, B. There is a chemical reaction as described by

$$A + B \xrightarrow{k} \text{Products} \qquad (6.9\text{-}37)$$

Figure 59 illustrates this situation. From Table XIV we can write immediately the governing equations for components A and B:

$$\frac{\partial C_A}{\partial t} = D_A \frac{\partial^2 C_A}{\partial x^2} - kC_A C_B \qquad (6.9\text{-}38)$$

$$\frac{\partial C_B}{\partial t} = D_B \frac{\partial^2 C_B}{\partial x^2} - kC_A C_B \qquad (6.9\text{-}39)$$

where R_A, the rate of consumption of component A, is given as $R_A = R_B = kC_A C_B$. The initial and boundary conditions for this situation might be

$$\left. \begin{array}{l} C_A = 0 \\ C_B = C_0 \end{array} \right\} \quad \text{for} \quad t = 0 \quad\quad x > 0 \qquad (6.9\text{-}40)$$

$$\left. \begin{array}{l} C_A = C^* \\ \dfrac{\partial C_B}{\partial x} = 0 \end{array} \right\} \quad \text{for} \quad x = 0 \quad\quad t > 0 \qquad (6.9\text{-}41)$$

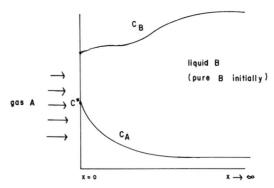

Fig. 59. Diffusion of a gas into a liquid with chemical reaction A + B → products.

Equations (6.9-41) state that the concentration of component A is an equilibrium or saturation value at $x = 0$ and that the flux of component B, $N_B = D_B \partial C_B / \partial x$ through the surface at $x = 0$ is zero. Another boundary condition is given by

$$C_A \to 0 \quad \text{and} \quad C_B \to 0 \quad \text{as } x \to \infty \qquad (6.9\text{-}42)$$

which states that because of the chemical reaction, all of A is consumed as it diffuses in the x-direction. The nonlinear, coupled equations (6.9-38) and (6.9-39) can be solved by an asymptotic approximation.[14] We begin the solution by introducing a set of dimensionless variables:

$$\alpha = \frac{C_A}{C^*} \quad \beta = \frac{C_B}{C_0} \quad \theta = kC^*t \quad \xi = \left(\frac{kC^*}{D_A}\right)^{1/2} x \qquad (6.9\text{-}43)$$

and parameters

$$\varDelta = \frac{D_B}{D_A} \quad \varGamma = \frac{C_0}{C^*} \qquad (6.9\text{-}44)$$

Substituting equations (6.9-43) and (6.9-44) into (6.9-38) and (6.9-39), we get the simplified governing equations

$$\frac{\partial \alpha}{\partial \theta} = \frac{\partial^2 \alpha}{\partial \xi^2} - \varGamma \alpha \beta \qquad (6.9\text{-}45)$$

and

$$\frac{\partial \beta}{\partial \theta} = \varDelta \frac{\partial^2 \beta}{\partial \xi^2} - \alpha \beta \qquad (6.9\text{-}46)$$

with dimensionless boundary and initial conditions now given by

$$\alpha = 0 \qquad \text{for } \theta = 0 \quad \xi > 0 \qquad (6.9\text{-}47)$$

$$\beta = 1 \qquad \text{for } \theta = 0 \quad \xi > 0 \qquad (6.9\text{-}48)$$

$$\alpha = 1 \quad \frac{\partial \beta}{\partial \xi} = 0 \qquad \text{for } \xi = 0 \quad \theta > 0 \qquad (6.9\text{-}49)$$

$$\alpha = 0 \quad \beta = 0 \qquad \text{as } \xi \to \infty \quad \theta > 0 \qquad (6.9\text{-}50)$$

The solution of equations (6.9-45) through (6.9-50) proceeds as follows.

---

[14] W. F. Ames, *Non-linear Partial Differential Equations in Engineering*, Academic Press, New York (1965), p. 240.

Case I

For small values of time (θ), α is small, $\Gamma\alpha\beta$ is essentially zero, and $\beta \cong 1$. Equation (6.9-45) is therefore one of simple diffusion, resulting in

$$\frac{\partial\alpha}{\partial\theta} = \frac{\partial^2\alpha}{\partial\xi^2} \tag{6.9-51}$$

The solution of this approximate problem is given by[15]

$$\alpha = \text{erfc}\left(\frac{\eta}{2}\right) \tag{6.9-52}$$

where

$$\xi = \eta\theta^{1/2} \tag{6.9-53}$$

Case II

If diffusion is negligible compared to the rapid chemical reaction, new variables are introduced:

$$\tau = \Gamma\theta \qquad \eta = \Gamma^{1/2}\xi \qquad \xi' = \xi/\Delta^{1/2} \tag{6.9-54}$$

We can convert equations (6.9-45) and (6.9-46) with (6.9-54) into

$$\frac{\partial\alpha}{\partial\tau} = \frac{\partial^2\alpha}{\partial\eta^2} - \alpha \tag{6.9-55}$$

when $\tau \ll \Gamma$ and $\eta \ll (\tau\Delta)^{1/2}$. Equation (6.9-55) can be solved to get[16]

$$\alpha = \tfrac{1}{2}e^{-\eta}\,\text{erfc}\left[\frac{\eta}{(2\tau)^{1/2}} - \tau^{1/2}\right] + \tfrac{1}{2}e^{\eta}\,\text{erfc}\left[\frac{\eta}{(2\tau)^{1/2}} + \tau^{1/2}\right] \tag{6.9-56}$$

If a gas diffusing into an infinite column of solvent dimerizes according to the reaction $2A \xrightarrow{k} A_2$, the governing equation from Table XIV is

$$\frac{\partial C_A}{\partial t} = D_A\frac{\partial^2 C_A}{\partial x^2} - kC_A^2 \tag{6.9-57}$$

with boundary conditions

$$C_A(x, t) = 0 \qquad \text{at } t = 0 \tag{6.9-58}$$

$$C_A(x, t) = C_0 \qquad \text{at } x = 0 \tag{6.9-59}$$

$$C_A(x, t) = 0 \qquad \text{as } x \to \infty \tag{6.9-60}$$

[15] *Ibid.*
[16] *Ibid.*

The steady-state solution of equation (6.9-57) ($\partial C_A/\partial t = 0$) with the boundary conditions is given by[17]

$$C_A = \frac{C_0}{[1 + (kC_0/6D_A)^{1/2}x]^2} \qquad (6.9-61)$$

The flux at $x = 0$ is

$$N_A \big|_{x=0} = -D_A \frac{\partial C_A}{\partial x}\bigg|_{x=0} = (\tfrac{2}{3}kD_AC_0^3)^{1/2} \qquad (6.9-62)$$

For unsteady-state conditions with no chemical reaction ($k = 0$), equation (6.9-57) yields

$$C_A = C_0 \operatorname{erfc}\left(\frac{x}{2\sqrt{D_At}}\right) \qquad (6.9-63)$$

and

$$N_A \big|_{x=0} = -D_A \frac{\partial C_A}{\partial x}\bigg|_{x=0} = \left(\frac{D_AC_0^2}{\pi t}\right)^{1/2} \qquad (6.9-64)$$

Another approximate method for the solution of equation (6.9-57) is the method of moments.[18] By the method of moments we operate upon equation (6.9-57) by multiplying it by dx and integrating between 0 and ∞. The result is

$$\frac{\partial}{\partial t} \int_0^\infty C_A dx = -D_A \frac{\partial C_A}{\partial x}\bigg|_0^\infty - k \int_0^\infty C_A^2 dx \qquad (6.9-65)$$

We next choose an appropriate function of $C_A(x, t)$ that conforms to the boundary and initial conditions:

$$C_A(x, t) = \begin{cases} C_0\left(1 - \dfrac{x}{x_0(t)}\right)^2 & 0 < x < x_0(t) \qquad (6.9-66) \\ 0 & x_0(t) < x < \infty \qquad (6.9-67) \end{cases}$$

$$x_0(t) = 0 \qquad \text{at } t = 0 \qquad (6.9-68)$$

where $x_0(t)$ is the penetration distance (or moving-front location.) Figure 60 defines the phenomenon. Substituting equation (6.9-66) into (6.9-65) leads to[19]

$$\frac{1}{3}\frac{dx_0(t)}{dt} = \frac{2D_A}{x_0(t)} - \frac{kC_0}{5}x_0(t) \qquad (6.9-69)$$

[17] D. M. Himmelblau and K. B. Bischoff, *Process Analysis and Simulation*, John Wiley, New York (1968), p. 53.

[18] *Ibid.*, p. 54.

[19] *Ibid.*, p. 54.

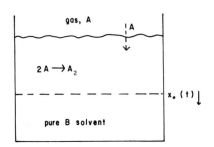

Fig. 60. A gas dissolving into a liquid and dimerizing along a moving front.

or

$$x_0(t) = \left[\frac{10D_A}{kC_0} \left(1 - e^{-6kC_0 t/5} \right) \right]^{1/2} \tag{6.9-70}$$

and

$$N_A \big|_{x=0} = -D_A \frac{\partial C_A}{\partial x} \bigg|_{x=0} = \left(\tfrac{2}{5} k D_A C_0^3 \right)^{1/2} \left(1 - e^{-6kC_0 t/5} \right)^{-1/2} \tag{6.9-71}$$

For steady-state conditions, let $t \to \infty$ in equation (6.9-71). The result is

$$N_A \big|_{x=0} = \left(\tfrac{2}{5} k D_A C_0^3 \right)^{1/2} \quad \text{(method of moments—steady state)} \tag{6.9-72}$$

which can be compared with equation (6.9-62). For the case where there is no chemical reaction, let $k \to 0$ in (6.9-71), which yields[20]

$$N_A \big|_{x=0} = \left(\frac{D_A C_0^2}{3t} \right)^{1/2} \quad \text{(method of moments—no chemical reaction)} \tag{6.9-73}$$

which can be compared with equation (6.9-64). Another choice for the trial function, replacing equation (6.9-66), is

$$C_A(x, t) = C_0 e^{-a(t)x} \tag{6.9-74}$$

with

$$a(t) \to \infty \quad \text{at } t = 0 \tag{6.9-75}$$

Substituting equation (6.9-74) into (6.9-65) yields

$$\frac{d}{dt} \left(\frac{1}{a(t)} \right) = D_A a(t) - \frac{kC_0}{2a(t)} \tag{6.9-76}$$

or

$$a(t) = \left(\frac{kC_0}{2D_A} \right)^{1/2} \left(1 - \frac{kC_0}{2D_A} e^{-kC_0 t} \right)^{-1/2} \tag{6.9-77}$$

[20] *Ibid.*, p. 54.

and

$$N_A \big|_{x=0} = (\tfrac{1}{2}kD_AC_0^3)^{1/2}\left(1 - \frac{kC_0}{2D_A} e^{-kC_0t}\right)^{-1/2} \qquad (6.9\text{-}78)$$

For unsteady-state diffusion in one direction with reversible kinetics, as given by

$$A \underset{k_2}{\overset{k_1}{\rightleftharpoons}} B \qquad (6.9\text{-}79)$$

the governing equations, from Table XIV, are

$$\frac{\partial C_A}{\partial t} = D_{AS} \frac{\partial^2 C_A}{\partial z^2} - k_1C_A + k_2C_B \qquad (6.9\text{-}80)$$

and

$$\frac{\partial C_B}{\partial t} = D_{BS} \frac{\partial^2 C_B}{\partial z^2} + k_1C_A - k_2C_B \qquad (6.9\text{-}81)$$

where $K = k_1/k_2$ is the equilibrium constant. This model is applicable for the diffusion of A in a solvent S containing reactant B. Figure 61 illustrates the behavior of C_A and C_B. Some initial conditions are

$$\frac{C_B}{C_A} = K \quad \text{and} \quad C_A + C_B = C_0 \quad \text{at } t = 0 \qquad (6.9\text{-}82)$$

where C_0 is the total concentration of A and B. From equation (6.9-82), we get an expression for C_A:

$$C_A = \frac{C_0}{1 + K} \quad \text{at } t = 0 \qquad (6.9\text{-}83)$$

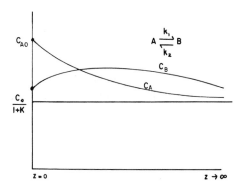

Fig. 61. Diffusion and reversible chemical reaction.

which is applicable also at $z = \infty$. (At $z = \infty$ it requires an infinite amount of time to bring about changes from the initial condition):

$$C_A = \frac{C_0}{1 + K} \qquad \text{at } z = \infty \qquad (6.9\text{-}84)$$

Other boundary conditions applicable to this phenomenon are

$$C_A = C_{A0} \qquad \text{at } z = 0 \qquad (6.9\text{-}85)$$

$$\frac{\partial C_B}{\partial z} = 0 \qquad \text{at } z = 0 \quad \text{(no flux of B into the system)} \qquad (6.9\text{-}86)$$

In some systems with simultaneous diffusion and chemical reaction of component A, it is observed that the chemical reaction involving A is very rapid and immobilizing. One may for this case consider that local equilibrium can exist between component A and the immobilized components. If S is the concentration of the immobilized substance (reaction product), then we may write from Table XIV the governing relationship

$$\frac{\partial C_A}{\partial t} = D_A \frac{\partial^2 C_A}{\partial x^2} - R_A \qquad (6.9\text{-}87)$$

where R_A = reaction term = $\partial S / \partial t$. If we assume $S = RC_A$, where R is a constant, then equation (6.9-87) becomes

$$\frac{\partial C_A}{\partial t} = \left(\frac{D_A}{R + 1} \right) \frac{\partial^2 C_A}{\partial x^2} \qquad (6.9\text{-}88)$$

The effect of the chemical reaction is to decrease the apparent diffusion coefficient.

Some nonlinear transport equations may be solved approximately by the Galerkin method.[21] Suppose that a solution of

$$L\{u\} = \frac{\partial^2 u}{\partial x^2} + \frac{\partial^2 u}{\partial y^2} - (1 - x^2) = 0 \qquad (6.9\text{-}89)$$

is required, where L is a differential operator and u may be velocity. The Galerkin method involves the approximation of the velocity function u by $\bar{u}$, as given by

$$\bar{u} = \sum_{i=1}^{N} C_i v_i(x, y) = C_1 \underbrace{(1 - x^2)(1 - y^2)}_{v_1} + C_2 \underbrace{(1 - x^2)(1 - y^2)x^2}_{v_2} \qquad (6.9\text{-}90)$$

[21] R. Schechter, *The Variational Method in Engineering*, McGraw-Hill, New York (1967), p. 230.

where

$$v_1 = (1 - x^2)(1 - y^2) \tag{6.9-91}$$

$$v_2 = (1 - x^2)(1 - y^2)x^2 \tag{6.9-92}$$

and C_1 and C_2 are arbitrary constants. The forms of v_1 and v_2 are chosen to satisfy the boundary conditions

$$u = 0 \quad \text{for } x = \pm 1 \text{ and } y = \pm 1 \tag{6.9-93}$$

The Galerkin method generally requires that the constants be selected so that the N algebraic equations

$$\int L\{\bar{u}\}v_i dx dy = 0 \quad i = 1, 2, \ldots, N \tag{6.9-94}$$

are satisfied. Substituting equations (6.9-90) into (6.9-89) with (6.9-94), we get

$$L\{\bar{u}\} = -2(1 - y^2)(C_1 - C_2 + 6C_2x^2)$$
$$-2(C_1 + C_2x^2 - C_1x^2 - C_2x^4) - (1 - x^2) \tag{6.9-95}$$

$$4C_1 + \frac{24}{35} C_2 = -1 \tag{6.9-96}$$

$$\frac{24}{5} C_1 + \frac{76}{15} C_2 = -1 \tag{6.9-97}$$

and

$$C_1 = \frac{-575}{2228} \quad \text{and} \quad C_2 = \frac{105}{2228} \tag{6.9-98}$$

Thus equation (6.9-90) is the approximate solution with constants given by equation (6.9-98).

6.10. Complex Chemical Reactions in Reactors (Diffusion is Negligible)

As an illustration of some interesting though complicated kinetics, we cite the problem of chlorinating liquid benzene by sparging chlorine gas into it. The reactions are written as follows:

$$C_6H_6 + Cl_2 \xrightarrow{k_1} C_6H_5Cl + HCl \tag{6.10-1}$$

$$C_6H_5Cl + Cl_2 \xrightarrow{k_2} C_6H_4Cl_2 + HCl \tag{6.10-2}$$

$$C_6H_4Cl_2 + Cl_2 \xrightarrow{k_3} C_6H_3Cl_3 + HCl \tag{6.10-3}$$

If all of the chlorine which enters the reactor undergoes chemical reaction, the kinetics equations corresponding to equations (6.10-1) to (6.10-3) are

$$-V\frac{dQ}{dt} = k_1PQ \tag{6.10-4}$$

$$-V\frac{dR}{dt} = k_2PR - k_1PQ \tag{6.10-5}$$

$$-V\frac{dS}{dt} = -k_2PR + k_3PS \tag{6.10-6}$$

$$-V\frac{dT}{dt} = -k_3PS \tag{6.10-7}$$

where

P = mole fraction of chlorine present

Q = mole fraction of benzene present

R = mole fraction of monochlorobenzene

S = mole fraction of dichlorobenzene

T = mole fraction of trichlorobenzene

$k_1/k_2 = 8.0$

$k_2/k_3 = 30.0$

The material balance is

$$Q + R + S + T = 1 \tag{6.10-8}$$

In equations (6.10-4) through (6.10-7) the terms on the right-hand side are the reaction terms, R_i, defined as the rate of consumption of component i. The left-hand side of the equations represents the rate of accumulation of component i in a constant volume V. We can also use initial conditions such as

$$Q = 1 \quad \text{and} \quad R = 0 \quad \text{at } t = 0 \tag{6.10-9}$$

Equations (6.10-4) through (6.10-7) are nonlinear first-order differential equations. By dividing equations (6.10-5), (6.10-6), and (6.10-7) by (6.10-4) (which can be done if we have total derivatives), we can get[22]

$$\frac{dR}{dQ} = \frac{k_2R}{k_1Q} - 1 = \frac{R - 8Q}{8Q} \tag{6.10-10}$$

[22] V. G. Jenson and G. V. Jeffreys, *op. cit.*, p. 27.

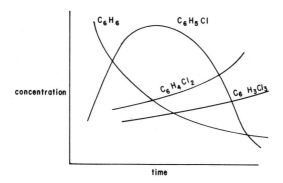

Fig. 62. Solution of complex reaction equations
(6.10-4) through (6.10-7).

which is a first-order homogeneous differential equation. Its solution is

$$Q = K\left(\frac{7R}{Q} + 8\right)^{-8/7} \tag{6.10-11}$$

With equation (6.10-9) substituted into (6.10-11) we can evaluate K and, with some rearrangement, get

$$R = 8(Q^{1/8} - Q)/7 \tag{6.10-12}$$

In a similar manner S can be determined in terms of Q and R from equations (6.10-6) and (6.10-4). If equation (6.10-12) is then used to replace R, we finally arrive at[23]

$$S = \frac{240}{(7)(29)(239)} (29Q - 239Q^{1/8} + 210Q^{1/240}) \tag{6.10-13}$$

Thus, all the quantities, Q, R, S, and T, can be evaluated (relative to P). It is therefore possible to construct graphs showing the consumption of the material of interest, as illustrated in Figure 62.

There are occasionally very complex reactions occurring that may be represented as

$$A_1 \underset{k_{12}}{\overset{k_{21}}{\rightleftharpoons}} A_2$$

$$\begin{array}{cc} k_{31} & k_{32} \\ k_{13} & k_{23} \end{array} \tag{6.10-14}$$

$$A_3$$

where k_{ij} are the rate constants and $\sum_{j=1}^{3} k_{ij}A_j$ is the rate of formation of A_i.

[23] *Ibid.*, p. 28.

For the reaction scheme given by equation (6.10-14) in a batch reactor (no diffusion) we can get from Table XIV the governing equations for each component:

$$\frac{dA_1}{dt} = -(k_{21} + k_{31})A_1 + k_{12}A_2 + k_{13}A_3 \tag{6.10-15}$$

$$\frac{dA_2}{dt} = k_{21}A_1 - (k_{12} + k_{32})A_2 + k_{23}A_3 \tag{6.10-16}$$

$$\frac{dA_3}{dt} = k_{31}A_1 + k_{32}A_2 - (k_{13} + k_{23})A_3 \tag{6.10-17}$$

This system may be generalized to n chemical species, each of which is coupled to every other specie by chemical reaction:

$$\frac{dA_1}{dt} = -\sum_{j=1}^{n} k_{j1}A_1 + k_{12}A_2 + k_{13}A_3 + \cdots + k_{1n}A_n$$

$$\frac{dA_2}{dt} = k_{21}A_1 - \sum_{j=1}^{n} k_{j2}A_2 + k_{23}A_3 + \cdots + k_{2n}A_n \tag{6.10-18}$$

$$\dotfill$$

$$\frac{dA_n}{dt} = k_{n1}A_1 + k_{n2}A_2 + \cdots + k_{n,n-1}A_{n-1} - \sum_{j=1}^{n} k_{jn}A_n$$

Equation (6.10-18) may be written in matrix form as

$$\frac{d\bar{\bar{A}}}{dt} = \bar{\bar{K}}\bar{\bar{A}} \tag{6.10-19}$$

where the double bar indicates matrix notation and $\bar{\bar{K}}$ is the matrix of reaction rate constants.

One can sometimes postulate a four-component reaction system such as

$$\begin{array}{ccc} A_1 & \rightleftharpoons & A_2 \\ \updownarrow & & \updownarrow \\ A_4 & \rightleftharpoons & A_3 \end{array} \tag{6.10-20}$$

where there is no direct coupling between A_1 and A_3 or between A_2 and A_4. A simpler, less complicated kinetics situation is

$$A_1 \xrightarrow{k_1} A_2 \xrightarrow{k_2} A_3 \tag{6.10-21}$$

For this consecutive chemical reaction, we can write

$$\frac{dA_1}{dt} = -k_1 A_1$$

$$\frac{dA_2}{dt} = k_1 A_1 - k_2 A_2 \qquad (6.10\text{-}22)$$

$$\frac{dA_3}{dt} = k_2 A_2$$

or, in more compact matrix form,

$$\frac{d\bar{A}}{dt} = \bar{\bar{K}}\bar{A} \qquad (6.10\text{-}23)$$

with

$$\bar{\bar{K}} = \begin{pmatrix} -k_1 & 0 & 0 \\ k_1 & -k_2 & 0 \\ 0 & k_2 & 0 \end{pmatrix} \qquad (6.10\text{-}24)$$

6.11. Diffusion, Convection, and Chemical Reaction in Thin Films (Wetted-Wall Columns)

The flow of fluids in thin films, accompanied by mass transfer from a gas phase into the liquid film, is sketched in Figure 63. If the liquid film is thin it may be acceptable to "flatten" the geometry of circular thin films

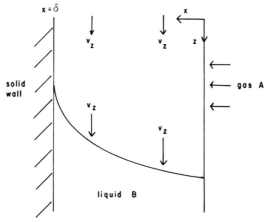

Fig. 63. Diffusion and convection in a thin film of liquid.

(in wetted-wall columns) to rectangular coordinates. From Table XIV we can get the governing equation for this situation,

$$v_z \frac{\partial C_A}{\partial z} = D_A \frac{\partial^2 C_A}{\partial x^2} \tag{6.11-1}$$

where v_z is the fluid velocity, no chemical reaction is occurring, and the gravitational contribution has been neglected. Equation (6.11-1) implies that v_x is negligible (laminar flow) and $v_y = 0$ (no variation of properties in the y-direction). If we knew the velocity profile, i.e., how v_z varies with x, we could solve equation (6.11-1). The velocity profile is found by the procedures presented in Chapters 3 and 4. If gas A dissolves in liquid B and reacts as $A + B \xrightarrow{k}$ Products, then the governing equation is

$$v_z \frac{\partial C_A}{\partial z} = D_A \frac{\partial^2 C_A}{\partial x^2} + kC_A C_B \tag{6.11-2}$$

Some typical boundary conditions for this steady-state problem (fluid acceleration is zero and entrance and exit effects are neglected) are

$$C_A = 0 \qquad\qquad \text{at } z = 0 \tag{6.11-3}$$

$$C_A = C_{A0} \qquad\qquad \text{at } x = 0 \tag{6.11-4}$$

$$\frac{\partial C_A}{\partial x} = 0 \quad (N_{AZ} = 0) \qquad \text{at } x = \delta \tag{6.11-5}$$

Equation (6.11-1) (no chemical reaction) could also have been derived by starting with a differential element as shown in Figure 64. From equation (5.1-8), applied to the differential element in Figure 64, we can get

$$(N_{Az} W \Delta x)\big|_z + (N_{Ax} W \Delta z)\big|_x - (N_{Az} W \Delta x)\big|_{z+\Delta z} - (N_{Ax} W \Delta z)\big|_{x+\Delta x} = 0 \tag{6.11-6}$$

Dividing by $W\Delta x \Delta z$ as usual, and performing the limiting process, we get

$$\frac{\partial N_{Az}}{\partial z} + \frac{\partial N_{Ax}}{\partial x} = 0 \tag{6.11-7}$$

which is shown in Table XV. From equation (5.1-19) we can get the flux equations

$$N_{Az} = -D_A \frac{\partial C_A}{\partial z}^{\,0} + C_A v_z \tag{6.11-8}$$

$$N_{Ax} = -D_A \frac{\partial C_A}{\partial x} + C_A v_x^{\,0} \tag{6.11-9}$$

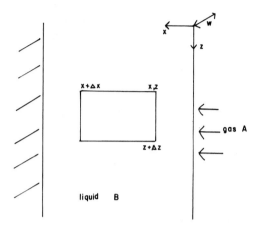

Fig. 64. Differential fluid element in a falling film.

It has been assumed in equations (6.11-8) and (6.11-9) that in the z-direction the mass is transferred primarily by convection and that in the x-direction the mass is transferred primarily by molecular diffusion. Substitution of equations (6.11-8) and (6.11-9) into (6.11-7) leads to (6.11-1). If the velocity profile, v_z, can be considered "flat," then $v_z = (v_z)_{\text{average}} = $ constant. It is common to invoke the concept of "short penetration," i.e., the assumption that component A does not go very far into the liquid film. Hence component A "sees" only a small portion of the velocity profile, mostly of the order of the maximum velocity. Thus $v_{\max}$ replaces v_z in equation (6.11-1). Since A does not penetrate very deeply into the liquid film, boundary condition (6.11-5) can be replaced by

$$C_A = 0 \qquad \text{at } x = \infty \qquad (6.11\text{-}10)$$

implying that since component A doesn't "see" much of the film, the film might as well be infinite in thickness. This is a mathematical convenience since it provides a more readily used boundary condition for the solution of equation (6.11-1). With equations (6.11-10), (6.11-4), (6.11-3), and (6.11-1), and with $v_z = v_{\max}$, the concentration profile can be expressed analytically as[24]

$$\frac{C_A}{C_{A0}} = \text{erfc} \frac{x}{\sqrt{4D_A z/v_{\max}}} \qquad (6.11\text{-}11)$$

The erfc form is the complementary error function and $\text{erfc}(\eta)$ is equal to

[24] R. B. Bird, W. E. Stewart, and E. N. Lightfoot, op. cit., p. 539.

$1 - \text{erf}(\eta)$, where $\text{erf}(\eta)$ is the error function of η. In this falling film example, the mass (or moles) of A transferred into the film is given by

$$N_{Ax}\big|_{x=0} = -D_A \frac{\partial C_A}{\partial x}\bigg|_{x=0} = C_{A0}\sqrt{\frac{D_A v_{\max}}{\pi z}} \qquad (6.11\text{-}12)$$

where equation (6.11-11) has been used in (6.11-12). The total mass (or moles) transferred is

$$W_A = \int_0^W \int_0^L N_{Ax}\big|_{x=0}\, dzdy = WLC_{A0}\sqrt{\frac{4D_A v_{\max}}{\pi L}} \qquad (6.11\text{-}13)$$

where W is the width of the film.

In the analysis of a wetted-wall column as described by equation (6.11-2), if the concentration of component B is not essentially changed by the reaction, then the rate of consumption of component A is given by

$$R_A = k_1 C_A C_B \cong k C_A \qquad (6.11\text{-}14)$$

so that the reaction is said to be pseudo-first-order. For this situation the governing equation becomes

$$v_z \frac{\partial C_A}{\partial z} = D_A \frac{\partial^2 C_A}{\partial x^2} - k C_A \qquad (6.11\text{-}15)$$

For a Newtonian fluid, from equation (4.1-7) the velocity profile in rectangular coordinates is given by

$$v_z = v_{\max}\left(1 - \frac{x^2}{\delta^2}\right) \qquad (6.11\text{-}16)$$

For short contact times (small columns or high fluid velocities) or short penetration of gas A into liquid B (sparingly soluble), $v_z \cong v_{\max}$ in equation (6.11-16). With this simplification, equation (6.11-15) can be solved by taking the Laplace transform $\mathscr{L}_{z \to s}$ of (6.11-15). The result is

$$D_A \frac{\partial^2 \bar{C}_A(x, s)}{\partial x^2} - (k + sv_{\max})\bar{C}_A(x, s) = 0 \qquad (6.11\text{-}17)$$

The solution of equation (6.11-17) is

$$\bar{C}_A(x, s) = A \exp\left(x\sqrt{\frac{k + sv_{\max}}{D_A}}\right) + B \exp\left(-x\sqrt{\frac{k + sv_{\max}}{D_A}}\right)$$

$$(6.11\text{-}18)$$

For short penetration of A into the liquid film a useful boundary condition is

$$C_A \text{ is finite as } x \to \infty \qquad (6.11\text{-}19)$$

Application of condition (6.11-19) in (6.11-18) yields $A = 0$. Another boundary constraint is

$$\frac{\partial C_A}{\partial x} = 0 \qquad \text{at } x = \delta \qquad (6.11\text{-}20)$$

Equation (6.11-20) relates to the requirement that the flux $N_{Ax} = D_A \partial C_A / \partial x = 0$ at $x = \delta$, i.e., there is no flux of A across the wall. Equation (6.11-19) is the choice we will make here simply because it leads to a less complex solution. Another boundary condition is

$$C_A = C_0 \text{ (equilibrium or saturation)} \qquad \text{at } x = 0 \quad \text{(gas–liquid interface)} \qquad (6.11\text{-}21)$$

In order to use equation (6.11-21) in (6.11-18) we need to transform (6.11-21), $\mathscr{L}_{z \to s}$:

$$\bar{C}_A(x, s) = \frac{C_0}{s} \qquad \text{at } x = 0 \qquad (6.11\text{-}22)$$

Substituting equation (6.11-22) into (6.11-18) allows us to evaluate B, yielding

$$\bar{C}_A(x, s) = \frac{C_0}{s} \exp\left[-x \sqrt{\frac{v_{\max}}{D_A} \left(s + \frac{k}{v_{\max}} \right)} \right] \qquad (6.11\text{-}23)$$

From the Laplace transform tables we get the inverse Laplace transform of equation (6.11-23), which is[25]

$$\frac{C_A}{C_0} = \tfrac{1}{2} \exp\left(-x \sqrt{\frac{k}{D_A}} \right) \operatorname{erfc}\left(\frac{x}{2} \sqrt{\frac{v_{\max}}{D_A z}} - \sqrt{\frac{kz}{v_{\max}}} \right)$$

$$+ \tfrac{1}{2} \exp\left(x \sqrt{\frac{k}{D_A}} \right) \operatorname{erfc}\left(\frac{x}{2} \sqrt{\frac{v_{\max}}{D_A z}} + \sqrt{\frac{kz}{v_{\max}}} \right) \qquad (6.11\text{-}24)$$

The absorption at the gas–liquid interface expressed as flux, N_{Ax}, is given by[26]

$$N_{Ax} = -D_A \left. \frac{\partial C_A}{\partial x} \right|_{x=0}$$

$$= C_0 \sqrt{k D_A} \left[\operatorname{erf} \sqrt{\frac{kz}{v_{\max}}} + \sqrt{\frac{v_{\max}}{\pi kz}} \exp\left(\frac{-kz}{v_{\max}} \right) \right] \qquad (6.11\text{-}25)$$

[25] G. E. Roberts and H. Kaufman, op. cit.

[26] V. G. Jenson and G. V. Jeffreys, op. cit., p. 300.

We can also obtain N_{Ax} by first determining $\bar{N}_{Ax}$:

$$\mathscr{L}_{z \to s} N_{Ax} = \bar{N}_{Ax} = \mathscr{L}_{z \to s}\left\{-D_A \frac{\partial C_A}{\partial x}\bigg|_{x=0}\right\} = -D_A \frac{\partial \bar{C}_A(x, s)}{\partial x}\bigg|_{x=0} \quad (6.11\text{-}26)$$

Combining equation (6.11-23) and (6.11-26), we can get

$$\bar{N}_{Ax} = C_0 \sqrt{D_A v_{max}} \frac{\sqrt{s + k/v_{max}}}{s} \quad (6.11\text{-}27)$$

which can be inverted to yield equation (6.11-25).

6.12. Tubular Chemical Reactors

Tubular chemical reactors combine some aspects of fluid flow, heat transport, and mass transport and are interesting pieces of equipment to study. In the simpler analyses we neglect the temperature gradients that may be produced by the heat of reaction. Suppose in this simplified analysis we consider the reaction

$$A \xrightarrow{k} B \quad (6.12\text{-}1)$$

and write the governing equation for the tubular reactor in rectangular coordinates from Table XIV. Figure 65 sketches this situation. The governing equation is

$$D_A \frac{d^2 C_A^{II}}{dx^2} - v_x \frac{d C_A^{II}}{dx} - k C_A^{I'} = 0 \quad (6.12\text{-}2)$$

It is assumed that there are no gradients in the y- or z-directions. Axial (x-direction) transport is governing. For the entrance region, where there is no chemical reaction occurring, we find an equation similar to (6.12-2)

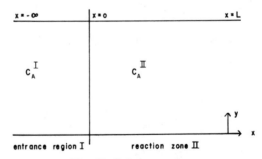

Fig. 65. Tubular reactor.

except that there is no chemical reaction term present:

$$D_A \frac{d^2 C_A{}^I}{dx^2} - v_x \frac{dC_A{}^I}{dx} = 0 \tag{6.12-3}$$

The boundary conditions for this steady-state problem are

$$C_A{}^I = C_A{}^0 \qquad \text{at } x = -\infty \tag{6.12-4}$$

$$C_A{}^I = C_A{}^{II} \qquad \text{at } x = 0 \tag{6.12-5}$$

$$\frac{dC_A{}^I}{dx} = \frac{dC_A{}^{II}}{dx} \qquad \text{at } x = 0 \tag{6.12-6}$$

$$\frac{dC_A{}^{II}}{dx} = 0 \qquad \text{at } x = L \tag{6.12-7}$$

We have assumed continuity of concentration between entrance zone and reactor zone [equation (6.12-5)]. Equation (6.12-6) states that the axial flux $N_{Ax}{}^I$ leaving zone I is equal to the axial flux into zone II, $N_{Ax}{}^{II}$, and $C_A v_x{}^I|_{x=0} = C_A v_x{}^{II}|_{x=0}$. Because there can be no further changes in concentration of A beyond the reactor zone, the concentration gradient beyond $x = L$ is zero $(dC_A/dx = 0)$. Therefore at $x = L$, it is also necessary that $dC_A{}^{II}/dx = 0$ to insure continuity. Equations (6.12-2) and (6.12-3) are readily solved when v_x is a constant and the results are

$$C_A{}^{II} = A \exp\left[\frac{v_x x}{2D_A}(1 + a)\right] + B \exp\left[\frac{-v_x x}{2D_A}(1 - a)\right] \tag{6.12-8}$$

and

$$C_A{}^I = \alpha + \beta \exp\left(\frac{v_x x}{D_A}\right) \tag{6.12-9}$$

where $a = \sqrt{1 + 4k D_A/v_x{}^2}$. Application of boundary equations (6.12-4) through (6.12-7) allows the evaluation of the constants, yielding finally $C_A{}^{II}$ in[27]

$$\frac{C_A{}^{II}}{C_A{}^0} = \frac{2}{k} \exp\left(\frac{v_x x}{2D_A}\right)\left[(a + 1)\exp\left\{\frac{v_x a}{2D_A}(L - x)\right\}\right.$$

$$\left. + (a - 1)\exp\left\{\frac{-v_x a}{2D_A}(L - x)\right\}\right] \tag{6.12-10}$$

In some designs a catalyst is coated onto the surface of a tubular reactor, so that a gas flowing through the reactor is reactive only at the

[27] *Ibid.*, p. 59.

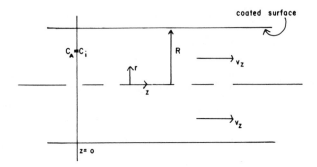

Fig. 66. Flow through a tubular reactor with coated walls.

solid surface. The governing equation for steady-state tube flow in cylindrical coordinates is obtained from Table XIV and is given by

$$v_z \frac{\partial C_A}{\partial z} = D_A\left(\frac{\partial^2 C_A}{\partial r^2} + \frac{1}{r}\frac{\partial C_A}{\partial r}\right) \qquad (6.12\text{-}11)$$

Figure 66 illustrates this case. The boundary conditions are

$$C_A = C_i \qquad \text{at } z = 0 \quad \text{(entrance region)} \quad (6.12\text{-}12)$$

$$\frac{\partial C_A}{\partial r} = 0 \qquad \text{at } r = 0 \quad \text{(symmetry)} \qquad (6.12\text{-}13)$$

$$N_{Ar} = D_A \frac{\partial C_A}{\partial r} = -kC_A \qquad \text{at } r = R \quad \text{(wall reaction)} \qquad (6.12\text{-}14)$$

Equation (6.12-14) states that component A diffuses to the wall surface at $r = R$ and then is consumed at the boundary by a first-order reaction. Thus, the chemical reaction shows up as a boundary condition, not as a term in the governing equation. We have assumed in equation (6.12-11) that there are no radial velocity components (laminar flow). For simplicity here, v_z is constant (plug flow). Equations (6.12-11) through (6.12-14) can be solved by Laplace transforms techniques. Taking the Laplace transform, $\mathscr{L}_{z \to p}$, of equation (6.12-11), we get

$$p\bar{C}_A(p, r) - C_i = \frac{D_A}{v_z}\left(\frac{d^2\bar{C}_A(p, r)}{dr^2} + \frac{1}{r}\frac{d\bar{C}_A(p, r)}{dr}\right) \qquad (6.12\text{-}15)$$

where $\mathscr{L}_{z \to p}C_A(z, r) = \bar{C}_A(p, r)$. Since the boundary conditions are expressed as $C_A(z, r)$, we must transform these boundary conditions before they can be used in equation (6.12-15). Equation (6.12-13) transforms to

$$\frac{d\bar{C}_A(p, r)}{dr} = 0 \qquad \text{at } r = 0 \qquad (6.12\text{-}16)$$

and equation (6.12-14) transforms to

$$D_A \frac{d\bar{C}_A(p, r)}{dr} = k\bar{C}_A(p, r) \qquad \text{at } r = R \qquad (6.12\text{-}17)$$

In equation (6.12-15) we have already used (6.12-12). Equation (6.12-15) is a form of Bessel's equation which has as its solution

$$\bar{C}_A(p, r) - \frac{C_i}{p} = AI_0\left(\sqrt{\frac{v_z p}{D_A}}\, r\right) + BK_0\left(\sqrt{\frac{v_z p}{D_A}}\, r\right) \qquad (6.12\text{-}18)$$

where $I_0(r)$ and $K_0(r)$ are tabulated Bessel functions. $K_0(r)$ goes to infinity as r goes to zero. Since C_A must be finite as r goes to zero this requires that B must be equal to zero in (6.12-18). In order to use boundary equations (6.12-16) and (6.12-17) we differentiate (6.12-18), yielding

$$\frac{d\bar{C}_A}{dr} = A\sqrt{\frac{v_z p}{D_A}}\, I_1\left(\sqrt{\frac{v_z p}{D_A}}\, r\right) \qquad (6.12\text{-}19)$$

where

$$\frac{d}{dr} I_0(\alpha r) = \alpha I_1(\alpha r) \qquad (6.12\text{-}20)$$

and $I_1(\alpha r)$ is another tabulated Bessel function. Applying equation (6.12-17) to (6.12-19) yields the value for the arbitrary constant A, so that equation (6.12-18) finally becomes

$$\bar{C}_A(p, r) = C_i\left[\frac{1}{p} - \frac{(k/D_A)p - I_0[\sqrt{(v_z p/D_A)}\, r]}{(k/D_A)I_0[\sqrt{(v_z p/D_A)}R] + (v_z p/D_A)I_1[\sqrt{(v_z p/D_A)}R]}\right]$$

$$(6.12\text{-}21)$$

If the inversion of equation (6.12-21) can be found,

$$\mathscr{L}_{p \to z}^{-1} \bar{C}_A(p, r) = C_A(z, r) \qquad (6.12\text{-}22)$$

then the mean concentration of C_A at a particular z location can be obtained in a straightforward manner by applying the mean value theorem,

$$\langle C_A \rangle = \frac{\displaystyle\int_0^{2\pi}\int_0^R C_A(z, r)r\,dr\,d\theta}{\displaystyle\int_0^{2\pi}\int_0^R r\,dr\,d\theta} \tag{6.12-23}$$

where $\langle C_A \rangle$ signifies a mean value of C_A with respect to r (at a particular location of z). It is this mean value of C_A that one measures experimentally.

If the tubular reactor with the coated reactive walls has the diffusion transport confined to a small zone adjacent to the pipe wall, we can analyze the "boundary layer" problem. If the velocity profile is originally given by

$$v_z = v_{max}\left(1 - \frac{r^2}{R^2}\right) \tag{6.12-24}$$

where v_{max} is the maximum velocity, we can define a direction $y = R - r$ as shown in Figure 67. The y distance is measured from the wall and is small. Thus,

$$1 - \frac{r^2}{R^2} = 1 - \frac{(R - y)^2}{R^2} = 1 - \left(\frac{R^2 + y^2 - 2Ry}{R^2}\right)$$

$$= \frac{-y^2 + 2Ry}{R^2} \cong \frac{2y}{R}$$

Equation (6.12-24) therefore becomes

$$v_z \cong v_{max}\frac{2y}{R} \tag{6.12-25}$$

The governing mass transport equation for this reactor (in rectangular

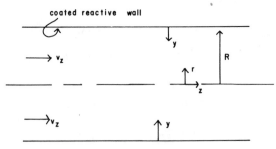

Fig. 67. Tubular reactor with reactive wall–boundary layer analysis.

coordinates) is again obtained from Table XIV and is

$$v_z \frac{\partial C_A}{\partial z} = D_A \frac{\partial^2 C_A}{\partial y^2} \qquad (6.12\text{-}26)$$

where v_z is given by equation (6.12-25). The boundary conditions may be

$$C_A = 0 \qquad \text{at } y = 0 \qquad (6.12\text{-}27)$$

$$C_A \rightarrow C_0 \qquad \text{as } y \rightarrow \infty \qquad (6.12\text{-}28)$$

Note that equation (6.12-27) states that the reaction at the wall instantaneously and irreversibly consumes the A component that diffuses to the wall. The region beyond the reaction zone ($y \rightarrow \infty$) remains unaltered. Equations (6.12-25) through (6.12-28) can be solved more readily by introducing a new variable:

$$\eta = \left(\frac{v_{\max}}{D_A R} \right)^{1/3} \frac{y}{x^{1/3}} \qquad (6.12\text{-}29)$$

which leads to

$$\frac{d^2 C_A}{d\eta^2} + \tfrac{2}{3}\eta^2 \frac{dC_A}{d\eta} = 0 \qquad (6.12\text{-}30)$$

By another variable substitution, $P = dC/d\eta$ and $dP/d\eta = d^2C/d\eta^2$, we can eventually solve equation (6.12-30), yielding

$$C_A = \frac{C_0 \displaystyle\int_0^{(y/z^{1/3})(v_{\max}/D_A R)^{1/3}} \exp(-\tfrac{2}{9}\eta^3)\,d\eta}{\displaystyle\int_0^{\infty} \exp(-\tfrac{2}{9}\eta^3)\,d\eta} \qquad (6.12\text{-}31)$$

It is possible from (6.12-31) to obtain an expression for the diffusional flux to the wall of the tube:

$$N_{Ay}\big|_{y=0} = D_A \frac{\partial C_A}{\partial y}\bigg|_{y=0} = \frac{D_A C_0}{z^{1/3}} \left(\frac{v_{\max}}{D_A R} \right)^{1/3} \frac{1}{\displaystyle\int_0^{\infty} \exp(-\tfrac{2}{9}\eta^3)\,d\eta} \qquad (6.12\text{-}32)$$

where the differentiation was performed by the Leibnitz rule. If the integral in equation (6.12-32) is evaluated, the final result is[28]

$$N_{Ay}\big|_{y=0} = 0.67 C_0 D_A \left(\frac{v_{\max}}{D_A R z} \right)^{1/3} \qquad (6.12\text{-}33)$$

[28] V. G. Levich, *Physiochemical Hydrodynamics*, Prentice-Hall, Englewood Cliffs, N. J. (1962), p. 87.

In an isothermal plug flow chemical reactor the chemical kinetics may be given by an empirical relationship such as

$$R_A = kC_A{}^s \tag{6.12-34}$$

where k is the rate constant and s is a constant. The governing equation in rectangular coordinates) is, from Table XIV,

$$v_x \frac{\partial C_A}{\partial x} = D_A \frac{\partial^2 C_A}{\partial x^2} - kC_A{}^s \tag{6.12-35}$$

Figure 68 shows the situation. The boundary conditions for this plug flow reactor may be given as

$$C_A = C_1 \qquad \text{at } x = 0 \tag{6.12-36}$$

$$\frac{dC_A}{dx} = 0 \qquad \text{at } x = L \tag{6.12-37}$$

For tubular reactors with $A \xrightarrow{k}$ Products, if we ignore diffusional gradients, from Table XIV

$$v_z \frac{\partial C_A}{\partial z} = -kC_A \tag{6.12-38}$$

where z is the length of the reactor. If this reactor has plug flow conditions ($v_z = $ constant) and a boundary condition such as

$$C_A = C_{A0} \qquad \text{at } z = 0 \tag{6.12-39}$$

we can solve for C_A, yielding

$$\frac{C_A}{C_{A0}} = e^{-kz/v_z} \tag{6.12-40}$$

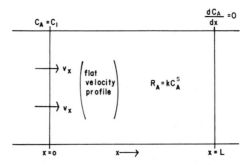

Fig. 68. Plug flow chemical reactor.

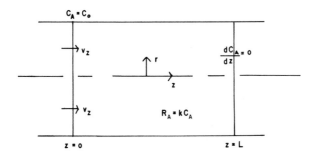

Fig. 69. Plug flow tubular reactor with internal kinetics.

To attempt to assess the importance of neglecting axial dispersion (or diffusion), the dispersion term should be included in the governing equation, so that (6.12-38) becomes

$$v_z \frac{\partial C_A}{\partial z} = D_z \frac{\partial^2 C_A}{\partial z^2} - kC_A \qquad (6.12\text{-}41)$$

where D_z is a dispersion or diffusion coefficient. We then compare predicted results for C_A using solutions to the two models. Figure 69 sketches the situation.

Another interesting mass transport phenomenon is shown in Figure 70, where a fluid flowing through a circular tube has material diffusing into it from the wall. From Table XIV we get the governing equation (in cylindrical coordinates):

$$v_z \frac{\partial C_A}{\partial z} = D_A \frac{1}{r} \frac{\partial}{\partial r} \left(r \frac{\partial C_A}{\partial r} \right) \qquad (6.12\text{-}42)$$

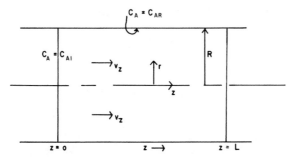

Fig. 70. Tubular plug flow reactor with wall dissolving into the main stream.

where v_z is a constant. Some boundary conditions are

$$C_A = C_{A1} \quad \text{for } z = 0 \qquad (6.12\text{-}43)$$

$$C_A = C_{AR} \quad \text{at } r = R \qquad (6.12\text{-}44)$$

The solution of equations (6.12-42), (6.12-43), and (6.12-44) is[29]

$$\frac{C_A - C_{A1}}{C_{AR} - C_{A1}} = 1 - 2 \sum_{n=1}^{\infty} \frac{1}{\alpha_n} \frac{J_0(\alpha_n r/R)}{J_1(\alpha_n)} \exp\left(\frac{-D_A}{v_z R^2} \alpha_n^2 z\right) \qquad (6.12\text{-}45)$$

where α_n are the roots of $J_0(\alpha_n) = 0$. Equation (6.12-45) can be obtained by a Laplace transform technique. In research experiments on this type of equipment, average concentrations are usually measured, so that by using the mean value theorem at $z = L$,

$$\langle C_A |_{z=L} \rangle = \frac{\int_0^R C_A |_{z=L} \, dr}{\int_0^R dr} \qquad (6.12\text{-}46)$$

we can get

$$\frac{\langle C_A |_{z=L} \rangle - C_{A1}}{C_{AR} - C_{A1}} = 1 - 4 \sum_{n=1}^{\infty} \frac{1}{\alpha_n} \exp\left(\frac{-D_A \alpha_n^2}{v_z R^2} L\right) \qquad (6.12\text{-}47)$$

which presents concentration C_A in terms of its average value. If the velocity distribution is not constant but parabolic, equation (6.12-42) becomes

$$v_{\max}\left[1 - \left(\frac{r}{R}\right)^2\right] \frac{\partial C_A}{\partial z} = D_A \frac{1}{r} \frac{\partial}{\partial r}\left(r \frac{\partial C_A}{\partial r}\right) \qquad (6.12\text{-}48)$$

No rigorous analytic solution of (6.12-48) is available, but an approximate solution is given by[30]

$$\frac{\langle C_A |_{z=L} \rangle - C_{A1}}{C_{AR} - C_{A1}} = 1 - \sum_{n=1}^{\infty} a_n \exp\left\{-b_n\left(\frac{D_A}{v_{av} R}\right) L\right\} \qquad (6.12\text{-}49)$$

where $v_{av} = \frac{1}{2} v_{\max}$ and a_n and b_n are given as

$$a_1 = 0.820 \qquad b_1 = 3.658$$
$$a_2 = 0.972 \qquad b_2 = 22.178$$
$$a_3 = 0.0135 \qquad b_3 = 53.05$$

---

[29] T. B. Drew and J. W. Hoopes, ed., *Advances in Chemical Engineering*, Vol. 1, Academic Press, New York (1956), p. 216.
[30] *Ibid.*, p. 217.

If an incompressible non-Newtonian fluid (power law) is flowing in a horizontal circular tubular reactor, then from Table XIV, equation (5.1-34), and Table III we can write the governing mass transport, heat transport, and momentum transport equations. These are

$$v_z \frac{\partial C_A}{\partial z} = D_A \left(\frac{\partial^2 C_A}{\partial r^2} + \frac{1}{r} \frac{\partial C_A}{\partial r} \right) + R_A \qquad (6.12\text{-}50)$$

$$\varrho C_p v_z \frac{\partial T}{\partial z} = k \frac{1}{r} \frac{\partial}{\partial r} \left(r \frac{\partial T}{\partial r} \right) + S_r \qquad (6.12\text{-}51)$$

$$\frac{\partial P}{\partial z} + \frac{1}{r} \frac{\partial}{\partial r} (r \tau_{rz}) = 0 \qquad (6.15\text{-}52)$$

where R_A is the chemical reaction term, S_r the heat generation term caused by R_A, C_p the heat capacity, k the thermal conductivity, and τ_{rz} the shear stress term. The solution of equations (6.12-50) through (6.12-52) will be coupled and be quite complex.

For turbulent flow situations, or if there are transport gradients in more than one direction, it is necessary to define more than one transport coefficient. For example, suppose we have a square duct in which a fluid is in turbulent flow. We could define D_{Ax}, D_{Ay}, and D_{Az} to account for diffusion in the three rectangular directions. From Table XIV we could then arrive, for flow in the z-direction, at the governing equation

$$\frac{\partial C_A}{\partial t} + v_z \frac{\partial C_A}{\partial z} = D_{Az} \frac{\partial^2 C_A}{\partial z^2} \qquad (6.12\text{-}53)$$

where v_z is the turbulent velocity profile and D_{Az} the "effective" diffusivity. Similarly, in tubular packed reactors at steady state, as shown in Figure

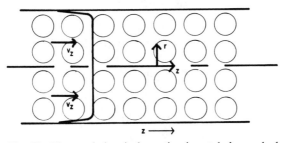

Fig. 71. Flow and chemical reaction in a tubular packed column.

71, the heat flow equation is obtained from equation (5.1-34) and is

$$\varrho C_p \bar{v}_z \frac{\partial T}{\partial z} = \bar{k}\left(\frac{\partial^2 T}{\partial r^2} + \frac{1}{r}\frac{\partial T}{\partial r}\right) + R_A \varDelta H \qquad (6.12\text{-}54)$$

where $\bar{v}_z$ is the assumed uniform fluid velocity. The heat generation term S_r arising from the chemical reaction has been expressed as a chemical rate-of-production term R_A multiplied by the partial molal enthalpy of the reaction. For a Newtonian fluid under turbulent flow conditions we can get from Table IX the governing fluid flow equation

$$\frac{\partial P}{\partial z} = \frac{1}{r}\frac{\partial}{\partial r}\left(r\bar{\mu}\frac{\partial \bar{v}_z}{\partial r}\right) \qquad (6.12\text{-}55)$$

where

$$\bar{\mu} = \mu + \mu^{(t)}$$

μ = laminar flow viscosity

$\mu^{(t)}$ = turbulent contribution to momentum transport

$\bar{v}_z$ = time-averaged velocity

For a multicomponent fluid flowing in a plug flow tubular reactor we can sometimes neglect diffusional effects. The reactions which take place can be written as

$$\sum_{i=1}^{n} a_{ij}A_i = 0 \qquad j = 1, 2, \ldots, m \qquad (6.12\text{-}56)$$

where a_{ij} is positive for a product and negative for a reactant. There are m reactions and n chemical species. For example, with a reaction system as given by

$$\alpha R + \beta B \underset{k_2}{\overset{k_1}{\rightleftharpoons}} \gamma P \qquad (6.12\text{-}57)$$

we can rearrange (6.12-57) to conform to (6.12-56). The result is

$$\gamma P - \alpha R - \beta B = 0 \qquad j = 1 \qquad (6.12\text{-}58)$$

and

$$A_1 = P \qquad a_{11} = \gamma$$
$$A_2 = R \qquad a_{21} = \alpha$$
$$A_3 = B \qquad a_{31} = \beta$$

For a plug flow reactor (neglecting diffusion), the governing transport equation for component i (rectangular coordinates) from Table XIV is[31]

$$G \frac{dg_i}{dx} = a \sum_{j=1}^{m} f_{ij} \qquad (6.12\text{-}59)$$

where

G = total mass flow rate in the x-direction

g_i = moles of species i per unit mass of the mixture

a = cross-sectional area of the reactor

f_{ij} = rate of generation (or depletion) of species i in reaction j, moles per unit volume per time

x = axial direction

In equation (6.12-57) f_{11} is the rate of production of component P

$$f_{11} = k_1 C_R^\alpha C_B^\beta - k_2 C_P^\gamma \qquad (6.12\text{-}60)$$

where C is the concentration in moles per unit volume. The reaction terms f_{ij} are not independent but are related by an equation such as

$$\frac{f_{ij}}{a_{ij}} = \frac{f_{kj}}{a_{kj}} \qquad (6.12\text{-}61)$$

If equation (6.12-58) is applied to (6.12-61), the result is

$$\frac{f_{11}}{\gamma} = \frac{f_{21}}{\alpha} = \frac{f_{31}}{\beta} \qquad (6.12\text{-}62)$$

where

f_{11} = rate of production of P

f_{21} = rate of production of R

f_{31} = rate of production of B

Thus, from equation (6.12-62), $f_{11} = (\gamma/\alpha)f_{21}$. If we define an intrinsic production rate f_j by

$$\frac{f_{ij}}{a_{ij}} = \frac{f_{kj}}{a_{kj}} = f_j \qquad (6.12\text{-}63)$$

[31] N. R. Amundson, *Mathematical Methods in Chemical Engineering*, Prentice-Hall, Englewood Cliffs, N. J. (1966), p. 62.

we can get

$$f_{ij} = a_{ij} f_j \qquad (6.12\text{-}64)$$

If equation (6.12-64) is substituted into the governing transport equation (6.12-59) the result is

$$G \frac{dg_i}{dx} = a \sum_{j=1}^{m} a_{ij} f_j \qquad (6.12\text{-}65)$$

or, dividing by a, we can get for the ith component

$$w \frac{dg_i}{dx} = \sum_{j=1}^{m} a_{ij} f_j \qquad (6.12\text{-}66)$$

where w is the mass flux of the mixture. Equation (6.12-66) may be expressed in matrix form as

$$w \frac{d\bar{g}}{dx} = \bar{A}\bar{f} \qquad (6.12\text{-}67)$$

where

$$\bar{g} = \begin{pmatrix} g_1 \\ g_2 \\ \cdot \\ \cdot \\ \cdot \\ g_n \end{pmatrix} = \text{column vector}$$

$$\bar{f} = \begin{pmatrix} f_1 \\ f_2 \\ \cdot \\ \cdot \\ \cdot \\ f_n \end{pmatrix} = \text{column vector}$$

$$\bar{A} = \begin{pmatrix} a_{11} & \cdots & a_{1n} \\ \vdots & & \vdots \\ a_{n1} & \cdots & a_{nn} \end{pmatrix} = n \times n \text{ matrix}$$

For this plug flow reactor with diffusion neglected a heat transport equation (neglecting heat conduction) is given by the following equation (obtained from an energy balance around a differential element):

$$\left(G \sum_{i=1}^{n} g_i h_i \right)\bigg|_x - \left(G \sum_{i=1}^{n} g_i h_i \right)\bigg|_{x+\Delta x} + [ph(t - T)]\Delta x = 0 \qquad (6.12\text{-}68)$$

where

$$h_i = \text{partial molal enthalpy of species } i$$

$$p = \text{surface area of the reactor}$$

$$h = \text{heat transfer coefficient at the tube wall}$$

$$t = \text{ambient temperature}$$

$$T = \text{temperature of the reaction mixture}$$

If we take the usual limiting process, $\Delta x \to 0$, after dividing by Δx, equation (6.12-68) becomes

$$G \frac{d}{dx} \sum_{i=1}^{n} g_i h_i - ph(t - T) = 0 \qquad (6.12\text{-}69)$$

or

$$G \sum_{i=1}^{n} h_i \frac{dg_i}{dx} + G \sum_{i=1}^{n} g_i \frac{dh_i}{dx} - ph(t - T) = 0 \qquad (6.12\text{-}70)$$

From the definition of enthalpy, $dh_i = C_{pi}dT$, we can get $dh_i/dx = C_{pi}dT/dx$, and with this relationship substituted into equation (6.12-70), and using (6.12-67), the result, after some rearrangement, is

$$a\bar{h}^T \bar{\bar{A}} \bar{f} + G \frac{dT}{dx} \sum_{i=1}^{n} g_i C_{pi} - ph(t - T) = 0 \qquad (6.12\text{-}71)$$

where

$$C_{pi} = \text{molar heat capacity of species } i$$

$$\bar{h}^T = \text{transpose of } \bar{h} = (h_1, h_2, \ldots, h_n)$$

$$\bar{h} = \begin{pmatrix} h_1 \\ h_2 \\ \cdot \\ \cdot \\ \cdot \\ h_n \end{pmatrix} = \text{column vector}$$

The quantity

$$\sum_{i=1}^{n} g_i C_{pi} = C_{pa}$$

is the molar heat capacity per unit mass of the reaction mixture. Using matrix operations we can get[32]

$$\bar{h}^T \bar{\bar{A}} \bar{f} = \bar{f}^T \bar{\bar{A}}^T \bar{h} \qquad (6.12\text{-}72)$$

---

[32] *Ibid.*, p. 65.

and

$$\bar{A}^T \bar{h} = \begin{pmatrix} \Delta H_1 \\ \Delta H_2 \\ \vdots \\ \Delta H_n \end{pmatrix} = \overline{\overline{\Delta H}}$$

where $\overline{\overline{\Delta H}}$ is the heat of reaction. With equations (6.12-67) and (6.12-72) in (6.12-71), a more useful energy transport equation is obtained:

$$a \bar{f}^T \overline{\overline{\Delta H}} + GC_{pa} \frac{dT}{dx} - ph(t - T) = 0 \qquad (6.12\text{-}73)$$

where C_{pa} is the average heat capacity of the mixture. For the plug flow reactor we also need a momentum (fluid flow) balance about the differential volume element. A macroscopic balance with viscous effects implicitly included in the resistance at the wall can be obtained from Table III and is given by[33]

$$\frac{dP}{dx} + \varrho v_x \frac{dv_x}{dx} + \tau_R \frac{S}{a} = 0 \qquad (6.12\text{-}74)$$

where

$x = $ the axial direction

$\varrho = $ density

$P = $ pressure

$\tau_R = $ shear stress at the wall $= (\Delta P/L)R/2$

$R = $ radius of the tube

$S = $ perimeter of the reactor

$v_x = $ linear velocity

$a = $ cross-sectional area

The shear stress at the wall may be written in terms of a Fanning friction factor by combining the τ_R relationship with equation (4.6-3). The result, after some rearrangement, is

$$\tau_R = f \frac{w^2}{8\varrho} \qquad (6.12\text{-}75)$$

where w is the mass velocity per square foot of the reaction mixture. With

---

[33] *Ibid.*, p. 66.

equation (6.12-75) in (6.12-74), we get

$$\frac{dP}{dx} + \varrho v_x \frac{dv_x}{dx} + fS \frac{w^2}{8\varrho a} = 0 \qquad (6.12\text{-}76)$$

We now have enough relationships for the plug flow reactor to compute dg_i/dx from equation (6.12-66), dT/dx from (6.12-73), and, dP/dx from (6.12-76).

6.13. Macroscopic Analysis of Plug Flow Tubular Reactors

Instead of writing a microscopic mass transport relationship from Table XIV, we could write a macroscopic mass transport equation[34]

$$\underbrace{\frac{\partial C_A}{\partial t}}_{\text{accumulation}} + \underbrace{\frac{\partial (v_z C_A)}{\partial z}}_{\substack{\text{bulk} \\ \text{transport}}} = \underbrace{R_i}_{\substack{\text{generation} \\ \text{by chemical} \\ \text{reaction}}} + \underbrace{m_i}_{\substack{\text{transport} \\ \text{through the} \\ \text{system} \\ \text{surfaces} \\ \text{(boundaries)}}} \qquad (6.13\text{-}1)$$

where v_z is the plug flow (bulk flow). Equation (6.13-1) is similar to the microscopic relationship, equation (5.1-25), except that in the macroscopic approach we ignore diffusional transport or incorporate it into the expression for transport through the boundaries. The macroscopic energy balance for the plug flow reactor is handled similarly, as given by[35]

$$\varrho C_P \underbrace{\left(\frac{\partial T}{\partial t} + v_z \frac{\partial T}{\partial z} \right)}_{\substack{\text{accumulation} \qquad \text{bulk} \\ \text{transport}}} = \underbrace{S_R}_{\text{generation}} + \underbrace{E^{(t)}}_{\substack{\text{transport} \\ \text{through} \\ \text{system} \\ \text{surfaces} \\ \text{(boundaries)}}} \qquad (6.13\text{-}2)$$

where $E^{(t)}$ is the interphase transfer of energy. Many times it is of interest to determine the required size of a plug flow reactor. From equation (6.13-1) we can get, at steady state,

$$\frac{d(v_z C_A)}{dz} = R_A \qquad (6.13\text{-}3)$$

or

$$v_z \frac{dC_A}{dz} = R_A \qquad (6.13\text{-}4)$$

[34] D. M. Himmelblau and K. B. Bischoff, op. cit., p. 24.
[35] Ibid.

With

$$F = \text{feed rate}$$

$$C = \text{density of the feed}$$

$$S = \text{cross-sectional area of the reactor}$$

we can get $v_z = F/CS$, and if this is substituted into equation (6.13-4), the result, after some rearrangement, is

$$\int_{C_{A0}}^{C_{Af}} \frac{dC_A}{R_A} = \frac{1}{v_z} \int_0^L dz = \frac{L}{v_z} = \frac{CSL}{F} = \frac{CV}{F} \qquad (6.13\text{-}5)$$

where V is the volume of the reactor that is to be determined.

For a tubular reactor with a diffusion coefficient in the z-direction which is different from the the diffusion coefficient in the r-direction, a transport equation can be obtained from Table XIV and is

$$v_z(r) \frac{\partial C_A}{\partial z} = \bar{D}_z(r) \frac{\partial^2 C_A}{\partial z^2} + \frac{1}{r} \left(\frac{\partial}{\partial r} r \bar{D}_r(r) \frac{\partial C_A}{\partial r} \right) + R_A \qquad (6.13\text{-}6)$$

Some typical boundary conditions might be

$$\frac{\partial C_A}{\partial z} = 0 \qquad \text{at } Z = L \quad \text{(end of reactor)} \qquad (6.13\text{-}7)$$

$$\frac{\partial C_A}{\partial r} = 0 \qquad \text{at } r = 0 \quad \text{(symmetry)} \qquad (6.13\text{-}8)$$

$$\frac{\partial C_A}{\partial r} = 0 \qquad \text{at } r = R \quad \text{(no flux through the wall)} \qquad (6.13\text{-}9)$$

An additional condition expresses the concept that all of component A entering the reactor by convection becomes a flux at $z = 0$:

$$v_z(r)[C_A(0, r) - C_0] = \bar{D}_z(r) \frac{\partial C_A}{\partial z}\bigg|_{z=0} \qquad (6.13\text{-}10)$$

where C_0 is the entering (feed) concentration of component A and $C_A(0,r)$ is the concentration of A just inside the reactor. A general energy balance can be obtained from equation (5.1-34) as shown:[36]

$$\varrho C_P \left[v_z(r) \frac{\partial T}{\partial z} \right] = k_z(r) \frac{\partial^2 T}{\partial z^2} + \frac{1}{r} \left(\frac{\partial}{\partial r} r k_r(r) \frac{\partial T}{\partial r} \right) + \Delta H R_A \qquad (6.13\text{-}11)$$

[36] *Ibid.*, p. 48.

where the thermal conductivity varies in the r- and z-directions. In equation (6.13-11), ΔH is the heat of the reaction R_A. Some typical energy transport boundary conditions are

$$\frac{\partial T}{\partial z} = 0 \quad \text{at } z = L \quad \text{(end of reactor)} \qquad (6.13\text{-}12)$$

$$\frac{\partial T}{\partial r} = 0 \quad \text{at } r = 0 \quad \text{(symmetry)} \qquad (6.13\text{-}13)$$

$$k_r(r)\left.\frac{\partial T}{\partial r}\right|_{r=R} = U[T_S - T]|_{r=R} \quad \text{(heat flow through the reactor wall is dissipated into the surroundings by convection)} \qquad (6.13\text{-}14)$$

where U is the overall heat transfer coefficient at the reactor wall outside surface. One other boundary condition, which equates the energy coming into the reactor by convection with the resulting energy flow by convection and heat conduction at $z = 0$, is

$$\varrho C_P v_z(r)(T_0 - T^0) = \varrho C_P v_z(r)[T|_{z=0} - T^0] - k_z(r)\left.\frac{\partial T}{\partial z}\right|_{z=0} \qquad (6.13\text{-}15)$$

T^0 is the reference temperature for the enthalpy terms and T_0 is the feed temperature. If radial gradients are ignored, the derivatives with respect to r are zero, i.e., $\partial(\;\;)/\partial r = 0$. This macroscopic approach alters equations (6.13-11) and (6.13-14) so that equation (6.13-11) becomes

$$\varrho C_P\left[v_z\,\frac{\partial T}{\partial z}\right] = k\,\frac{\partial^2 T}{\partial z^2} + \Delta H R_A + U\,\frac{A}{V}\,(T_s - T) \qquad (6.13\text{-}16)$$

and (6.13-14) is not applicable. In (6.13-16) we now have included an energy term due to heat flow through the boundaries, $U(A/V)(T_s - T)$. The ratio A/V is surface area/volume, and T_s is the temperature of the surroundings. Macroscopic balances between the inlet and outlet positions of the reactor yield

$$\Delta[C_A v_z S] = R_{A_{av}} V_{tot} \qquad (6.13\text{-}17)$$

and

$$\Delta[\varrho C_P(T - T^0)v_z S] = \Delta H_{tot} R_{A_{av}} V_{tot} + U A_{tot}(T_s - T) \qquad (6.13\text{-}18)$$

where

$$\Delta = (\quad)\big|_{\text{inlet}} - (\quad)\big|_{\text{outlet}}$$

$S = $ cross-sectional area of the reactor

$V_{\text{tot}} = $ total volume of the reactor

$A_{\text{tot}} = $ total surface area of the reactor

$\Delta H_{\text{tot}} = $ net heat of reaction $= \Delta H_{\text{products}} - \Delta H_{\text{reactants}}$

For macroscopic balances such as (6.13-17) and (6.13-18) there are no boundary conditions since they are all accounted for in the macroscopic governing equations.

6.14. Unsteady-State Response of Chemical Reactors

Once a mathematical model (governing equation) has been written, it is often of interest to examine the response of the system to certain inputs or driving forces. Typical driving forces are shown in Figure 72. Suppose we have a governing equation such as

$$\frac{dy}{dt} + ay = x(t) \tag{6.14-1}$$

where $x(t)$ is the input or forcing function. The solution of equation (6.14-1) can be shown to be

$$y = e^{-at} \int_0^t x(\lambda)e^{a\lambda}d\lambda = \int_0^t e^{a(\lambda-t)}x(\lambda)d\lambda \tag{6.14-2}$$

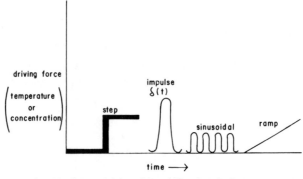

Fig. 72. Some driving forces for chemical systems.

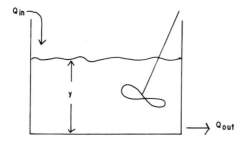

Fig. 73. Illustration for impulse forcing function.

where y is the response to $x(t)$. If $x(t)$ is a step input given by

$$x(t) = a \qquad \text{at } t = 0 \qquad (6.14\text{-}3)$$

and

$$y(t) = 0 \qquad \text{at } t = 0 \qquad (6.14\text{-}4)$$

then we get from equation (6.14-2) the result

$$y = \int_0^t e^{a(\lambda - t)} a\, dy = 1 - e^{-at} \qquad (6.14\text{-}5)$$

The impulse or Dirac delta function, $x(t) = \delta(t)$, describes a pulse introduced into the system. In using the delta function we make use of the equation

$$\int_0^t f(t')\delta(t' - \alpha)dt' = f(\alpha) \qquad (6.14\text{-}6)$$

where $\delta(t) = 0$ for $t \neq 0$ and $\delta(0) = 1$. As an example of the use of the impulse, consider the case of water flowing into an empty tank as sketched in Figure 73. A macroscopic material balance for this system is given from equation (5.1-8) as

$$A\frac{dy}{dt} = Q_{\text{in}} - Q_{\text{out}} \qquad (6.14\text{-}7)$$

where y is the height of water in the tank at time t, A the cross-sectional area of the tank, and Q the volumetric flow rate. If we assume that Q_{out} is proportional to the height of liquid in the tank, then $Q_{\text{out}} = Ky$, and equation (6.14-7) becomes

$$A\frac{dy}{dt} = Q_{\text{in}} - Ky \qquad (6.14\text{-}8)$$

or

$$\frac{dy}{dt} + \frac{Ky}{A} = \frac{Q_{in}}{A} = x(t) \quad \text{(forcing function)} \quad (6.14\text{-}9)$$

with

$$y(t) = 0 \quad \text{at } t = 0 \quad\quad\quad (6.14\text{-}10)$$

If $x(t) = \delta(t)$ we can solve equation (6.14-9) by Laplace transform techniques, with

$$\mathscr{L}_{t \to s}\delta(t) = 1 \quad\quad\quad (6.14\text{-}11)$$

The result is

$$s\bar{y}(s) + \frac{K}{A}\,\bar{y}(s) = 1 \quad\quad\quad (6.14\text{-}12)$$

where $\mathscr{L}_{t \to s}\,y(t) = \bar{y}(s)$. Equation (6.14-12) can be written explicitly for $\bar{y}(s)$ and then inverted to yield

$$y(t) = e^{-(K/A)t} \quad\quad\quad (6.14\text{-}13)$$

If a pulse of tracer (or dye) is introduced into a flow system governed by a partial differential equation (distributed-parameter system), this is described by[37]

$$\frac{\partial C_A}{\partial t} + v_z\frac{\partial C_A}{\partial z} = \text{Source term} \quad\quad (6.14\text{-}14)$$

where equation (6.14-14) is obtained from Table XIV by lumping the radial effects into a source term. The boundary and initial conditions for this situation might be

$$C_A(t, z) = 0 \quad \text{at } z = 0 \quad \text{(inlet condition)} \quad (6.14\text{-}15)$$

$$C_A(t, z) = 0 \quad \text{at } t = 0 \quad \text{(initial condition)} \quad (6.14\text{-}16)$$

If a fixed quantity of material is introduced at $z = 0$ and $t = 0$, the source term can be expressed as

$$\text{Source term} = \frac{I}{A}\,\delta(t)\delta(z) \quad\quad\quad (6.14\text{-}17)$$

where I is the amount of material introduced, and A is the cross-sectional

----

[37] *Ibid.*, p. 113.

area of the reactor. From the definition of the Dirac delta function,

$$\int_{-\infty}^{+\infty} \delta(\alpha)d\alpha = 1 \tag{6.14-18}$$

and with equation (6.14-17) substituted into (6.14-14), we can solve the resulting equation by the Laplace transform technique. First taking the Laplace transform with respect to t, $\mathscr{L}_{t \to s}$, and then with respect to z, $\mathscr{L}_{z \to p}$, we get from equations (6.14-14) through (6.14-17) the result

$$s\bar{C}_A(s, z) + v_z \frac{d\bar{C}_A(s, z)}{dz} = \frac{I}{A} \delta(z) \tag{6.14-19}$$

and then

$$s\bar{\bar{C}}_A(s, p) + v_z[p\bar{\bar{C}}_A(s, p)] = \frac{I}{A} \tag{6.14-20}$$

or

$$\bar{\bar{C}}_A(s, p) = \frac{I/Av_z}{(s/v_z) + p} \tag{6.14-21}$$

In the above operations, v_z has been considered constant. The final solution ts obtained by sequentially inverting equation (6.14-21) twice, $\mathscr{L}_{p \to z}^{-1}$ and ihen $\mathscr{L}_{s \to t}^{-1}$. The results are

$$\bar{C}_A(s, z) = \frac{I}{F} e^{-sz/v_z} \tag{6.14-22}$$

$$C_A(t, z) = \frac{I}{F} \delta\left(t - \frac{z}{v_z}\right) \tag{6.14-23}$$

where $F = Av_z$. In equation (6.14-23) C_A is zero until $t = z/v_z$. This implies that the impulse proceeds as a front down the reactor. Another slightly more complicated illustration of the pulse concept is the equation

$$\text{Source term} = I\delta(z - z_0)\delta(t)f(r) \tag{6.14-24}$$

where

$$I = \text{amount of tracer injected}$$

$$\delta(z - z_0) = \text{Dirac delta or impulse function introduced at } z = z_0$$

$$\delta(t) = \text{impulse activated at } t = 0$$

$$f(r) = \text{radial distribution of the tracer}$$

In some situations there is both axial and radial dispersion. For these cases, a tracer injected at the axis of the tube (a point source) will move down the reactor and spread rapidly by dispersion. From Table XIV we can write the governing equation

$$\frac{\partial C_A}{\partial t} + v_z \frac{\partial C_A}{\partial z} = \bar{D}_z \frac{\partial^2 C_A}{\partial z^2} + \bar{D}_r \frac{1}{r} \frac{\partial}{\partial r} \left(r \frac{\partial C_A}{\partial r} \right) + I\delta(z - z_0)f(r)$$

$$(6.14\text{-}25)$$

where

$$\text{Source} = I\delta(z - z_0)f(r)$$

$$I = \pi R^2 v_z (C_A)_{av} = \text{amount of A injected per unit time}$$

Another method of analyzing a system is by a frequency response technique. The response of a linear system to a sinusoidal input is a sinusoidal output. There may be a difference between input and output, showing up as a displacement (phase angle shift) or a change in amplitude (gain). Figure 74 illustrates this. For example, suppose we have the governing equation

$$\frac{dy}{dt} + ay = x(t) \tag{6.14-26}$$

with

$$y(t) = 0 \qquad \text{at } t = 0 \tag{6.14-27}$$

and

$$x(t) = ab \sin wt \tag{6.14-28}$$

Equations (6.14-26), (6.14-27), and (6.14-28) can be solved by Laplace

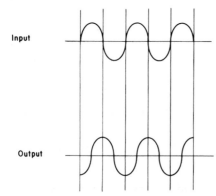

Input

Output

Fig. 74. Illustration of phase angle shift in a linear system.

transform techniques with the result[38]

$$y(t) = \left[y_0 + \frac{ba \sin \phi}{\sqrt{a^2 + w^2}} \right] e^{-at} + \underbrace{\frac{ba}{\sqrt{a^2 + w^2}} \sin(wt + \phi)}_{\text{frequency response}} \quad (6.14\text{-}29)$$

where

$$y_0 = y(0)$$

and

$$\phi = -\tan^{-1}\left(\frac{w}{a} \right)$$

After a sufficient time, e^{-at} reduces the first term in equation (6.14-29) to zero, and we can compare the output $y(t)$ to the input $x(t)$.

If a first-order reaction is carried out isothermally in a plug flow tubular reactor with a sinusoidal input, we can get from Table XIV a governing equation such as

$$\frac{\partial C_A}{\partial t} + v_z \frac{\partial C_A}{\partial z} = -kC_A \quad (6.14\text{-}30)$$

with boundary equations

$$C_A(t, z) = C_0 \sin wt \qquad \text{at } z = 0 \quad (6.14\text{-}31)$$

$$C_A(t, z) = 0 \qquad \text{at } t = 0 \quad (6.14\text{-}32)$$

By Laplace transforms, $\mathscr{L}_{t \to s}$, we can get from (6.14-30) and (6.14-32)

$$\frac{d\bar{C}_A(s, z)}{dz} + \left(\frac{s + k}{v_z} \right) \bar{C}_A(s, z) = 0 \quad (6.14\text{-}33)$$

where $\mathscr{L}_{t \to s} C_A(t, z) = \bar{C}_A(s, z)$. By transforming equation (6.14-31) we can convert it to an appropriate form for use with (6.14-33):

$$\bar{C}_A(s, z) = C_0 \left(\frac{w}{s^2 + w^2} \right) \qquad \text{at } z = 0 \quad (6.14\text{-}34)$$

Thus we get

$$\bar{C}_A(s, z) = \left(C_0 \frac{w}{s^2 + w^2} \right) e^{-(s+k)z/v_z} \quad (6.14\text{-}35)$$

The ratio

$$\frac{\text{output (Laplace transformed)}}{\text{input (Laplace transformed)}} = \frac{\bar{C}_A(s, z)}{\bar{C}_A(s, z)\big|_{z=0}} = e^{-(s+k)z/v_z} \quad (6.14\text{-}36)$$

[38] *Ibid.*, p. 118.

produces the definition of a transfer function, $G(s, z)$:

$$G(s, z) = \frac{\text{output (transformed)}}{\text{input (transformed)}} = e^{-(s+k)z/v_z} \qquad (6.14\text{-}37)$$

The transfer function is useful in characterizing the behavior of systems which are described by differential equations. We can take the Laplace transform of the nth-order ordinary differential equation

$$a_n \frac{d^n y}{dx^n} + a_{n-1} \frac{d^{n-1} y}{dx^{n-1}} + \cdots + a_1 \frac{dy}{dx} + a_0 y = x(t) \qquad (6.14\text{-}38)$$

and get

$$a_n s^n \bar{y}(s) + a_{n-1} s^{n-1} \bar{y}(s) + \cdots + a_1 \bar{y}(s) + a_0 \bar{y}(s) = \bar{x}(s) \qquad (6.14\text{-}39)$$

with

$$y(0) = y'(0) = y''(0) = \cdots = y^{(n-1)}(0) = 0$$

and define a transfer function, $G(s)$, by

$$G(s) = \frac{\bar{y}(s)}{\bar{x}(s)} = \frac{1}{a_n s^n + a_{n-1} s^{n-1} + \cdots + a_1 s + a_0} \qquad (6.14\text{-}40)$$

In handling nonlinear systems, sometimes the original equation is linearized by a redefinition of the dependent variable in terms of a perturbation. For example, a well-mixed reactor, sketched in Figure 75, has as its governing equation

$$V \frac{dC_A}{dt} = FC_0 - FC_A - VkC_A \qquad (6.14\text{-}41)$$

with

$$V = \text{volume} = \text{constant}$$

$$kC_A = \text{reaction term}$$

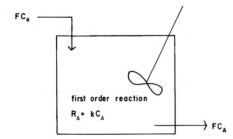

Fig. 75. Stirred tank reactor.

For this case, we define the concentration C_A as

$$C_A = C_{AS} + \bar{C}_A \tag{6.14-42}$$

where C_{AS} is the steady-state concentration and $\bar{C}_A$ a small concentration perturbation, and

$$C_0 = C_{0S} + \bar{C}_0 \tag{6.14-43}$$

where C_0 is the inlet concentration and C_{0S} and $\bar{C}_0$ correspond to the C_A perturbation terms. Substituting equations (6.14-42) and (6.14-43) into (6.14-41), we get

$$V \frac{d\bar{C}_A}{dt} = F(C_{0S} + \bar{C}_0) - F(C_{AS} + \bar{C}_A) - Vk(C_S + \bar{C}_A) \tag{6.14-44}$$

At steady state

$$0 = FC_{0S} - FC_{AS} - VkC_{AS} \tag{6.14-45}$$

is obtained from (6.14-41). Equation (6.14-45) can be substituted into (6.14-44) to get

$$\frac{d\bar{C}_A}{dt} = \frac{F\bar{C}_0}{V} - \frac{F\bar{C}_A}{V} - k\bar{C}_A \tag{6.14-46}$$

We can take the Laplace transform of equation (6.14-46) and rearrange terms to yield

$$\frac{\bar{\bar{C}}_A}{(F/V)\bar{\bar{C}}_0} = \frac{1}{s + (F/V) + k} = G(s) = \text{transfer function} \tag{6.14-47}$$

where

$$\mathscr{L}_{t \to s}\bar{C}_A = \bar{\bar{C}}_A \qquad \mathscr{L}_{t \to s}\bar{C}_0 = \bar{\bar{C}}_0$$

Nonlinear equations may be approximated by linearizing the nonlinear terms, as described below:

1. Truncated Taylor series expansions (retain the first-order term)
2. Perturbation methods involving small deviations about the steady-state values (or other reference values)
3. Sectional linearization, i.e., the nonlinear term is split into different linear functions over different regions.

As another linearization example, suppose we write the governing equation for a well-stirred chemical reactor in which a second-order reaction occurs:

$$V \frac{dC_A}{dt} = QC_{Ai} - QC_A - VkC_A^2 \tag{6.14-48}$$

where

$$V = \text{volume of the reactor} = \text{constant}$$

$$r_A = kC_A{}^2 = \text{rate of consumption of component A}$$

$$Q = \text{volumetric flow rate}$$

The nonlinear equation is linearized by use of the equation

$$C_{Ai} = C_{Ai}^0 + C_{Ai}' \qquad C_{Ai}' \ll C_{Ai}^0 \qquad (6.14\text{-}49)$$

where C_{Ai}' is a small perturbation around the inlet average C_{Ai}^0, and $C_A{}^0$ is the mean value for the exiting stream,

$$C_A = C_A{}^0 + C_A' \qquad (6.14\text{-}50)$$

From equation (6.14-50) we can get

$$C_A{}^2 = (C_A{}^0)^2 + (C_A')^2 + 2C_A{}^0 C_A' = (C_A{}^0)^2 + 2C_A{}^0[C_A - C_A{}^0] \qquad (6.14\text{-}51)$$

where $(C_A')^2$ is neglected. If equations (6.14-49), (6.14-50), and (6.14-51) are substituted into (6.14-48), a linear equation is produced:

$$V\frac{dC_A}{dt} + [Q + 2kVC_A{}^0]C_A = [QC_{Ai} + kV(C_A{}^0)^2] \qquad (6.14\text{-}52)$$

6.15. Miscellaneous Reactor Models

One of the most complicated physical situations to analyze is the fluidized bed, which is illustrated in Figure 76. There seems to be two distinct regions within the reactor: (a) a dense region with a large concentration of particles in the fluid and (b) a dilute or bubble phase consisting almost entirely of fluid. Those who attempt to describe the reactor behavior

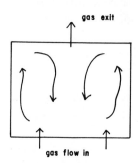

Fig. 76. Fluidized bed flow patterns.

with one dispersion (diffusion) coefficient have had difficulty matching theory and experiment. It has been proposed that the governing equations for this fluidized bed reactor are given by[39]

$$V_2 \frac{\partial C_2}{\partial t} + Q_2 \frac{\partial C_2}{\partial z^*} = K(C_1 - C_2) \tag{6.15-1}$$

and

$$V_1 \frac{\partial C_1}{\partial t} + Q_1 \frac{\partial C_1}{\partial z^*} = \frac{\bar{D}_1 V_1}{L^2} \frac{\partial^2 C_1}{\partial z^{*2}} + K(C_2 - C_1) \tag{6.15-2}$$

where

C_1 = concentration of reactant in phase 1 (dense phase)

C_2 = concentration of reactant in phase 2 (bubble phase)

$z^* = z/l$ = dimensionless position

L = total bed length

K = interphase mass transfer coefficient

$\bar{D}_1$ = dispersion coefficient in the dense phase

V_1, V_2 = volumes of phases 1 and 2

Q_1, Q_2 = volumetric flow rates of phases 1 and 2

With perfect mixing, $\bar{D}_1 \to \infty$, or, alternatively, if we assume plug flow, $\bar{D}_1 \to 0$.

Unsteady-state absorption by a fixed bed absorber can be modeled in simplified terms as shown in Figure 77. The governing equations for this situation are given by[40]

$$\frac{\partial C_L}{\partial t} + V_L \frac{\partial C_L}{\partial z} = -(k_a)(C_L - C_{Le}) \qquad \text{liquid phase} \tag{6.15-3}$$

and

$$m \frac{\partial C_s}{\partial t} = (k_a)(C_L - C_{Le}) \qquad \text{solid stationary phase} \tag{6.15-4}$$

where the subscript L and S refer to the liquid and solid phases. The term k_a is the mass transport coefficient. As the liquid initially progresses up through the solid bed with velocity, V_L, the interphase mass transport will not start until the fluid front reaches the solids at location z. By a slight

[39] *Ibid.*, p. 216.
[40] *Ibid.*, p. 27.

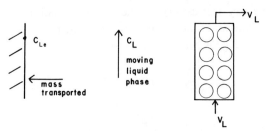

Fig. 77. Simple model of fixed bed absorber.

change of variables, we can redefine the clock time, t, to bed time, t', so that $t' = 0$ corresponds to the time the front first reaches the location z. This is shown in the equation

$$t' = t - \left(\frac{z}{V_L}\right) \tag{6.15-5}$$

By the chain rule for differentiation, we can convert equations (6.15-3) and (6.15-4) from t to t':

$$\left(\frac{\partial C_L}{\partial t}\right)_z = \left(\frac{\partial C_L}{\partial t'}\right)_z \tag{6.15-6}$$

$$\left(\frac{\partial C_S}{\partial t}\right)_z = \left(\frac{\partial C_S}{\partial t'}\right)_z \tag{6.15-7}$$

$$\left(\frac{\partial C_L}{\partial z}\right)_t = \left(\frac{\partial C_L}{\partial t'}\right)_z \left(\frac{\partial t'}{\partial z}\right)_t + \left(\frac{\partial C_L}{\partial z}\right)_{t'} \left(\frac{\partial z}{\partial z}\right)_t$$
$$= \frac{1}{V_L} \left(\frac{\partial C_L}{\partial t'}\right)_z + \left(\frac{\partial C_L}{\partial z}\right)_{t'} \tag{6.15-8}$$

and hence equations (6.15-3) and (6.15-4) become

$$\left(\frac{\partial C_L}{\partial t'}\right)_z + \left(\frac{\partial C_L}{\partial t'}\right)_z + V_L \left(\frac{\partial C_L}{\partial z}\right)_{t'} = -(k_a)(C_L - C_{Le}) \tag{6.15-9}$$

$$\left(\frac{\partial C_S}{\partial t'}\right)_z = \left(\frac{k_a}{m}\right)(C_L - C_{Le}) \tag{6.15-10}$$

The boundary conditions are

$$C_S = 0 \qquad \text{at } t' = 0 \tag{6.15-11}$$

$$C_L = C_{L0} \qquad \text{at } z = 0 \tag{6.15-12}$$

In gas absorption equipment the usual design assumptions simplify the transport analysis as shown in Figure 78. If we postulate that both the liquid and gas phases are moving, then we can write as macroscopic models for the physical situation,

$$V_g \frac{\partial C_{Ag}}{\partial z} = m_{Ag} \qquad (6.15\text{-}13)$$

and

$$V_L \frac{\partial C_{AL}}{\partial z} = m_{AL} \qquad (6.15\text{-}14)$$

where m_{Ag} and m_{AL} are interphase mass transfer terms such that $m_{Ag} = -m_{AL}$ and V_g and V_L are gas and liquid flow rates. A mass transport coefficient is defined by

$$m_{Ag} = -k_{Ag}a(C_{Ag} - C_{Agi}) \qquad (6.15\text{-}15)$$

where

$a = $ interfacial contact area per unit volume

$k_{Ag} = $ gas mass transport coefficient

$C_{Ag} = $ bulk gas phase concentration

$C_{Agi} = $ interfacial gas concentration of A

We can also define an overall mass transfer coefficient by

$$m_{Ag} = -K_{Ag}a(C_{Ag} - C_{Ag}^{*}) \qquad (6.15\text{-}16)$$

where

$C_{Ag}^{*} = $ concentration in the gas phase which would be in equilibrium with the bulk liquid phase, C_{AL}

$K_{Ag} = $ overall mass transfer coefficient

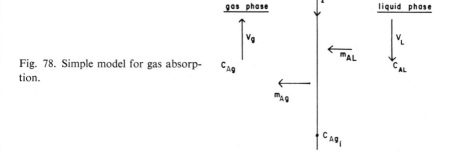

Fig. 78. Simple model for gas absorption.

Combining equation (6.15-13) and (6.15-16), we get the governing equation

$$V_g \frac{\partial C_{Ag}}{\partial z} = -K_{Ag}a(C_{Ag} - C_{Ag}^*) \qquad (6.15\text{-}17)$$

If we introduce the definitions

$$y_A = \text{mole fraction of A in the gas phase} = C_{Ag}/C_g$$
$$C_g = \text{total moles per unit volume in the gas phase}$$
$$\varrho_g = \text{mass density of the gas}$$
$$G_g = \text{mass velocity of the gas} = v_g \varrho_g$$
$$\bar{M}_g = \text{average molecular weight of the gas}$$

then equation (6.15-17) becomes

$$\frac{G_g}{\varrho_g} \frac{dy_A}{dz} = -K_{Ag}a(y_A - y_A^*) \qquad (6.15\text{-}18)$$

Separating the variables in equation (6.15-18) and integrating, we get

$$z = \underbrace{\frac{G_g}{K_{Ag}a\varrho_g}}_{= (HTU)_{og}} \underbrace{\int_{y_A}^{y_{A0}} \frac{dy_A}{y_A - y_A^*}}_{(NTU)_g} \qquad (6.15\text{-}19)$$

where

$$y_A = y_{A0} \qquad \text{at } z = 0$$

and

$$(HTU)_{og} = \text{overall height of a "transfer unit" in the gas phase}$$
$$(NTU)_g = \text{number of "transfer units" in the gas phase}$$

Assignments in Chapter 6

6.1. Verify equations (6.1-8), (6.1-10), (6.2-6), (6.2-11), (6.2-12), (6.2-15), (6.2-20), (6.2-24), (6.2-31), (6.2-36), (6.2-37), (6.2-38), (6.2-39), (6.2-47), (6.3-7), (6.3-8), (6.3-13), (6.3-15), (6.3-36), (6.3-37), (6.4-5), (6.4-9), (6.4-31), (6.4-32), (6.4-33), (6.4-34), (6.5-11), (6.5-12), (6.5-13), (6.5-14), (6.5-15), (6.7-3), (6.7-4), (6.7-5), (6.7-7), (6.7-16), (6.7-17), (6.7-22), (6.7-24), (6.7-27), (6.7-36), (6.9-5), (6.9-8), (6.9-9), (6.9-11), (6.9-12), (6.9-14), (6.9-21), (6.9-22), (6.9-25), (6.9-27), (6.9-35), (6.9-36), (6.9-45), (6.9-46), (6.9-55),

(6.9-56), (6.9-62), (6.9-63), (6.9-64), (6.9-69), (6.9-70), (6.9-71), (6.9-76), (6.9-77), (6.9-78), (6.10-13), (6.10-19), (6.10-23), (6.10-24), (6.11-7), (6.11-11), (6.11-12), (6.11-13), (6.11-17), (6.11-25), (6.11-27), (6.12-8), (6.12-9), (6.12-20), (6.12-30), (6.12-31), (6.12-32), (6.12-45), (6.12-47), (6.12-54), (6.12-68), (6.12-71), (6.12-74), (6.12-75), (6.14-2), (6.14-19), (6.14-20), (6.14-33), (6.14-35), (6.14-41), (6.14-46), (6.14-47), (6.14-51), and (6.14-52).

6.2.[41] For the flow of heat in a long circular cylinder of radius a, the governing equation and initial and boundary conditions are

$$\frac{\partial^2 v}{\partial r^2} + \frac{1}{r}\frac{\partial v}{\partial r} = \frac{1}{K}\frac{\partial v}{\partial t}$$

$$v(r, t) = v_0 \qquad \text{at } r = a$$

$$v(r, t) = 0 \qquad \text{at } t = 0$$

(a) Break down this equation by $\mathscr{L}_{t\to p}$ to get

$$\bar{v}(r, p) = \frac{v_0}{p}\frac{I_0(qr)}{I_0(qa)}$$

where

$$q^2 = p/K$$

and finally

$$v(r, t) = v_0\left[1 - \frac{2}{a}\sum_{n=1}^{\infty}\frac{e^{-K\alpha_n^2 t}J_0(r\alpha_n)}{\alpha_n J_1(a\alpha_n)}\right]$$

where α_n are the roots of $I_0(\alpha a) = 0$. This form of solution is often unsuitable for use with small values of the time. We can avoid this problem by expanding $\bar{v}(r, p)$ in a Taylor series and then inverting the result term by term.

(b) Use

$$I_0(p) \cong \frac{e^p}{2\pi p}\left[1 + \left(\frac{1}{8}p\right) + \left(\frac{9}{128}\right)p^2 + \cdots\right]$$

to get $\bar{v}(r, p)$ and then $v(r, t)$. Compare this result with that in part (a).

6.3. Rework equation (6.9-9) for $nA \to A_n$.

[41] C. J. Tranter, *Integral Transforms in Mathematical Physics*, Methuen, London (1966), p. 28.

6.4.[42] For one-dimensional unsteady-state diffusion with first-order kinetics, show that

(a)
$$\frac{\partial C_A}{\partial t} = D \frac{\partial^2 C_A}{\partial x^2} - kC_A$$

With

$$t = 0 \qquad C_A = 0$$
$$x = 0 \qquad C_A = C_{A0}$$
$$x = \infty \qquad C_A = 0$$

show that

(b)
$$\frac{C_A}{C_{A0}} = \tfrac{1}{2} \exp\left(-x\sqrt{\frac{k}{D}}\right) \operatorname{erfc}\left(\frac{x}{\sqrt{4Dt}} - \sqrt{kt}\right)$$
$$+ \tfrac{1}{2} \exp\left(x\sqrt{\frac{k}{D}}\right) \operatorname{erfc}\left(\frac{x}{\sqrt{4Dt}} + \sqrt{kt}\right)$$

(c)
$$N_A\big|_{x=0} = C_{A0}\sqrt{Dk}\left(\operatorname{erf}\sqrt{kt} + \frac{e^{-kt}}{\sqrt{\pi kt}}\right)$$

(d) W_A = total moles absorbed in the time interval $0 \le t \le t_0$

$$W_A = C_{A0}\sqrt{Dt_0}\left[\left(\sqrt{kt_0} + \frac{1}{2\sqrt{kt_0}}\right)\operatorname{erf}\sqrt{kt_0} + \frac{1}{\sqrt{\pi}} e^{-kt_0}\right]$$

(e) Describe a typical phenomenon for which (a) to (d) apply.

6.5.[43] For two immiscible solvents, the transfer of component A from one phase to the other may be written as

$$\frac{\partial C_1}{\partial t} = D_1 \frac{\partial^2 C_1}{\partial z^2} \qquad -\infty < z < 0$$

$$\frac{\partial C_2}{\partial t} = D_2 \frac{\partial^2 C_2}{\partial z^2} \qquad 0 < z < \infty$$

where C_1 is the concentration of A in phase 1 and C_2 is the concentration

---

[42] R. B. Bird, W. E. Stewart, and E. N. Lightfoot, *Transport Phenomena*, John Wiley, New York (1960), p. 621.
[43] *Ibid.*, p. 624.

of A in phase 2. The initial and boundary conditions may be

$$t = 0 \qquad C_1 = C_1{}^0 \qquad -\infty < z < 0$$

$$t = 0 \qquad C_2 = C_2{}^0 \qquad 0 < z < \infty$$

$$z = 0 \qquad C_2 = mC_1 \qquad t > 0$$

$$z = 0 \qquad D_1 \frac{\partial C_1}{\partial z} = D_2 \frac{\partial C_2}{\partial z}$$

$$z = -\infty \qquad C_1 = C_1{}^0$$

$$z = +\infty \qquad C_2 = C_2{}^0$$

(a) Explain the initial and boundary conditions.
(b) Show that

$$\frac{C_1 - C_1{}^0}{C_2{}^0 - mC_1{}^0} = \frac{1 + \mathrm{erf}\, z/\sqrt{4D_1 t}}{m + \sqrt{D_1/D_2}}$$

$$\frac{C_2 - C_2{}^0}{C_1{}^0 - (1/m)C_2{}^0} = \frac{1 - \mathrm{erf}\, z/\sqrt{4D_2 t}}{(1/m) + \sqrt{D_2/D_1}}$$

6.6.[44] In a well-stirred reactor, the kinetics are

$$A \underset{k_A}{\overset{k_B}{\rightleftharpoons}} B \xrightarrow{k_C} C$$

and the fill rate is Q (with concentration C_{A0}).
(a) Explain how we get

$$\frac{dm_A}{dt} = QC_{A0} - k_B m_A + k_A m_B$$

$$\frac{dm_B}{dt} = -(k_A + k_C)m_B + k_B m_A$$

(b) With $t = 0$, $m_B = 0$ and $dm_B/dx = 0$, solve for m_B.

6.7.[45] A fluid containing component A is pumped with volumetric flow rate Q into an empty holding tank where A decomposes by first-order

[44] Ibid., p. 701.
[45] Ibid., p. 707.

kinetics. After the tank is full, the outlet valve is opened and Q flows out as Q flows in.

(a) Find $C_A(t)$ for the filling period.
(b) Find $C_A(t)$ after the outlet valve is opened.

6.8.[46] Suppose we have absorption of a gas into a liquid which contacts a solid catalytic surface where the absorbed gas reacts.

(a) Show that the governing equation may be written as

$$D \frac{\partial^2 C}{\partial x^2} = \left(\frac{\partial v}{\partial x} x \right) \frac{\partial C}{\partial z}$$

where $[(\partial v/\partial x)x]$ is the fluid velocity at the solid–liquid interface.
(b) Explain the significance of the boundary conditions

$$C = C_0 \qquad \text{at } z = 0$$

$$D \frac{\partial C}{\partial x} = kC \qquad \text{at } x = 0$$

$$C = \text{finite} \qquad \text{as } x \to \infty$$

(c) Solve by the Laplace transform technique.

6.9. When a chemical reactor is operated continuously, mixing becomes a strong factor in performance. According to some mixing models, it is assumed that this situation is equivalent to a diffusion phenomenon governed by

$$D_z \frac{\partial^2 C}{\partial z^2} = v_z \frac{\partial C}{\partial z} + R$$

$$C = C_0 \qquad \text{at } z = 0$$

$$D_z \frac{\partial C}{\partial z} = 0 \qquad \text{at } z = z \big|_{\text{outlet}}$$

Explain this.

6.10. Consider a solute being leached from spherical particles.

(a) The rate of diffusion of the solute to the surface is given by $(N \cdot 4\pi a^2)$, where N is the flux. Explain this.

---

[46] G. Astarita, *Mass Transfer with Chemical Reaction*, Elsevier, Amsterdam (1967), p. 126.

(b) A boundary condition for this phenomenon is given by

$$V \frac{\partial C}{\partial t} = \left(-4\pi a^2 D \left. \frac{\partial C}{\partial r} \right|_{r=a} \right) (m)$$

where V is the volume of the solvent. Explain this.

6.11.[47] Suppose we have diffusion of component A (injected as a point source) into fluid B, which is flowing in the z-direction with constant velocity v_0.

(a) Explain how we get the following equations:

$$v_0 \frac{\partial C_A}{\partial z} = D \nabla^2 C_A$$

$$C_A = 0 \qquad \text{as } r \to \infty$$

$$4\pi r^2 D \frac{\partial C_A}{\partial r} = Q_A \qquad \text{as } r \to 0$$

where $r = \sqrt{x^2 + y^2 + z^2}$ is the distance from the source and Q_A is the rate at which A enters the system.

(b) Show that the solution is

$$C_A = \frac{Q_A}{4\pi Dr} e^{-(v_0/2D)(r-z)}$$

6.12.[48] Consider a plug flow tubular reactor where the flowing fluid contains a small quantity of A. The inside of the cylindrical tube is coated with a catalyst which converts A to B by a first-order reaction.

(a) Show that the governing equations are

$$v_0 \frac{\partial C_A}{\partial z} = D \frac{1}{r} \frac{\partial}{\partial r} \left(r \frac{\partial C_A}{\partial r} \right)$$

$$C_A = C_{A0} \qquad \text{at } z = 0$$

$$-D \frac{\partial C_A}{\partial r} = k C_A \qquad \text{at } r = R$$

(b) Obtain the solution

$$\frac{C_A}{C_{A0}} = \sum_{n=1}^{\infty} \frac{2K}{(K^2 + \beta_n^2)} \frac{J_0(\beta_n r/R)}{J_0(\beta_n)} \exp(-\beta_n^2 DZ/v_0 R^2)$$

where $K = Rk/D$, and β_n are the roots of $r J_1(r)/[J_0(r)] = K$.

[47] T. B. Drew, J. W. Hoopes, and T. Vermeulen, eds., *Advances in Chemical Engineering*, Vol. 5, Academic Press, New York (1964), p. 218.
[48] *Ibid.*, p. 214.

6.13.[49] A mixing tank with an inlet stream containing three components 1, 2, 3 and a well-stirred outlet can be represented by

$$\frac{d}{dt}(V\rho) = F_0\rho_0 - P\rho$$

$$\frac{d}{dt}(V\rho_1) = F_0\rho_{01} - P\rho_1$$

$$\frac{d}{dt}(V\rho_2) = F_0\rho_{02} - P\rho_2$$

$$\frac{d}{dt}(V\rho_3) = F_0\rho_{03} - P\rho_3$$

(a) Sketch a figure showing this equipment and describe how the equations arise. (Let ρ signify density.)
(b) Show how we can get

$$\frac{d\rho_1}{dt} + \frac{\rho_1}{\tau} = \frac{\rho_{01}}{\tau}$$

$$\frac{d\rho_2}{dt} + \frac{\rho_2}{\tau} = \frac{\rho_{02}}{\tau}$$

where $T = V/F_0$.
(c) With

$$\bar{y} = \begin{pmatrix} \rho_1 \\ \rho_2 \end{pmatrix} \qquad \bar{A} = \begin{pmatrix} \dfrac{1}{\tau} & 0 \\ 0 & \dfrac{1}{\tau} \end{pmatrix} = \frac{1}{\tau}\bar{I} \qquad \bar{x} = \begin{pmatrix} \rho_{01}/\tau \\ \rho_{02}/\tau \end{pmatrix}$$

show how we can get the matrix equation

$$\frac{d\bar{y}}{dt} = \bar{A}\bar{y} = \bar{x}$$

(d) The solution to part (c) is

$$\bar{y} = e^{-\bar{A}t} \int_0^t e^{\bar{A}\lambda}\bar{x}d\lambda$$

For a step input of $\bar{x}$,

$$\rho = 0 \left[\overline{}^{\rho = \rho_0}_{t=0} \right.$$

[49] D. M. Himmelblau and K. B. Bischoff, *Process Analysis and Simulation*, John Wiley, New York (1968), p. 102.

show that the solution becomes

$$\bar{y} = e^{-It/\tau}\left[\int_0^t e^{It'/\tau}dt'\right]\bar{x}$$

(e) By performing the integration indicated in (d) and using

$$e^{It'/\tau} = I + \frac{It'}{\tau} + \frac{It'^2}{2!\tau^2} + \cdots$$

show that we can get

$$\bar{y} = \tau[I(1 - e^{-t/\tau})][x] = \tau(1 - e^{-t/\tau})\bar{x}$$

or

$$\begin{pmatrix} \varrho_1 \\ \varrho_2 \end{pmatrix} = \tau(1 - e^{-t/\tau})\begin{pmatrix} \varrho_{01} \\ \varrho_{02} \end{pmatrix} = \begin{pmatrix} \varrho_{01}(1 - e^{-t/\tau}) \\ \varrho_{02}(1 - e^{-t/\tau}) \end{pmatrix}$$

6.14.[50] In the study of the dynamics of a countercurrent packed gas absorber, the governing equations may be written as

$$\frac{\partial y}{\partial t} + v_g \frac{dy}{dz} = -k_L(x^* - x)$$

$$\frac{\partial x}{\partial t} - v_L \frac{\partial x}{\partial z} = k_L(x^* - x)$$

and

$$y(t, 0) = y_0(t)$$

$$x(t, \bar{z}) = 0, \quad \text{where } \bar{z} \text{ is one end of the absorber}$$

$$y = mx^*$$

where y and x are gas and liquid compositions.

(a) Show how these equations are obtained.

(b) By a Laplace transform technique, $\mathcal{L}_{t \to s}$, show that we can get

$$\bar{y}(s, z) = A_1 e^{r_1 z} + A_2 e^{r_2 z}$$

where

$$r_{1,2} = \frac{(a_{11} - a_{22}) \pm \sqrt{(a_{11} - a_{22})^2 + 4(a_{11}a_{12} - a_{12}a_{21})}}{2}$$

and

$$a_{11}, a_{12}, a_{21}, \text{ and } a_{22}$$

50 *Ibid.*, p. 134.

are found from

$$\frac{d\bar{y}}{dz} = -a_{11}\bar{y} + a_{12}\bar{x}$$

$$\frac{d\bar{x}}{dz} = -a_{21}\bar{y} + a_{22}\bar{x}$$

(c) Show that

$$A_1 = \frac{\bar{y}_0(s)e^{r_2z}(a_{11} + r_2)}{B}$$

$$A_2 = \frac{-\bar{y}_0(s)e^{r_1z}(a_{11} + r_1)}{B}$$

$$B = (a_{11} + r_2)e^{-r_1z} - (a_{11} + r_1)e^{-r_2z}$$

$$\bar{y}_0(s) = \mathcal{L}_{t\to s} y(t, 0)$$

6.15.[51] In a well-stirred tank which has a steam heat transfer surface proportional to the amount of liquid in the tank, the governing equation is

$$\frac{d}{d\theta}(C_p wt) = GH + UA(T - t)$$

where

$$C_p = \text{heat capacity of the liquid}$$
$$w = \text{amount of water in the tank}$$
$$t = \text{liquid temperature}$$
$$G = \text{liquid feed rate}$$
$$H = \text{enthalpy of entering liquid}$$
$$U = \text{overall heat transfer coefficient}$$
$$A = \text{heating area}$$
$$T = \text{heating steam temperature}$$
$$\theta = \text{time}$$

(a) Explain how we get this equation.
(b) Show how the equation is transformed to

$$f\frac{dt}{df} + (Bf + 1)t = BfT + t_0$$

[51] W. R. Marshall and R. L. Pigford, *Applications of Differential Equations to Chemical Engineering Problems*, University of Delaware, Dover, Delaware (1947), p. 16.

where

$$H = C_p(t_0 - t^0)$$
$$t^0 = 0 = \text{reference temperature}$$
$$A = wf = \text{exposed area}$$
$$w = \text{weight of liquid in the filled tank}$$
$$f = \text{fraction of the tank which is filled}$$
$$B = UA/GC_p$$

(c) If we define $\varphi = (t - t_0)/(T - t_0)$, show that the final solution is

$$\varphi = 1 - \frac{1 - e^{-Bf}}{Bf}$$

with $\varphi = 0$ at $f = 0$.

6.16. Solve equation (6.9-57) by an approximate Laplace transform technique and compare the result with equation (6.9-61).

For Further Reading

Transport Phenomena, by R. B. Bird, W. E. Stewart, and E. N. Lightfoot, John Wiley, New York, 1960.

Mathematical Methods in Chemical Engineering, by V. G. Jenson and G. V. Jeffreys, Academic Press, New York, 1963.

Applied Mathematics in Chemical Engineering, by H. S. Mickley, T. K. Sherwood, and C. E. Reed, McGraw-Hill, New York, 1957.

Mathematics of Engineering Systems, by F. H. Raven, McGraw-Hill, New York, 1966.

Similarity Analyses of Boundary Value Problems in Engineering, by A. G. Hansen, Prentice-Hall, Englewood Cliffs, N. J., 1964.

Process Analysis and Simulation, by D. M. Himmelblau and K. B. Bischoff, John Wiley, New York, 1968.

The Mathematics of Diffusion, by J. Crank, Oxford University Press, Oxford, 1956.

Applications of Differential Equations to Chemical Engineering Problems, by W. R. Marshall and R. L. Pigford, University of Delaware, Dover, Deiaware, 1947.

Advances in Chemical Engineering, T. B. Drew, J. W. Hoopes, and T. Vermeulen, eds., Vol. 5, Academic Press, New York, 1964.

Non-linear Partial Differential Equations in Engineering, by W. F. Ames, Academic Press, New York, 1965.

Mass Transfer with Chemical Reaction, by G. Astarita, Elsevier, Amsterdam, 1967.

Mathematical Methods in Chemical Engineering, by N. R. Amundson, Prentice-Hall, Englewood Cliffs, N. J., 1966.

Physiochemical Hydrodynamics, by V. G. Levich, Prentice-Hall, Englewood Cliffs, N. J., 1962.

Integral Transforms in Mathematical Physics, by C. J. Tranter, Methuen, London, 1966.

Part III

TRANSPORT ANALYSIS
IN DISCRETE PROCESSES

Chapter 7

Finite Difference Calculus

In this chapter we introduce the calculus of finite differences, with applications in difference equations, interpolation and extrapolation and solutions to simple difference equations. Finite difference integration methods are discussed. More complex finite difference equations are then solved by operator techniques. Finally simultaneous difference equations and non-linear equations are solved.

7.1. Elementary Difference Operations

In dealing with finite difference or staged processes, where there are incremental variations, it is useful to go to the calculus of finite differences in order to set up and solve the governing equations. First we introduce the notation

$$U_x = f(x) \qquad (7.1\text{-}1)$$

and

$$U_{x+\Delta x} = f(x + \Delta x) \qquad (7.1\text{-}2)$$

where U_x is any function of the variable x. We now define a difference ΔU_x by

$$\Delta U_x = U_{x+1} - U_x \qquad (7.1\text{-}3)$$

where $\Delta x = 1$. Some specific examples of equation (7.1-3) are

$$\Delta x^2 = (x + 1)^2 - x^2 \qquad (7.1\text{-}4)$$

$$\Delta \log x = \log(x + 1) - \log x \qquad (7.1\text{-}5)$$

$$\Delta 2^x = 2^{x+1} - 2^x \qquad (7.1\text{-}6)$$

299

TABLE XVI

Difference Table for the Function $U_x = x^3$

x	$U_x = x^3$	ΔU_x	$\Delta^2 U_x$	$\Delta^3 U_x$
0	0			
		1		
1	1		6	
		7		6
2	8		12	
		19		
3	27			

If data are collected incrementally, with x the independent variable and, for instance, $U_x = x^3$ the dependent variable, we can construct a difference table as shown in Table XVI, where $U_1 = 1$, $U_2 = 8$, $U_3 = 27$,

$$\Delta U_0 = U_1 - U_0 = 1$$
$$\Delta U_1 = U_2 - U_1 = 8 - 1 = 7$$
$$\Delta U_2 = U_3 - U_2 = 27 - 8 = 19$$

and

$$\Delta^2 U_1 = \Delta(\Delta U_1) = \Delta(U_2 - U_1) = \Delta U_2 - \Delta U_1 = 19 - 7 = 12$$

We can develop a series of formulas from these elementary operations:

$$\Delta U_0 = U_1 - U_0 \tag{7.1-7}$$
$$\Delta^2 U_0 = \Delta(\Delta U_0) = \Delta(U_1 - U_0) = \Delta U_1 - \Delta U_0 \tag{7.1-8}$$
$$= (U_2 - U_1) - (U_1 - U_0) \tag{7.1-9}$$
$$= U_2 - 2U_1 + U_0 \tag{7.1-10}$$

Similarly, it can be shown that the third difference can be written as

$$\Delta^3 U_0 = U_3 - 3U_2 + 3U_1 - U_0 \tag{7.1-11}$$

and

$$\Delta^3 U_1 = U_4 - 3U_3 + 3U_2 - U_1 \tag{7.1-12}$$

Note that the coefficients in equations (7.1-11) and (7.1-12) correspond to the coefficients found in the binomial expansion formula

$$(a - b)^3 = a^3 - 3a^2b + 3ab^2 - ab^3 \tag{7.1-13}$$

Some more elementary difference operations are

$$\Delta C = 0 \qquad C = \text{constant} \tag{7.1-14}$$

$$\Delta^n(U_x + V_x) = \Delta^n U_x + \Delta^n V_x \tag{7.1-15}$$

$$\Delta^n C U_x = C \Delta^n U_x \tag{7.1-16}$$

The difference operation symbolized by Δ is the finite difference analogy of the differential derivative d and obtains many similar results.

The expression $x(x - 1)(x - 2) \cdots (x - n + 1)$ is frequently encountered and is symbolized by $x^{(n)}$:

$$x^{(n)} = x(x - 1)(x - 2) \ldots (x - n + 1) \tag{7.1-17}$$

Similarly, we define $x^{|n|}$ as

$$x^{|n|} = x(x + 1)(x + 2) \ldots (x + n - 1) \tag{7.1-18}$$

The utility of the forms $x^{(n)}$ and $x^{|n|}$, particularly $x^{(n)}$, is that they allow us to use the difference operation Δ in a manner parallel to the derivative operator d. For example, it can be shown that[1]

$$\Delta(a + bx)^{(n)} = nb(a + bx)^{(n-1)} \tag{7.1-19}$$

and also

$$\Delta^n x^{(n)} = n! \tag{7.1-20}$$

$$\Delta x^{(n)} = nx^{(n-1)} \tag{7.1-21}$$

The greatest utility of equation (7.1-18) rests in forms such as[2]

$$\Delta \frac{1}{x^{|n|}} = \frac{-n}{x^{|n+1|}} \tag{7.1-22}$$

which is analogous to the differential operation

$$\frac{d}{dx}\left(\frac{1}{x^n}\right) = \frac{-nx^{n-1}}{x^{2n}} = \frac{-n}{x^{n+1}} \tag{7.1-23}$$

---

[1] C. H. Richardson, *The Calculus of Finite Differences*, Van Nostrand, New York (1954), p. 9.

[2] *Ibid.*

TABLE XVII

x	U_x	ΔU_x	$\Delta^2 U_x$	$\Delta^3 U_x$	$\Delta^4 U_x$
0	1	3	3	1	0
1	4	6	4	1	0
2	10	10	5	1	
3	20	15	6		
4	35	21			
5	56				

7.2. Interpolation and Extrapolation with Finite Differences

Where data are collected incrementally, finite difference approaches can be useful in the analysis. For example, if we collected data as shown in Table XVII, it is possible to express the data in an infinite finite difference series using Newton's formula[3]

$$U_x = U_0 + x^{(1)}\Delta U_0 + \frac{x^{(2)}}{2!}\Delta^2 U_0 + \cdots + \frac{x^{(n)}}{n!}\Delta^{(n)}U_0 \qquad (7.2\text{-}1)$$

Note the similarity between equation (7.2-1) and the Taylor series expansion of $f(x)$. For the data given above, and with equation (7.2-1), we can immediately write down the general term for U_x:

$$U_x = 1 + x^{(1)} \cdot 3 + \frac{x(x-1)}{2!} \cdot 3 + \frac{x(x-1)(x-2)}{3!} \cdot 1$$
$$= \tfrac{1}{6}(x^3 + 6x^2 + 11x + 6) \qquad (7.2\text{-}2)$$

The final results in equation (7.2-2) are expressed in the usual exponential forms, such as x^3, whereas we need forms such as $x^{(3)}$ when performing finite difference operations. Thus a facile technique is needed for interchanging forms such as x^3 and $x^{(3)}$. One direct method is to write out the equivalent forms with undetermined coefficients:

$$2x^3 - 3x^2 + 3x - 10 = Ax^{(3)} + Bx^{(2)} + Cx^{(1)} + D \qquad (7.2\text{-}3)$$

and evaluate A, B, C, and D. Multiplying out the terms in equation (7.2-3) yields

$$2x^3 - 3x^2 + 3x - 10 = A(x)(x-1)(x-2) + B(x)(x-1) + C(x) + D \qquad (7.2\text{-}4)$$

[3] *Ibid.*, p. 10.

By matching coefficients of x^3, x^2, x^1, and x^0 we get the values for A, B, C, and D, and hence the final result

$$2x^3 - 3x^2 + 3x - 10 = 2x^{(3)} + 3x^{(2)} + 2x^{(1)} - 10 \qquad (7.2\text{-}5)$$

The procedure outlined here is general and can be used without resort to formulas that require memorization.

With the data of Table XVII and with equation (7.2-1), we can extrapolate to U_8, which is U_x at $x = 8$. The result is

$$U_8 = U_0 + 8\Delta U_0 + \frac{8 \cdot 7}{2!} \Delta^2 U_0 + \frac{8 \cdot 7 \cdot 6}{3!} \Delta^3 U_0 \qquad (7.2\text{-}6)$$

and from the table we get

$$U_0 = 1 \qquad \Delta U_0 = 3 \qquad \Delta^2 U_0 = 3 \qquad \Delta^3 U_0 = 1$$

so that finally

$$U_8 = 1 + (8)(3) + (28)(3) + (56)(1) = 165 \qquad (7.2\text{-}7)$$

We can also interpolate, so that if $U_{6.2}$ is required, it can be obtained from equation (7.2-1) as follows:

$$U_{6.2} = 1 + (6.2)(3) + \frac{(6.2)(5.2)}{2!} (3) + \frac{(6.2)(5.2)(4.2)}{3!} (1) \quad (7.2\text{-}8)$$

Newton's finite difference formula, equation (7.2-1), can also be used for interpolation of data such as shown in Table XVIII. Given values of

TABLE XVIII

Interpolation of Trigonometric Data[a]

$$U_x = 10^4 \sin(45 + 5x)$$
$$\theta = 45 + 5x$$

θ	x	U_x	ΔU_x	$\Delta^2 U_x$	$\Delta^3 U_x$
45	0	7071	589	-57	-7
50	1	7660	532	-64	
55	2	8192	468		
60	3	8660			

[a] C. H. Richardson, *The Calculus of Finite Differences*, Van Nostrand, New York (1954), p. 65.

$\sin \theta$ for $\theta = 45°$, $50°$, $55°$, $60°$, find $\sin 52°$. To accomplish this, we set up a forward difference table as given in Table XVIII. Decimals are avoided by multiplying the sine function by 10^4 and by using a transformation, $\theta = 45 + 5x$. Thus $52°$ corresponds to $x = 1.4$. Using equation (7.2-1) with Table XVIII, we get the interpolated value for $\sin 52°$ as

$$U_{1.4} = 7071 + (1.4)(589) + \frac{(1.4)(1.4 - 1)}{2!} (-57)$$

$$+ \frac{(1.4)(1.4 - 1)(1.4 - 2)}{3!} (-7) = 7880.032 \qquad (7.2-9)$$

Newton's formula, equation (7.2-1), is called the forward interpolation formula. The value of x was chosen near the beginning of the tabulated values. If the value of x is near the end of the data table, we may not have all the required differences. In these cases we use Newton's backward interpolation formula.[4] There also are other formulas for interpolation given by Gauss, Sterling, and Bessel.[5]

7.3. The Difference Operation E

We now introduce a difference operator E which works like the D operator in ordinary differential equations. From

$$\Delta U_x = U_{x+1} - U_x \qquad (7.3-1)$$

we can write

$$U_{x+1} = U_x + \Delta U_x = (1 + \Delta)U_x \qquad (7.3-2)$$

and define the E operator as

$$E = 1 + \Delta \qquad (7.3-3)$$

so that equation (7.3-2) becomes

$$U_{x+1} = EU_x \qquad (7.3-4)$$

This indicates that E operating on U_x gives the same function but with the index raised one integer. It can be shown that

$$E^2 U_x = E[EU_x] = EU_{x+1} = U_{x+2} \qquad (7.3-5)$$

$$E^3 U_x = U_{x+3} \qquad (7.3-6)$$

---

[4] *Ibid.*, p. 67.
[5] *Ibid.*, p. 69.

and, in general,

$$E^n U_x = U_{x+n} \tag{7.3-7}$$

One sample operation is

$$E(x^2 + \sin x) = (x + 1)^2 + \sin(x + 1) \tag{7.3-8}$$

and from another, if

$$EU_x = x + \sin x \tag{7.3-9}$$

then

$$U_x = (x - 1) + \sin(x - 1) \tag{7.3-10}$$

Later in the chapter we will use the E operator to solve the difference equations which are analogous to ordinary differential equations solved by D operator techniques.

7.4. Finite Difference Integration

Finite difference integration is defined by the inverse difference operator, $1/\Delta$ or Δ^{-1}, which has similarities to integration operations in infinitesimal calculus. For example, given the relationship

$$\Delta V_x = U_x \tag{7.4-1}$$

then, by definition,

$$V_x = \frac{1}{\Delta} U_x = \Delta^{-1} U_x \tag{7.4-2}$$

Specifically, if we have, for instance,

$$\Delta \frac{x^{(3)}}{3} = x^{(2)} \tag{7.4-3}$$

then by equation (7.4-2) we get

$$\Delta^{-1} x^{(2)} = \frac{x^{(3)}}{3} + C \tag{7.4-4}$$

where C is an integration constant. Note the similarity of equation (7.4-4) to

$$\int x^2 dx = \frac{x^3}{3} + C \tag{7.4-5}$$

TABLE XIX

Table of Finite Difference Integral Operations[a]

Finite differences	Finite integrals
1. $\Delta(U_x + V_x - W_z)$ $= \Delta U_x + \Delta V_x - \Delta W_x$	$\Delta^{-1}(U_x + V_x - W_x)$ $= \Delta^{-1}U_x + \Delta^{-1}V_x - \Delta^{-1}W_x$
2. $\Delta k U_x = k\Delta U_x$	$\Delta^{-1}k U_x = k\Delta^{-1}U_x$
3. $\Delta a^x = (a-1)a^x$	$\Delta^{-1}a^x = \dfrac{a^x}{a-1}, \quad a \neq 1$
4. $\Delta x^{(n)} = nx^{(n-1)}$	$\Delta^{-1}x^{(n)} = \dfrac{x^{(n+1)}}{n+1}$
5. $\Delta\dfrac{1}{x^{\lvert n\rvert}} = \dfrac{-n}{x^{\lvert n+1\rvert}}$	$\Delta^{-1}\dfrac{1}{x^{\lvert n\rvert}} = \dfrac{1}{(1-n)x^{\lvert n-1\rvert}}, \quad n \neq 1$
6. $\Delta(a+bx)^{(n)} = bn(a+bx)^{(n-1)}$	$\Delta^{-1}(a+bx)^{(n)} = \dfrac{(a+bx)^{(n+1)}}{b(n+1)}$
7. $\Delta\dfrac{1}{(a+bx)^{\lvert n\rvert}} = \dfrac{-bn}{(a+bx)^{\lvert n-\ldots\rvert}}$	$\Delta^{-1}\dfrac{1}{(a+bx)^{\lvert n\rvert}} = \dfrac{1}{b(1-n)(a+bx)^{\lvert n-1\rvert}}$
8. $\Delta \sin(a+bx)$ $= 2\sin\dfrac{b}{2}\cos\left(a + \dfrac{b}{2} + bx\right)$	$\Delta^{-1}\cos(a+bx)$ $= \dfrac{1}{2\sin(b/2)}\sin\left(a - \dfrac{b}{2} + bx\right)$
9. $\Delta \cos(a+bx)$ $= -2\sin\dfrac{b}{2}\sin\left(a + \dfrac{b}{2} + bx\right)$	$\Delta^{-1}\sin(a+bx)$ $= \dfrac{-1}{2\sin(b/2)}\cos\left(a - \dfrac{b}{2} + bx\right)$
10. $\Delta\dbinom{x}{n} = \dbinom{x}{n-1}$	$\Delta^{-1}\dbinom{x}{n} = \dbinom{x}{n+1}$
11. $\Delta U_x V_x = U_x\Delta V_x + V_{x+1}\Delta U_x$	$\Delta^{-1}[U_x\Delta V_x] = U_x V_x - \Delta^{-1}[V_{x+1}\Delta U_x]$
12. $\Delta x! = x(x!)$	$\Delta^{-1}x(x!) = x!$
13. $\Delta \tan(a+bx)$ $= \dfrac{\sin b}{\cos(a+bx)\cos(a+b+bx)}$	$\Delta^{-1}\sec(a+bx)\sec(a+b+bx)$ $= \csc b \tan(a+bx)$
14. $\Delta \cot(a+bx)$ $= \dfrac{-\sin b}{\sin(a+bx)\sin(a+b+bx)}$	$\Delta^{-1}\csc(a+bx)\csc(a+b+bx)$ $= -\csc b \cot(a+bx)$

TABLE XIX (*continued*)

Finite differences	Finite Integrals
15. $\Delta \sec(a + bx)$ $= \dfrac{2 \sin(b/2) \sin(a + b/2 + bx)}{\cos(a + bx) \cos(a + b + bx)}$	$\Delta^{-1} \dfrac{\sin(a + b/2 + bx)}{\cos(a + bx) \cos(a + b + bx)}$ $= \dfrac{\sec(a + bx)}{2 \sin(b/2)}$
16. $\Delta \csc(a + bx)$ $= \dfrac{-2 \sin(b/2) \cos(a + b/2 + bx)}{\sin(a + bx) \sin(a + b + bx)}$	$\Delta^{-1} \dfrac{\cos(a + b/2 + bx)}{\sin(a + bx) \sin(a + b + bx)}$ $= \dfrac{-\csc(a + bx)}{2 \sin(b/2)}$
17. $\Delta\left(\dfrac{U_x}{V_x}\right) = \dfrac{V_x \Delta U_x - U_x \Delta V_x}{V_x V_{x+1}}$	
18. $\Delta^n X^n = n!$	

[a] C. H. Richardson, *The Calculus of Finite Differences*, Van Nostrand, New York (1954), p. 22.

It should be reemphasized here that we operate on $x^{(n)}$, not x^n or $x^{|n|}$. Many finite difference integrals have already been worked out, as given in Table XIX. Thus, to integrate $\Delta^{-1} 3^x$, we simply look up the result in the tables and obtain

$$\Delta^{-1} 3^x = \frac{3^x}{3 - 1} + C = \frac{3^x}{2} + C \qquad (7.4\text{-}6)$$

Another example is

$$\Delta^{-1}(x^3 - 2x^2 + 7x - 12) = \Delta^{-1}(x^{(3)} + x^{(2)} + 6x^{(1)} - 12)$$

$$= \frac{x^{(4)}}{4} + \frac{x^{(3)}}{3} + 3x^{(2)} - 12x^{(1)} + C \qquad (7.4\text{-}7)$$

where x^n is converted to $x^{(n)}$ by the method of Section 7.2.

7.5. Summation of Infinite Series by Finite Integration

Finite difference calculus can also be used to find the summation of infinite series. Given

$$\Delta V_x = U_x \qquad (7.5\text{-}1)$$

or

$$V_x = \Delta^{-1}U_x \tag{7.5-2}$$

we can write from equation (7.5-1) the sequence of equations

$$
\begin{aligned}
V_1 &- V_0 &= U_0 \\
V_2 &- V_1 &= U_1 \\
V_3 &- V_2 &= U_2 \\
V_n &- V_{n-1} &= U_{n-1} \\
V_{n+1} &- V_n &= U_n
\end{aligned} \tag{7.5-3}
$$

Adding the terms in equation (7.5-3), we get

$$V_{n+1} - V_0 = U_0 + U_1 + U_2 + \cdots + U_n = \sum_0^n U_x \tag{7.5-4}$$

In equation (7.5-2) we can put definite limits on the integration, giving

$$V_x \Big|_0^{n+1} = V_{n+1} - V_0 = \Delta^{-1}U_x \Big|_0^{n+1} \tag{7.5-5}$$

Equations (7.5-4) and (7.5-5) yield

$$\sum_0^n U_x = \Delta^{-1}U_x \Big|_0^{n+1} \tag{7.5-6}$$

which is a convenient formula for finding an infinite sum by integrating the general term. For example, the series $2 + 4 + 6 + \cdots + 2n$ is summed ($U_x = 2x$), and from equation (7.5-6) we get the final summation result

$$\sum_1^n U_x = \Delta^{-1}2x \Big|_0^{n+1} = x^{(2)} \Big|_1^{n+1} = (n+1)^{(2)} - (1)^{(2)}$$
$$= (n+1)(n) - (1)(0) = (n+1)(n) \tag{7.5-7}$$

(note that $n = 1$ was the lower limit). As another example, find the sum of $(1)(2) + (2)(3) + (3)(4) + \cdots + n(n+1)$. For this case, $U_x = x(x+1) = (1+x)^{(2)} = (a+bx)^{(m)}$, where $a = 1$, $b = 1$, and $m = 2$. From equation (7.5-6) we get the summed result as

$$\sum_1^n U_x = \sum_1^n (1+x)^{(2)} = \Delta^{-1}(1+x)^{(2)} \Big|_1^{n+1}$$
$$= \frac{(1+x)^{(3)}}{3} \Big|_1^{n+1} = \frac{(n+2)^{(3)}}{3} - \frac{2^{(3)}}{3}$$
$$= \tfrac{1}{3}(n+2)(n+1)(n) \tag{7.5-8}$$

As a final example, we find the sum of $1^2 + 2^2 + 3^2 + \cdots + n^2$ by noting that $U_x = x^2 = (x)(x-1) + x = x^{(2)} + x^{(1)}$, so that

$$\sum_{1}^{n} U_x = \Delta^{-1}[x^{(2)} + x^{(1)}]\Big|_{1}^{n+1} = \left[\frac{x^{(3)}}{3} + \frac{x^{(2)}}{2}\right]\Big|_{1}^{n+1}$$

$$= \frac{(n+1)^{(3)}}{3} + \frac{(n+1)^{(2)}}{2} = \frac{(n+1)(n)(n-1)}{3} + \frac{(n+1)(n)}{2}$$

$$= \frac{(n)(n+1)(2n+1)}{6} \qquad\qquad\qquad (7.5\text{-}9)$$

where

$$1^{(2)} = (1)(0) = 0 \qquad \text{and} \qquad 1^{(3)} = 0$$

7.6. Finite Difference Integration Techniques

If the integration to be performed is more complex than that which is covered in the integration tables, it is possible to integrate by parts in a manner similar to the procedure in infinitesimal calculus. It can be shown[6] that

$$\Delta U_x V_x = U_x \Delta V_x + V_{x+1}\Delta U_x \qquad\qquad (7.6\text{-}1)$$

is true, so that from equation (7.6-1) we get

$$\Delta^{-1}[U_x \Delta V_x] = U_x V_x - \Delta^{-1}[V_{x+1}\Delta U_x] \qquad\qquad (7.6\text{-}2)$$

where $\Delta^{-1}[\Delta U_x V_x] = U_x V_x$. Note the analogy between equation (7.6-2) and the usual infinitesimal integration by parts.

As an illustration of equation (7.6-2), we perform the finite integration $\Delta^{-1}x3^x$. We proceed as follows:

$$U_x = x \qquad \Delta V_x = 3^x$$

$$\Delta U_x = 1 \qquad V_x = \Delta^{-1}3^x = \frac{3^x}{2}$$

$$V_{x+1} = \frac{3^{x+1}}{2} = \frac{3}{2}3^x$$

From equation (7.6-2) the results are

$$\Delta^{-1}x3^x = \frac{x3^x}{2} - \Delta^{-1}\left[\left(\frac{3}{2}3^x\right)(1)\right] + C \qquad\qquad (7.6\text{-}3)$$

[6] *Ibid.*, p. 29.

where C is an integration constant. Performing the integration indicated in equation (7.6-3), we finally get

$$\Delta^{-1}x3^x = x\,\frac{3^x}{2}\left(\frac{3}{2}\right)\left(\frac{3^x}{2}\right) + C = 3^x\left[\frac{x}{2} - \frac{3}{4}\right] + C \qquad (7.6\text{-}4)$$

Other integration techniques include a method of undetermined coefficients and undetermined functions. The integration method is based on the observation that the Δ difference operation converts polynomials into polynomials, exponential functions into exponential functions, trigonometric functions into trigonometric functions, etc. We can shown the method by a few examples.

(a) Find the finite integral

$$\Delta^{-1}\,\frac{x2^x}{(x+1)(x+2)}$$

Let

$$U_x = \frac{x2^x}{(x+1)(x+2)}$$

and look for a function V_x such that

$$\Delta V_x = U_x = \frac{x2^x}{(x+1)(x+2)} \qquad (7.6\text{-}5)$$

or

$$V_x = \Delta^{-1}\,\frac{x2^x}{(x+1)(x+2)} \qquad (7.6\text{-}6)$$

From *a priori* experience with finite difference operations, we assume a form for V_x such as

$$V_x = \frac{f(x)}{(x+1)}\,2^x \qquad (7.6\text{-}7)$$

where $f(x)$ is an undetermined function. Since

$$\Delta V_x = V_{x+1} - V_x = U_x \qquad (7.6\text{-}8)$$

we observe that the choice of the form of V_x in equation (7.6-7) will produce the results we seek, i.e., equation (7.6-8) becomes

$$\frac{f(x+1)2^{x+1}}{(x+2)} - \frac{f(x)2^x}{(x+1)} = \frac{x2^x}{(x+1)(x+2)} \qquad (7.6\text{-}9)$$

If we equate numerator terms after clearing fractions, the result is

$$2(x+1)f(x+1) - (x+2)f(x) = x \qquad (7.6\text{-}10)$$

The right side of Equation (7.6-10) is linear in x, which requires that the left side also be linear in x. This implies that the undetermined function $f(x)$ is a constant. Let $f(x) = k$, so that $f(x + 1) = k$, and equation (7.6-10) becomes

$$2(x + 1)k - (x + 2)k = x \qquad (7.6\text{-}11)$$

which yields

$$k = f(x) = 1 \qquad (7.6\text{-}12)$$

and thus from equations (7.6-6) and (7.6-7) we finally get

$$\Delta^{-1} \frac{x2^x}{(x + 1)(x + 2)} = \frac{1}{(x + 1)} 2^x + C \qquad (7.6\text{-}13)$$

(b) Find

$$\Delta^{-1} \frac{2x - 1}{2^{x-1}}$$

Proceeding as in the previous example, we get

$$\Delta V_x = U_x = \frac{2x - 1}{2^{x-1}} \qquad (7.6\text{-}14)$$

Let

$$V_x = \frac{f(x)}{2^{x-1}} \qquad (7.6\text{-}15)$$

so that from equations (7.6-14) and (7.6-15) we get

$$\frac{f(x + 1)}{2^x} - \frac{f(x)}{2^{x-1}} = \frac{2x - 1}{2^{x-1}} \qquad (7.6\text{-}16)$$

or

$$f(x + 1) - 2f(x) = 4x - 2 \qquad (7.6\text{-}17)$$

Evidently $f(x)$ must be a linear function of x. Let

$$f(x) = ax + b \qquad (7.6\text{-}18)$$

so that

$$f(x + 1) = a(x + 1) + b = ax + a + b \qquad (7.6\text{-}19)$$

From equations (7.6-17) and (7.6-19) we get

$$(ax + a + b) - 2(ax + b) = 4x - 2 \qquad (7.6\text{-}20)$$

Equating coefficients in (7.6-20) gives

$$a = -4 \qquad b = -2$$

and equation (7.6-18) becomes

$$f(x) = -4x - 2 \qquad (7.6\text{-}21)$$

Thus from equations (7.6-21) and (7.6-15) the final result is

$$V_x = \frac{-4x - 2}{2^{x-2}} = -\frac{2x + 1}{2^{x-2}} \qquad (7.6\text{-}22)$$

and hence

$$\varDelta^{-1} \frac{2x - 1}{2^{x-1}} = -\frac{2x + 1}{2^{x-2}} + C \qquad (7.6\text{-}23)$$

7.7. Solutions of First-Order Finite Difference Equations

We are now in a position to solve finite difference equations. For example, suppose a difference equation is given by

$$\varDelta^3 U_x + \varDelta^2 U_x - \varDelta U_x - U_x = 0 \qquad (7.7\text{-}1)$$

We can convert (7.7-1) to

$$U_{x+3} - 2U_{x+2} = 0 \qquad (7.7\text{-}2)$$

by noting that the first, second, and third differences can be written as

$$\varDelta U_x = U_{x+1} - U_x \qquad (7.7\text{-}3)$$

$$\varDelta^2 U_x = U_{x+2} - 2U_{x+1} + U_x \qquad (7.7\text{-}4)$$

$$\varDelta^3 U_x = U_{x+3} - 3U_{x+2} + 3U_{x+1} - U_x \qquad (7.7\text{-}5)$$

By a substitution of variable,

$$Z_x = U_{x+2}$$

combined with (7.7-2) we can get

$$Z_{x+1} - 2Z_x = 0 \qquad (7.6\text{-}6)$$

which is recognized as a first-order equation. Before showing a systematic approach to the solution of equations such as (7.6-6) we first apply a bit of intuition and "guess" at a solution:

$$Z_x = C_x 2^x \tag{7.7-7}$$

where C_x is an arbitrary constant or arbitrary periodic function of period 1. Substituting equation (7.7-7) into (7.7-6) yields the identity

$$C_{x+1} 2^{x+1} - 2C_x 2^x = 0 \tag{7.7-8}$$

indicating that equation (7.7-7) is the solution of (7.7-6). Thus we can solve some simple difference equations almost by inspection. For example, given the difference equation

$$U_{x+1} - U_x = a^x \qquad a \neq 1 \tag{7.7-9}$$

it is easy to see that equation (7.7-9) can be written as

$$\varDelta U_x = a^x \tag{7.7-10}$$

and hence

$$U_x = \varDelta^{-1} a^x = \frac{a^x}{a-1} + C \tag{7.7-11}$$

Another example is

$$\varDelta^2 U_x = 2^x + x \tag{7.7-12}$$

which can be integrated once to get

$$\varDelta U_x = \varDelta^{-1}(2^x + x) = 2^x + \frac{x^{(2)}}{2} + C_1 \tag{7.7-13}$$

and then once more to give

$$U_x = 2^x + \frac{x^{(3)}}{6} + C_1 x + C_2 \tag{7.7-14}$$

As a final example, consider the equation

$$U_{x+1} - aU_x = 0 \tag{7.7-15}$$

Multiplying equation (7.7-15) by an integrating factor, a^{-x-1}, we get

$$a^{-(x+1)} U_{x+1} - a^{-x} U_x = 0 \tag{7.7-16}$$

or

$$\varDelta a^{-x} U_x = 0 \tag{7.7-17}$$

From equation (7.7-17) we obtain

$$a^{-x}U_x = C \tag{7.7-18}$$

or

$$U_x = Ca^x \tag{7.7-19}$$

Now we set up procedures for systematically solving first-order linear equations of the form

$$U_{x+1} - A_x U_x = B_x \tag{7.7-20}$$

where A_x and B_x are functions of x. First consider the homogeneous equation, i.e., $B_x = 0$:

$$U_{x+1} - A_x U_x = 0 \tag{7.7-21}$$

From equation (7.7-21) it follows that

$$U_1 = U_0 A_0$$
$$U_2 = U_1 A_1 = U_0 A_0 A_1$$
$$U_3 = U_2 A_2 = U_0 A_0 A_1 A_2$$
$$\cdots\cdots\cdots\cdots\cdots\cdots$$
$$U_x = U_0 A_0 A_1 A_2 \cdots A_{x-1}$$

If we let $U_0 = C = $ an arbitrary constant, then in general the solution of equation (7.7-12) is

$$U_x = CA_0 A_1 A_2 \ldots A_{x-1} = C \prod_{x=0}^{x-1} A_x \tag{7.7-22}$$

where the symbol $\prod_{x=0}^{x-1}$ is the product symbol:

$$\prod_{x=0}^{x-1} A_x = A_0 A_1 \ldots A_{x-1}$$

Thus, the solution of a problem such as

$$U_{x+1} - 3U_x = 0 \tag{7.7-23}$$

can be written almost by inspection. Comparing equations (7.7-21) and (7.7-23), we note $A_x = 3$, so that $A_0 = A_1 = A_2 = \cdots = A_{x-1} = 3$, and the solution from equation (7.7-22) is

$$U_x = C \prod_{x=0}^{x-1} 3 = C3^x \tag{7.7-24}$$

Another common problem is illustrated by the equation

$$U_{x+1} - xU_x = 0 \qquad x > 0 \tag{7.7-25}$$

Here $A_x = x$ when we compare equations (7.7-25) and (7.7-21). Thus $A_1 = 1, A_2 = 2, A_3 = 3, \cdots, A_{x-1} = x - 1$, which, from equation (7.7-22), yields as a final solution

$$U_x = C \prod_{x=1}^{x-1} x = C(x - 1)! \tag{7.7-26}$$

where $C = U_1$ and $0! = 1$.

Now we turn to the more general solution of equation (7.7-20), where $B_x \neq 0$. To obtain the solution of equation (7.7-20), we let

$$U_x = Z_x V_x \tag{7.7-27}$$

where Z_x is the solution of equation (7.7-21) and V_x is a function which is to be determined. Substituting equation (7.7-27) into (7.7-20), we obtain

$$Z_{x+1} V_{x+1} - A_x Z_x V_x = B_x \tag{7.7-28}$$

If now we substitute for V_{x+1} in equation (7.7-28) by the definition

$$\Delta V_x = V_{x+1} - V_x \tag{7.7-29}$$

the result is

$$V_x(Z_{x+1} - A_x Z_x) + Z_{x+1} \Delta V_x = B_x \tag{7.7-30}$$

Since $Z_{x+1} - A_x Z_x = 0$ [Z_x is a solution of equation (7.7-21)], we simplify (7.7-30) to get

$$Z_{x+1} \Delta V_x = B_x \tag{7.7-31}$$

where

$$Z_x = \prod_{x=0}^{x-1} A_x \tag{7.7-32}$$

Thus from equation (7.7-31) we get the solution we seek for V_x:

$$V_x = \Delta^{-1} \frac{B_x}{Z_{x+1}} + C \tag{7.7-33}$$

Now we have a complete solution of equation (7.7-20) as given by (7.7-27):

$$U_x = Z_x V_x = CZ_x + Z_x \Delta^{-1} \frac{B_x}{Z_{x+1}} \tag{7.7-34}$$

where Z_x is given by equation (7.7-32), V_x is given by (7.7-33), and the integration constant is attached to the homogeneous solution Z_x. This is analogous to the procedure in differential equations. As an example of the above procedure, we solve the problem

$$U_{x+1} - 3U_x = 2^x \tag{7.7-35}$$

Here $A_x = 3$, $B_x = 2^x$, and

$$Z_x = \prod_{x=0}^{x-1} 3 = 3^x \tag{7.7-36}$$

From equation (7.7-33),

$$V_x = \Delta^{-1} \frac{2^x}{3^{x+1}} + C = \frac{1}{3} \Delta^{-1} \left(\frac{2}{3}\right)^x + C = -\left(\frac{2}{3}\right)^x + C \tag{7.7-37}$$

The final result is obtained from equation (7.7-34):

$$U_x = Z_x V_x = C3^x - 2^x \tag{7.7-38}$$

7.8. Solutions of Higher-Order Linear Difference Equations

A linear difference equation of higher order is given, in general, by

$$U_{x+n} + A_1 U_{x+n-1} + \cdots + A_n U_x = X(x) \tag{7.8-1}$$

where $A_1, A_2, \ldots, A_n$ are constants. The method of solution of equation (7.8-1) will be by the E operator technique, as introduced in Section 7.3. The E operator technique parallels the D operator solution of ordinary differential equations. Recalling from equation (7.3-7) that $E^n U_x = U_{x+n}$, we can convert equation (7.8-1) to an E operator form,

$$(E^n + A_1 E^{n-1} + \cdots + A_n)U_x = X(x) \tag{7.8-2}$$

or, symbolically,

$$f(E)U_x = X(x) \tag{7.8-3}$$

Equation (7.8-3) has a solution

$$U_x = U_{xc} + U_{xp} \tag{7.8-4}$$

where U_{xc} is the complementary solution ($X = 0$) and U_{xp} is the particular solution ($X \neq 0$). In ordinary differential equations the complementary solution is found by letting $y = e^{mx}$ in the homogeneous equation and then solving the auxiliary equation in m. We do a similar thing for finite difference equations. The homogeneous form of equation (7.8-1) is

$$U_{x+n} + A_1 U_{x+n-1} + \cdots + A_n U_x = 0 \tag{7.8-5}$$

Substituting $U_x = a^x$ into equation (7.8-5), we get

$$a^x(a^n + A_1 a^{n-1} + \cdots + A_n) = 0 \tag{7.8-6}$$

which yields the auxiliary equation

$$a^n + A_1 a^{n-1} + \cdots + A_n = 0 \tag{7.8-7}$$

Thus the complementary solution U_{xc} is given by

$$U_{xc} = C_1 a_1{}^x + C_2 a_2{}^x + \cdots + C_n a_n{}^x \tag{7.8-8}$$

where $a_1, a_2, \ldots, a_n$ are the roots of equation (7.8-7). This result is illustrated by the following examples.

(a) Find the solution of

$$U_{x+2} - 5U_{x+1} + 6U_x = 0 \tag{7.8-9}$$

The auxiliary equation is

$$a^2 - 5a + 6 = 0 \tag{7.8-10}$$

with the roots

$$a = 3 \qquad a = 3 \tag{7.8-11}$$

Thus, we get the solution to the homogeneous equation (7.8-9) as

$$U_x = C_1 3^x + C_2 2^x \tag{7.8-12}$$

If the auxiliary equation has multiple roots, for example, if $a_1 = a_2$, the solution is given by $(C_1 + C_2 x)a_1{}^x$. (Note the parallel between this result and the procedure followed in ordinary differential equations.) If there are k equal roots, $a_1 = a_2 = a_3 = \cdots = a_k$, then the solution must contain the terms $(C_1 + C_2 x + \cdots + C_k x^{k-1})a_1{}^x$. The next example illustrates this procedure.

(b) Solve

$$U_{x+3} - 3U_{x+1} - 2U_x = 0 \tag{7.8-13}$$

The auxiliary equation is

$$a^3 - 3a - 2 = 0 \qquad (7.8\text{-}14)$$

or

$$(a + 1)^2(a - 2) = 0 \qquad (7.8\text{-}15)$$

which gives

$$a = -1, -1, 2 \qquad (7.8\text{-}16)$$

The solution of equation (7.8-13) is then

$$U_x = (C_1 + C_2 x)(-1)^x + C_3 2^x \qquad (7.8\text{-}17)$$

If the roots of the auxiliary equation are not real, we can show that if the roots are $A + iB$, $A - iB$, the solution is[7]

$$U_x = r^x(C_1 \cos x\theta + C_2 \sin x\theta) \qquad (7.8\text{-}18)$$

where

$$r = +\sqrt{A^2 + B^2} \qquad (7.8\text{-}19)$$

and

$$\tan \theta = \frac{B}{A} \qquad (7.8\text{-}20)$$

If the complex roots are repeated, it can be shown that the equation

$$U_x = r^x[(C_1 + C_2 x) \cos x\theta + (C_3 + C_4 x) \sin x\theta] \qquad (7.8\text{-}21)$$

applies.[8] An example of complex roots is given below.
  (c) Solve

$$U_{x+2} + U_{x+1} + U_x = 0 \qquad (7.8\text{-}22)$$

The auxiliary equation is

$$a^2 + a + 1 = 0 \qquad (7.8\text{-}23)$$

which gives

$$a = -\frac{1}{2} + \frac{\sqrt{3}}{2} i, \quad -\frac{1}{2} - \frac{\sqrt{3}}{2} i \qquad (7.8\text{-}24)$$

---

[7] *Ibid.*, p. 109.
[8] *Ibid.*

Thus

$$A = -\frac{1}{2} \qquad B = \frac{\sqrt{3}}{2} \qquad r = \sqrt{A^2 + B^2} = 1$$

$$\tan \theta = \frac{B}{A} = -\sqrt{3} \qquad \theta = \frac{2\pi}{3} \qquad (7.8\text{-}25)$$

With equation (7.8-25) and (7.8-18), the solution of (7.8-22) is available. It is

$$U_x = 1^x \left(C_1 \cos \frac{2\pi}{3} x + C_2 \sin \frac{2\pi}{3} x \right) \qquad (7.8\text{-}26)$$

or

$$U_x = C_1 \cos \frac{2\pi}{3} x + C_2 \sin \frac{2\pi}{3} x \qquad (7.8\text{-}27)$$

Thus far only the homogeneous solution, U_{xc}, of the general equation (7.8-1) has been obtained. Now concentrating on the particular solution U_{xp}, we need to solve equation (7.8-3), which is rewritten here to indicate that it is U_{xp} that is being sought:

$$f(E)U_{xp} = X(x) \qquad (7.8\text{-}28)$$

Note the similarity of this procedure to that practiced in differential equations. When U_{xp} is obtained then by equation (7.8-4) we have the complete solution. Before proceeding to the solution of equation (7.8-28) by the E operator technique, a few useful theorems are introduced.[9]

Theorem I

If $f(E)$ is a polynomial in E, then

$$f(E)a^x = a^x f(a) \qquad (7.8\text{-}29)$$

where $f(E)$ is said to operate upon a^x. The functional form $f(a)$ indicates that wherever in the function f there is E, we replace E by a.

Theorem II

If $f(E)$ is a polynomial in E, and $F(x)$ is a function of x, then

$$f(E)a^x F(x) = a^x f(aE)F(x) \qquad (7.8\text{-}30)$$

[9] *Ibid.*, p. 111.

Theorem III

As a corollary to Theorem II,

$$(E - a)^n a^x F(x) = a^x a^n (E - 1)^n F(x) \tag{7.8-31}$$

Equation (7.8-28) requires that the particular solution be given by

$$U_{xp} = \frac{1}{f(E)} X \tag{7.8-32}$$

Equation (7.8-32) implies the justification for the following theorems which enable us to obtain U_{xp} quickly.

Theorem IV

$$\frac{1}{(E - a)^n} a^x = \frac{x^{(n)} a^{x-n}}{n!} \tag{7.8-33}$$

where the operator is $1/(E - a)^n$.

Theorem V

$$\frac{1}{f(E)} a^x = \frac{1}{f(a)} a^x \qquad f(a) \neq 0 \tag{7.8-34}$$

Theorem VI

$$\frac{1}{f(E)} a^x F(x) = a^x \frac{1}{f(aE)} F(x) \qquad f(aE) \neq 0 \tag{7.8-35}$$

Theorem VII

If $f(\Delta)$ is a polynomial in Δ, then

$$f(\Delta) a^x = a^x f(a - 1) \tag{7.8-36}$$

Theorem VIII

$$\frac{1}{f(\Delta)} a^x = \frac{a^x}{f(a - 1)} \qquad f(a - 1) \neq 0 \tag{7.8-37}$$

Some illustrative examples will show the utility of these theorems.
   (d) Solve

$$U_{x+2} - 5U_{x+1} + 6U_x = 5^x \tag{7.8-38}$$

Equation (7.8-38) can also be written as

$$(E^2 - 5E + 6)U_x = 5^x \qquad (7.8\text{-}39)$$

The auxiliary equation is

$$a^2 - 5a + 6 = 0 \qquad (7.8\text{-}40)$$

or

$$a = 3, 2 \qquad (7.8\text{-}41)$$

so that the complementary solution is

$$U_{xc} = C_1 3^x + C_2 2^x \qquad (7.8\text{-}42)$$

To find the particular solution, we write equation (7.8-39) as

$$U_{xp} = \frac{1}{f(E)} 5^x = \frac{1}{(E^2 - 5E + 6)} 5^x \qquad (7.8\text{-}43)$$

Invoking Theorem V, we get immediately from equation (7.8-43) the result

$$U_{xp} = \frac{1}{[5^2 - (5)(5) + 6]} 5^x = \tfrac{1}{6} 5^x \qquad (7.8\text{-}44)$$

Thus the complete solution is

$$U_x = U_{xc} + U_{xp} = C_1 3^x + C_2 2^x + \tfrac{1}{6} 5^x \qquad (7.8\text{-}45)$$

(e) Solve

$$(E^2 - 5E + 6)U_x = 3^x \qquad (7.8\text{-}46)$$

As in example (d),

$$U_{xc} = C_1 3^x + C_2 2^x \qquad (7.8\text{-}47)$$

We find U_{xp} by rewriting equation (7.8-46) as

$$U_{xp} = \frac{1}{(E^2 - 5E + 6)} 3^x = \frac{1}{f(E)} 3^x \qquad (7.8\text{-}48)$$

We cannot invoke Theorem V directly since $f(a) = 0$. Thus, we write equation (7.8-48) in an altered way, as

$$U_{xp} = \left(\frac{1}{E - 3}\right)\left(\frac{1}{E - 2} 3^x\right) \qquad (7.8\text{-}49)$$

Now we use Theorem V on the form $(1/E - 2)3^x$:

$$U_{xp} = \frac{1}{E-3}\left\{\frac{1}{E-2}3^x\right\} = \frac{1}{E-3}\left\{\left(\frac{1}{3-2}\right)3^x\right\} = \frac{1}{E-3}3^x \tag{7.8-50}$$

We now use Theorem IV, so that equation (7.8-50) becomes

$$U_{xp} = \frac{x3^{x-1}}{1!} = x3^{x-1} \tag{7.8-51}$$

Thus the complete solution is

$$U_x = U_{xc} + U_{xp} = C_1 3^x + C_2 2^x + x3^{x-1} \tag{7.8-52}$$

(f) Solve

$$U_{x+2} + U_{x+1} + U_x = x^2 + x + 1 \tag{7.8-53}$$

or

$$(E^2 + E + 1)U_x = x^2 + x + 1 \tag{7.8-54}$$

The auxiliary equation is

$$a^2 + a + 1 = 0 \tag{7.8-55}$$

which yields

$$a = -\frac{1}{2} + \frac{\sqrt{3}}{2}i, \quad -\frac{1}{2} - \frac{\sqrt{3}}{2}i \tag{7.8-56}$$

Hence

$$A = -\frac{1}{2} \qquad B = \frac{\sqrt{3}}{2} \qquad r = \sqrt{A^2 + B^2} = 1$$

$$\tan\theta = \frac{B}{A} = -\sqrt{3} \qquad \theta = \frac{2\pi}{3} \tag{7.8-57}$$

and the complementary solution is

$$U_{xc} = C_1 \cos\frac{2\pi}{3}x + C_2 \sin\frac{2\pi}{3}x \tag{7.8-58}$$

In solving for the particular solution, U_{xp}, it is better to convert equation (7.8-54) to the Δ difference form by substituting in equation (7.3-3) to yield

$$(3 + 3\Delta + \Delta^2)U_{xp} = x^2 + x + 1 \tag{7.8-59}$$

Thus, the particular solution is from equation (7.8-59)

$$U_{xp} = \frac{1}{3 + 3\Delta + \Delta^2}(x^2 + x + 1) = \frac{1}{f(\Delta)}(x^2 + x + 1) \tag{7.8-60}$$

The term $1/(3 + 3\varDelta + \varDelta^2)$ can be written in an ascending series in $\varDelta$,

$$\frac{1}{3 + 3\varDelta + \varDelta^2} = \tfrac{1}{3} - \tfrac{1}{3}\varDelta + \tfrac{2}{9}\varDelta^2 - \cdots \tag{7.8-61}$$

which, substituted into equation (7.8-60), yields the result

$$\frac{1}{3 + 3\varDelta + \varDelta^2}\, (x^2 + x + 1) = \tfrac{1}{3}[1 - \varDelta + \tfrac{2}{3}\varDelta^2 - \cdots](x^2 + x + 1) \tag{7.8-62}$$

If we now convert $(x^2 + x + 1)$ to $(x^{(2)} + 2x^{(1)} + 1)$ by the method of Section 7.2, the final result is

$$\begin{aligned}
U_{xp} &= \tfrac{1}{3}[1 - \varDelta + \tfrac{2}{3}\varDelta^2 - \cdots](x^{(2)} + 2x^{(1)} + 1) \\
&= \tfrac{1}{3}[x^{(2)} + 2x^{(1)} + 1 - 2x^{(1)} - 2 + \tfrac{4}{3}] \\
&= \tfrac{1}{3}[x(x - 1) + \tfrac{1}{3}] = \frac{x^2}{3} - \frac{x}{3} + \frac{1}{9} \tag{7.8-63}
\end{aligned}$$

(g) Solve

$$U_{x+2} - 7U_{x+1} - 8U_x = x^{(2)}2^x \tag{7.8-64}$$

or

$$(E^2 - 7E - 8)U_x = x^{(2)}2^x \tag{7.8-65}$$

The auxiliary equation is

$$a^2 - 7a - 8 = 0 \tag{7.8-66}$$

which gives

$$a = -1, 8 \tag{7.8-67}$$

and the complementary solution

$$U_{xc} = C_1(-1)^x + C_2(8)^x \tag{7.8-68}$$

To find the particular solution, we proceed as usual from equation (7.8-65)

$$U_{xp} = \frac{1}{E^2 - 7E - 8}\, x^{(2)}2^x = \frac{1}{f(E)}\, x^{(2)}2^x \tag{7.8-69}$$

By employing Theorem VI in equation (7.8-69), with $F(x) = x^{(2)}$ and $a = 2$, we can write immediately

$$U_{xp} = 2^x\, \frac{1}{4E^2 - 14E - 8}\, x^{(2)} \tag{7.8-70}$$

which we solve by a method similar to the previous problem. This problem could also be solved by the substitution

$$U_x = 2^x V_x \tag{7.8-71}$$

which, introduced into (7.8-63), yields the result

$$4V_{x+2} - 14V_{x+1} - 8V_x = x^{(2)} \tag{7.8-72}$$

or

$$(4E^2 - 14E - 8)V_x = x^{(2)} \tag{7.8-73}$$

We can now solve equation (7.8-73) by the techniques given previously with

$$V_x = V_{xc} + V_{xp} \tag{7.8-74}$$

and

$$U_x = 2^x V_x \tag{7.8-75}$$

7.9. Simultaneous Difference Equations

Given simultaneous difference equations, such as

$$U_{x+1} - V_x = 0 \tag{7.9-1}$$

and

$$V_{x+1} - U_x = 0 \tag{7.9-2}$$

we alter equation (7.9-1) by raising the x index,

$$U_{x+2} - V_{x+1} = 0 \tag{7.9-3}$$

Substituting equation (7.9-3) into (7.9-2) yields

$$U_{x+2} - U_x = 0 \tag{7.9-4}$$

which has the auxiliary equation

$$a^2 - 1 = 0 \tag{7.9-5}$$

or

$$a = 1, -1 \tag{7.9-6}$$

and the solution

$$U_x = C_1 + C_2(-1)^x \tag{7.9-7}$$

Substituting equation (7.9-7) into (7.9-2), we get

$$V_{x+1} = C_1 + C_2(-1)^x \qquad (7.9-8)$$

or

$$V_x = C_1 + C_2(-1)^{x-1} = C_1 - C_2(-1)^x \qquad (7.9-9)$$

7.10. Nonlinear Difference Equations of Higher Order

In considering first-order equations we have generally been able to solve equations of the form

$$p_x U_{x+1} + q_x U_x = r_x \qquad (7.10-1)$$

We have also been able to solve linear equations of higher order with constant coefficients. If the coefficients are not constant, then, in general, there are few general methods of solution. Certain special forms, however, may be solved. For example,

$$p_x U_x U_{x+1} + q_x U_{x+1} + r_x U_x = 0 \qquad (7.10-2)$$

is one of these special forms. Dividing equation (7.10-2) by $U_x U_{x+1}$ we get

$$\frac{r_x}{U_{x+1}} + \frac{q_x}{U_x} + p_x = 0 \qquad (7.10-3)$$

By the substitution $V_x = 1/U_x$ in equation (7.10-3), a linear equation of first order results,

$$r_x V_{x+1} + q_x V_x + p_x = 0 \qquad (7.10-4)$$

which can be solved.

An equation of the form

$$U_{x+1}^2 - 5U_{x+1}U_x + 6U_x^2 = 0 \qquad (7.10-5)$$

can be solved if we use the substitution

$$U_{x+1} = Z_x U_x \qquad (7.10-6)$$

in equation (7.10-5). The result is

$$Z_x^2 - 5Z_x + 6 = 0 \qquad (7.10-7)$$

or

$$Z_x = 3, 2 \qquad (7.10-8)$$

Thus equation (7.10-6) becomes

$$U_{x+1} = 3U_x \qquad (7.10\text{-}9)$$

or

$$U_{x+1} = 2U_x \qquad (7.10\text{-}10)$$

The solutions of equations (7.10-9) and (7.10-10) are, respectively,

$$U_x = C_1 3^x \qquad (7.10\text{-}11)$$

$$U_x = C_2 2^x \qquad (7.10\text{-}12)$$

In general, an equation homogeneous in U_x which has the general form

$$f\left(\frac{U_{x+1}}{U_x}, x\right) = 0 \qquad (7.10\text{-}13)$$

can be reduced to a linear equation by substitution of equation (7.10-6) into (7.10-13).[10]

No general method has yet been found to solve

$$Z_{x+2} + A_x Z_{x+1} + B_x Z_x = C_x \qquad (7.10\text{-}14)$$

where A_x, B_x, and C_x are functions of x. However, if the homogeneous portion, with $C_x = 0$, can be solved, then we can use the substitution $Z_x = U_x V_x$, where V_x is the solution of equation (7.10-14) with $C_x = 0$.[11]

Assignments in Chapter 7

7.1. Verify equations (7.2-2), (7.2-5), (7.4-3), (7.4-7), (7.6-10), (7.6-17), (7.6-21), (7.7-2), (7.7-26), (7.7-37), (7.8-17), (7.8-27), (7.8-62), and (7.8-70).

For Further Reading

The Calculus of Finite Differences, by C. H. Richardson, Van Nostrand, New York, 1954.

[10] *Ibid.*, p. 126.
[11] *Ibid.*, p. 127.

Chapter 8

Transport Analysis in Cascaded Systems

In this chapter we illustrate the finite difference approach of Chapter 7 in stagewise transport operations such as extraction, absorption, stirred tank reactors, and distillation columns. Startup and unsteady-state operations are included. Finite difference techniques are applied with the emphasis on closed form analytical solutions.

8.1. Extraction

In mass transfer operations such as extraction, many times it is desirable to improve the single contact process by "cascading" or introducing a multiflow operation, as shown in Figure 79. Applying equation (5.1-8) (material balance) around stage n gives

$$(Rx_{n-1} + Sy_{n+1})|_{\text{in}} - (Rx_n + Sy_n)|_{\text{out}} + 0 = 0 \qquad (8.1-1)$$

where

R, S = flow rates of the two phases (units consistent with x_n and y_n)

x_n = mole fraction of A in one phase, coming from stage n

y_n = mole fraction of A in the other phase, coming from stage n

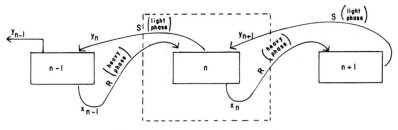

Fig. 79. Cascaded extraction system.

In this type of contact, there usually is a distribution coefficient m introduced to define the relationship between the concentrations of component A in the two phases:

$$y_n = mx_n \qquad (8.1\text{-}2)$$

Substitution of equation (8.1-2) into (8.1-1) yields

$$x_{n+1} - (\alpha + 1)x_n + \alpha x_{n-1} = 0 \qquad (8.1\text{-}3)$$

where $\alpha = R/mS$. Equation (8.1-3) is a second-order linear finite difference equation, whose solution by the methods of Chapter 7 is given by

$$y_n = mx_n = m(C_1 \alpha^n + C_2) \qquad (8.1\text{-}4)$$

where C_1 and C_2 are arbitrary constants to be determined.

8.2. Stripping Columns

Suppose we have a column containing contacting plates as shown in Figure 80. There are G_{N+1} moles per hour of a gas entering at the bottom of the column, where a solute is stripped from the gas by absorption in an oil flow of L_0 moles per hour. There is an equilibrium relationship

$$y_m = K_m x_m \qquad (8.2\text{-}1)$$

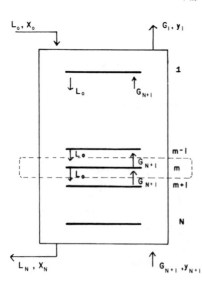

Fig. 80. Stripping column.

where K_m is the equilibrium constant and y_m and x_m are the mole fractions of the solute in the gas and liquid phases, respectively, on plate m. Usually the gas and liquid flow, G_{N+1} and L_0, are expressed as moles per hour of pure gas and pure liquid solvent. This makes it useful to define mole ratio units Y and X as the mole ratio in the gas and liquid phases respectively (moles of solute/mole of pure gas or liquid). From Figure 80 we see that a material balance over plate m gives

$$L_0(X_m - X_{m-1}) = G_{N+1}(Y_{m+1} - Y_m) \qquad (8.2\text{-}2)$$

From equations (8.2-1) and (8.2-2) we can get[1]

$$Y_m = \frac{Y_{m+1} + (L_0/K_m G_{N+1})Y_{m-1}}{1 + (L_0/K_m G_{N+1})} \qquad (8.2\text{-}3)$$

where the mole fractions x_m and y_m were converted to mole ratios X_m and Y_m by the relationship

$$y = \frac{Y}{Y+1} \qquad x = \frac{X}{X+1} \qquad (8.2\text{-}4)$$

Equation (8.2-4) is easily confirmed:

$$y = \frac{\alpha}{\alpha + \beta} = \frac{\alpha/\beta}{\alpha/\beta + 1} = \frac{Y}{Y+1}$$

where $Y = \alpha/\beta$. If we define

$$A = \frac{L_0}{K_m G_{N+1}} \qquad (8.2\text{-}5)$$

then equation (8.2-3) can be simplified to the governing finite difference equation

$$Y_{m+1} - (1 + A)Y_m + A Y_{m-1} = 0 \qquad (8.2\text{-}6)$$

Again, by the methods of Chapter 7, the solution of equation (8.2-6) is given by

$$Y_m = C_1 + C_2 A^m \qquad (8.2\text{-}7)$$

where C_1 and C_2 are arbitrary constants. By letting m become $N + 1$, 0, and 1, we can eliminate C_1 and C_2 in equation (8.2-7) to arrive at the

---

[1] V. G. Jenson and G. V. Jeffreys, *Mathematical Methods in Chemical Engineering*, Academic Press, New York (1963), p. 325.

Kremser–Brown equation[2]

$$\frac{Y_{N+1} - Y_1}{Y_{N+1} - Y_0} = \frac{A^{N+1} - A}{A^{N+1} - 1}$$

(8.2-8)

Note that $A^m \big|_{m=0} = A^0 = 1$.

8.3. Unsteady-State Finite Difference Analysis

The previous analysis was for steady-state situations. Consider now the startup and shutdown phases of the operation, or some upset that may be superimposed on the steady-state operation. Suppose we have N well stirred tanks arranged in a cascade as shown in Figure 81. Consider as a simple case a system where initially each tank contains V ft^3 of pure water. Then saltwater is fed to the first tank. The problem is to predict the output concentration from the last tank as a function of time. As usual, a material balance from equation (5.1-8) around tank n is the first step, yielding

$$RX_{n-1} - RX_n = \frac{d(VX_n)}{dt}$$

(8.3-1)

with the initial condition $X_n = 0$ at $t = 0$. Equation (8.3-1) readily lends itself to solution by a Laplace transform technique as described in Chapter 2. Taking the Laplace transform $\mathcal{L}_{t \to s}$ of equation (8.3-1), we get

$$R\bar{X}_{n-1}(s) - R\bar{X}_n(s) = V[s\bar{X}_n(s) - \overset{0}{\cancel{X_n(0)}}]$$

(8.3-2)

where $\mathcal{L}_{t \to s}X_n(t) = \bar{X}_n(s)$. Equation (8.3-2) is a linear first-order finite difference equation, which is solved by the methods of Chapter 7, giving

$$\bar{X}_n = A\left(\frac{R}{R + Vs}\right)^n$$

(8.3-3)

where A is an arbitrary constant. A boundary condition which can be invoked in order to evaluate the arbitrary constant is

$$X_n(t) = X_0 \quad \text{for } n = 0 \quad \text{(feed concentration)}$$

(8.3-4)

and

$$\mathcal{L}_{t \to s}\{X_n(t) = X_0\} = \left\{\bar{X}_n(s) = \frac{X_0}{s}\right\} \quad \text{for } n = 0$$

(8.3-5)

[2] *Ibid.*, p. 326.

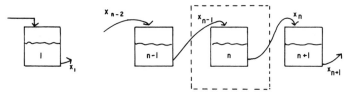

Fig. 81. Cascaded system in unsteady-state operation.

Combining equation (8.3-5) with (8.3-3), we evaluate A and get

$$\bar{X}_n = \frac{X_0}{s}\left(\frac{R}{R + Vs}\right)^n \tag{8.3-6}$$

Since the concentration at stage N is desired, we get this from equation (8.3-6) with $n = N$,

$$\bar{X}_N = \frac{X_0}{s}\left(\frac{R}{R + Vs}\right)^N \tag{8.3-7}$$

which can be inverted, $\mathscr{L}^{-1}_{s \to t}\bar{X}_N(s) = X_N(t)$, by resorting to the inversion tables to yield, finally,[3]

$$X_N = X_0 - X_0 e^{-Rt/V}\left[1 + \frac{Rt}{V} + \frac{(Rt/V)^2}{2!} + \cdots + \frac{(Rt/V)^{N-1}}{(N-1)!}\right] \tag{8.3-8}$$

Another situation which incorporates the previous ideas and in addition adds a generation or depletion term in the material balance is sketched in Figure 82. From equation (5.1-8) we get the following governing equation for a material balance around tank m, where a chemical reaction $r_A = kC_m$ is consuming the component of interest:

$$qC_{m-1} - qC_m - kC_m V = \frac{d(VC_m)}{dt} \tag{8.3-9}$$

with

$$q = \text{flow rate} = \text{constant}$$

$$V = \text{tank volume} = \text{constant}$$

$$kC_m V = r_A = \text{reaction term (depletion)}$$

$$C_m = \text{concentration in tank } m$$

An example of this situation is acetic anhydride being hydrolyzed in a stirred tank cascade.[4] Suppose initially 0.21 gram mole of acetic anhydride

[3] *Ibid.*, p. 339.
[4] *Ibid.*, p. 341.

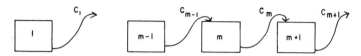

Fig. 82. Unsteady-state cascade with chemical reaction.

per liter is charged into each vessel. Then the feed stream of acetic anhydride to the first tank is begun. Dividing equation (8.3-9) by q and using a definition of residence time, $\theta = V/q = $ constant, we get

$$C_{m-1} - C_m - k\theta C_m = \theta \frac{dC_m}{dt} \tag{8.3-10}$$

which can be rearranged to give

$$\frac{1}{\theta} C_{m-1} - QC_m = \frac{dC_m}{dt} \tag{8.3-11}$$

where

$$Q = \frac{1 + k\theta}{\theta} = \text{constant} \tag{8.3-12}$$

Equation (8.3-11) is a linear differential difference equation which is amenable to a Laplace transform solution technique. Taking the Laplace transform of equation (8.3-11), $\mathscr{L}_{t \to s}$, we get

$$\frac{1}{\theta} \bar{C}_{m-1} - Q\bar{C}_m = s\bar{C}_m - C_m(0) \tag{8.3-13}$$

where $C_m(0)$ is the initial value of concentration, $C_m|_{t=0}$. Since in this case all the vessels are initially of the same composition, $C_m(0) = 0.21$. Rearrangement of equation (8.3-13) yields

$$\theta(s + Q)\bar{C}_m - \bar{C}_{m-1} = 0.21\theta \tag{8.3-14}$$

which can be solved by the methods of Chapter 7 to yield

$$\bar{C}_m = A[\theta(s + Q)]^{-m} + \frac{0.21\theta}{\theta(s + Q) - 1} \tag{8.3-15}$$

where A is an arbitrary constant. If the feed concentration is 0.137 gram mole of acetic anhydride per liter, the boundary condition for equation (8.3-15) is given by

$$C_m(t) = 0.137 \qquad \text{for } m = 0 \tag{8.3-16}$$

or

$$\bar{C}_m(s) = \frac{0.137}{s} \qquad \text{for } m = 0 \qquad (8.3\text{-}17)$$

where the Laplace transform of (8.3-16) has resulted in (8.3-17). With equation (8.3-17) substituted into (8.3-15), we can evaluate A and thus equation (8.3-15) yields

$$\bar{C}_m(s) = \frac{0.137}{s\theta^m(s+Q)^m} + \frac{0.210}{\theta(s+Q)-1}\left[1 - \frac{1}{\theta^m(s+Q)^m}\right] \qquad (8.3\text{-}18)$$

Equation (8.3-18) can be inverted, $\mathscr{L}^{-1}_{s \to t}$, to give the final solution,[5]

$$C_m(t) = \frac{0.137}{(\theta Q)^m} - \frac{0.137}{(\theta Q)^m}\left[1 + Qt + \frac{(Qt)^2}{2!} + \cdots + \frac{(Qt)^{m-1}}{(m-1)!}\right]e^{-Qt}$$

$$+ 0.21\left[1 + \frac{t}{\theta} + \frac{1}{2!}\left(\frac{t}{\theta}\right)^2 + \cdots + \frac{(t/\theta)^{m-1}}{(m-1)!}\right]e^{-Qt} \qquad (8.3\text{-}19)$$

Thus from equation (8.3-19) it is now possible to calculate the concentration of the effluent from a cascaded system starting with equal concentrations in each tank.

8.4. Cascaded Chemical Reactors at Steady State[6]

For example, suppose it is proposed to feed continuously 10,000 lb/hr of pure liquid A into the first of two identical stirred tank reactors operating in series as shown in Figure 83. Suppose the kinetics are given by

$$A \xrightarrow{k_1} B \xrightarrow{k_2} D \qquad (8.4\text{-}1)$$

A material balance for component A around tank n, from equation (5.1-8), is given by

$$C_{A,n-1} - C_{A,n} - k_1 C_{A,n}\theta = 0 \qquad (8.4\text{-}2)$$

where the rate of depletion of component A is $r_A = k_1 C_{A,n} V$, θ is the average residence time, $\theta = V/q$, q is the flow rate, and V is the volume of the tank. Equation (8.4-2) is solved by the methods of Chapter 7, and the result is

$$C_{A,n} = K_1 \varrho_1^n \qquad (8.4\text{-}3)$$

[5] Ibid., p. 342.
[6] Ibid., p. 327.

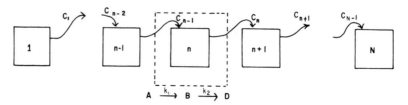

Fig. 83. Cascaded chemical reactors in series.

where

$$\varrho_1 = \frac{1}{1 + \alpha}$$

$$\alpha = k_1\theta$$

$$K_1 = \text{arbitrary constant}$$

For component B, a material balance around tank n yields

$$C_{B,n-1} - C_{B,n} = k_2 C_{B,n}\theta - k_1 C_{A,n}\theta \tag{8.4-4}$$

where the rate of depletion of B is $r_B = -k_2 C_{B,n} + k_1 C_{A,n}$. Substituting equation (8.4-3) into (8.4-4), and with $k_2\theta = \beta$, we can solve for $C_{B,n}$:

$$C_{B,n} = K_2 \varrho_2{}^n + \frac{\alpha K_1}{(\beta - \alpha)/(1 + \alpha)(1 + \beta)} \varrho_1^{n-1} \tag{8.4-5}$$

where K_2 is an arbitrary constant and $\varrho_2 = 1/(1 + \beta)$. The boundary conditions for this steady-state transport problem are given by

$$C_{A,n} = C_{A,0} \text{ (pure A)} \qquad \text{at } n = 0 \tag{8.4-6}$$

$$C_{B,n} = C_{B,0} = 0 \qquad \text{at } n = 0 \tag{8.4-7}$$

Application of (8.4-6) to equation (8.4-3) allows K_1 to be evaluated:

$$K_1 = C_{A,0} \tag{8.4-8}$$

Using equation (8.4-7) in (8.4-5), we find

$$K_2 = \frac{\alpha}{\alpha - \beta} C_{A,0} \tag{8.4-9}$$

where $\varrho_1{}^n \big|_{n=0} = \varrho_1{}^0 = 1$. With K_1 and K_2 evaluated, equations (8.4-3) and (8.4-5) finally become

$$C_{A,n} = C_{A,0}\varrho_1{}^n \tag{8.4-10}$$

$$C_{B,n} = \frac{\alpha C_{A,0}}{\beta - \alpha} (\varrho_1{}^n - \varrho_2{}^n) \tag{8.4-11}$$

Suppose now that it is necessary to find the optimal residence time, θ, which will maximize the amount of B coming from this cascade. One simple way of doing this is to differentiate equation (8.4-11) and then set the result equal to zero:

$$\frac{dC_{B,n}}{d\theta} = \frac{\alpha C_{A,0}}{\beta - \alpha} (\varrho_1{}^n \ln \varrho_1 - \varrho_2{}^n \ln \varrho_2) = 0 \tag{8.4-12}$$

or

$$\frac{\varrho_1{}^n}{\varrho_2{}^n} = \frac{\ln \varrho_2}{\ln \varrho_1} \tag{8.4-13}$$

Suppose that $k_1 = 0.1$ min^{-1}, $k_2 = 0.05$ min^{-1}, fluid density $= 60$ lb/ft^3, and $n = 2$. Then equation (8.4-13) becomes

$$\frac{(1 + 0.05\theta)^2}{(1 + 0.10\theta)^2} = \frac{\ln(1 + 0.05\theta)}{\ln(1 + 0.10\theta)} \tag{8.4-14}$$

which yields approximately $\theta = 0.456$ min. Since the mean residence time, θ, is

$$\theta = \frac{V}{q} = \frac{\text{reactor volume, ft}^3}{\text{volumetric feed rate, ft}^3/\text{min}}$$

we get for this problem,

$$\theta = \frac{V, \text{ ft}^3}{10{,}000 \text{ lbm/hr} \times (1 \text{ hr}/60 \text{ min}) \times (1 \text{ ft}^3/10 \text{ lbm})} = 0.456$$

or finally

$$V = \text{optimal volume of each reactor} = 1.27 \text{ ft}^3$$

8.5. Stirred Tank Reactors with Complex Chemical Reactions (Unsteady State)

Suppose N reactors initially contain pure solvent. Thereafter a feed rate q ft^3/min of a reactant A of concentration C_0 lb moles/ft^3 is fed to the first tank. Suppose further that the kinetics are given by

$$A \underset{k_1{}'}{\overset{k_1}{\rightleftharpoons}} B \underset{k_2{}'}{\overset{k_2}{\rightleftharpoons}} C \tag{8.5-1}$$

The governing equations for components A, B, and C are again obtained from equation (5.1-8) and, for material balances around tank n, are

$$qC_{A,n-1} - qC_{A,n} - Vk_1C_{A,n} + Vk_1'C_{B,n} = V\frac{dC_{A,n}}{dt} \qquad (8.5\text{-}2)$$

$$qC_{B,n-1} - qC_{B,n} - Vk_1'C_{B,n} - Vk_2C_{B,n} + Vk_1C_{A,n} + Vk_2'C_{C,n} = V\frac{dC_{B,n}}{dt}$$
$$(8.5\text{-}3)$$

$$qC_{C,n-1} - qC_{C,n} - Vk_2'C_{C,n} + Vk_2C_{B,n} = V\frac{dC_{C,n}}{dt} \qquad (8.5\text{-}4)$$

Defining a residence time as usual, $\theta = V/q =$ (volume of tank)/(flow rate), and rearranging equations (8.5-2), (8.5-3), and (8.5-4), we get the set of equations

$$\frac{dC_{A,n}}{dt} = \frac{C_{A,n-1}}{\theta} - \left(\frac{1}{\theta} + k_1\right)C_{A,n} + k_1'C_{B,n} \qquad (8.5\text{-}5)$$

$$\frac{dC_{B,n}}{dt} = \frac{C_{B,n-1}}{\theta} - \left(\frac{1}{\theta} + k_2 + k_1'\right)C_{B,n} + k_1C_{A,n} + k_2'C_{C,n} \qquad (8.5\text{-}6)$$

$$\frac{dC_{C,n}}{dt} = \frac{C_{C,n-1}}{\theta} - \left(\frac{1}{\theta} + k_2'\right)C_{C,n} + k_2C_{B,n} \qquad (8.5\text{-}7)$$

These equations may be solved by the Laplace transform technique shown in Section 8.3, or they may be solved by matrix algebra analysis. By the methods of Chapter 1, equations (8.5-5), (8.5-6), and 8.5-7 can be represented by the matrix equation[7]

$$\frac{d}{dt}\begin{bmatrix} C_{A,n} \\ C_{B,n} \\ C_{C,n} \end{bmatrix}$$
$$= \frac{1}{\theta}\begin{bmatrix} C_{A,n-1} \\ C_{B,n-1} \\ C_{C,n-1} \end{bmatrix} - \begin{bmatrix} \frac{1}{\theta} + k_1 & -k_1' & 0 \\ -k_1 & \left(\frac{1}{\theta} + k_2 + k_1'\right) & -k_2 \\ 0 & -k_2 & \left(\frac{1}{\theta} + k_2'\right) \end{bmatrix}\begin{bmatrix} C_{A,n} \\ C_{B,n} \\ C_{C,n} \end{bmatrix}$$
$$(8.5\text{-}8)$$

or

$$\frac{d\bar{\bar{C}}_n}{dt} = \frac{\bar{\bar{C}}_{n-1}}{\theta} - \bar{\bar{G}}\bar{\bar{C}}_n \qquad (8.5\text{-}9)$$

[7] *Ibid.*, p. 469.

where $\bar{C}_n$ and $\bar{C}_{n-1}$ are 3×1 matrices and $\bar{\bar{G}}$ is a 3×3 matrix. Equation (8.5-9) applies to each reaction vessel ($n = 1, 2, 3, \ldots, N$). Thus for the first reactor, $n = 1$, we get from (8.5-9)

$$\frac{d\bar{C}_1}{dt} = \frac{\bar{C}_0}{\theta} - \bar{\bar{G}}\bar{C}_1 \tag{8.5-10}$$

The solution of equation (8.5-10) is[8]

$$\bar{C}_1 = \frac{\bar{\bar{G}}^{-1}\bar{C}_0}{\theta} + e^{-\bar{\bar{G}}t}K_1 \tag{8.5-11}$$

where K_1 is an integration constant. Initially at $t = 0$, $C_A = C_B = C_C = 0$. Hence

$$\bar{C}_1 = \begin{bmatrix} C_{A,1} \\ C_{B,1} \\ C_{C,1} \end{bmatrix} = \begin{bmatrix} 0 \\ 0 \\ 0 \end{bmatrix} = 0$$

Substituting this initial condition into equation (8.5-11), we evaluate K_1:

$$K_1 = -\frac{\bar{\bar{G}}^{-1}\bar{C}_0}{\theta} \tag{8.5-12}$$

Thus equation (8.5-11) becomes

$$\bar{C}_1 = \frac{1}{\theta}(\bar{I} - e^{-\bar{\bar{G}}t})\bar{\bar{G}}^{-1}\bar{C}_0 \tag{8.5-13}$$

where $\bar{I}$ is the identity matrix. By letting $n = 2$ in equation (8.5-9) and substituting (8.5-13) into it, we reach the result by the same method used in getting $\bar{C}_1$. The solution for $\bar{C}_2$ is

$$\bar{C}_2 = \frac{\bar{\bar{G}}^{-2}\bar{C}_0}{\theta^2} - \frac{\bar{\bar{G}}^{-1}\bar{C}_0}{\theta^2} te^{-\bar{\bar{G}}t} + e^{-\bar{\bar{G}}t}K_2 \tag{8.5-14}$$

Applying the initial condition $\bar{C}_2 = 0$ at $t = 0$ allows the evaluation of K_2:

$$K_2 = -\frac{\bar{\bar{G}}^{-2}\bar{C}_0}{\theta^2} \tag{8.5-15}$$

and hence

$$\bar{C}_2 = \frac{1}{\theta^2}(\bar{I} - e^{-\bar{\bar{G}}t} - \bar{\bar{G}}te^{-\bar{\bar{G}}t})\bar{\bar{G}}^{-2}\bar{C}_0 \tag{8.5-16}$$

---

[8] *Ibid.*, p. 470.

By similar techniques[9] we can get $\bar{C}_3$ and then in general, $\bar{C}_N$:

$$\bar{C}_N = \frac{e^{-\bar{G}t}}{\theta^N}\left[\bar{I} - \frac{\bar{G}t}{1!} + \frac{(\bar{G}t)^2}{2!} + \cdots + \frac{(\bar{G}t)^{N-1}}{(N-1)!}\right]\bar{G}^{-N}\bar{C}_0 \quad (8.5\text{-}17)$$

8.6. Distillation Columns (Nonlinear Difference Equations)

The finite difference equations given in the previous sections were examples of linear difference equations. In the analysis of distillation column performance nonlinear difference equations arise from the mass balances. For example, suppose a benzene–toluene feed is introduced continuously to a distillation column as sketched in Figure 84. The feed enters the column at its boiling point, and the relative volatility, α, of benzene to toluene is 2.3, where relative volatility is defined by

$$\alpha = \frac{y_n/x_n}{(1 - y_n)/(1 - x_n)} \quad (8.6\text{-}1)$$

The terms y_n and x_n are mole fractions in the vapor and liquid phases (leaving plate n.) The distillation column also operates at a reflux ratio of 3.0, where reflux ratio is defined by

$$R = \text{reflux ratio} = \frac{\bar{L}}{D} \quad (8.6\text{-}2)$$

where $\bar{L}$ is the molar liquid downflow in the column (above the feed plate), and D is the product taken off the top of the column. A material balance around the entire column yields

$$F = D + W = 100 \quad \text{(basis: 100 lb moles of feed)} \quad (8.6\text{-}3)$$

A material balance on benzene around the entire column gives

$$Fx_F = Dx_D + Wx_W \quad (8.6\text{-}4)$$

where x_F, x_D, and x_W are mole fractions of benzene in the feed, product, and bottoms streams. From equation (8.6-4) we get

$$60 = 0.98D + 0.02W \quad (8.6\text{-}5)$$

Solving equations (8.6-3) and (8.6-5) simultaneously yields

$$D = 60.4 \qquad W = 39.6 \quad (8.6\text{-}6)$$

---

[9] *Ibid.*, p. 471.

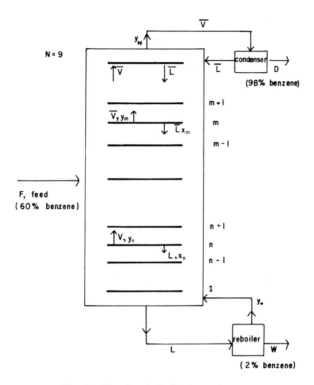

Fig. 84. Sketch of distillation column.

Another steady-state material balance between the bottom of the column and plate n in the lower (or stripping) section results in

$$\bar{L}x_{n+1} = \bar{V}y_n + Wx_W \qquad (8.6\text{-}7)$$

where $\bar{L}$ is the molar liquid flow down the column (below the feed plate), and $\bar{V}$ the molar vapor flow up the column (below the feed plate). Figure 85 shows the relationship between the flows in the sections above and below the feed plate when the feed is saturated liquid. From equation (8.6-1),

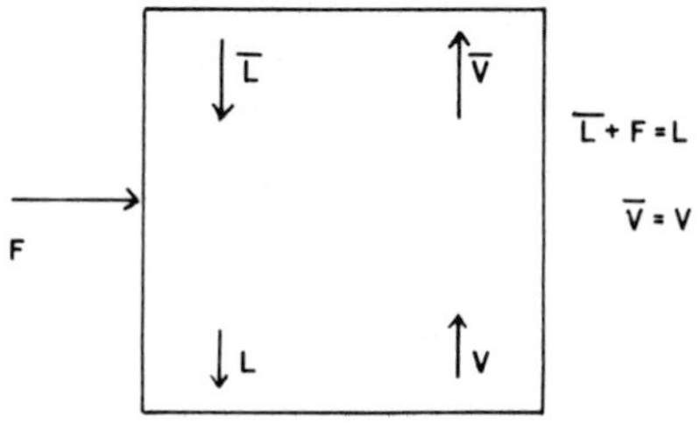

Fig. 85. Flow in the distillation column when the feed is a saturated liquid.

we can solve for y_n,

$$y_n = \frac{\alpha x_n}{1 + (\alpha - 1)x_n} \qquad (8.6\text{-}8)$$

which can be substituted into equation (8.6-7) and, with some rearrangement, yields[10]

$$x_{n+1}x_n + \frac{x_{n+1}}{\alpha - 1} - \left[\frac{\alpha V + (\alpha - 1)Wx_W}{L(\alpha - 1)}\right]x_n - \frac{Wx_W}{L(\alpha - 1)} = 0 \quad (8.6\text{-}9)$$

From equation (8.6-2), with $R = 3.0$ and a material balance around the feed plate, $F + \bar{L} = L$, we can get $L = F + RD = 100 + (3.0)(60.4) = 281$ lb moles/hr. Defining

$$A = \frac{1}{\alpha - 1} = \frac{1}{(2.3 - 1)} = 0.769$$

$$B = \frac{\alpha V + (\alpha - 1)Wx_W}{L(\alpha - 1)} = 1.523$$

$$C = \frac{Wx_W}{L(\alpha - 1)} = 0.0022$$

we can condense equation (8.6-9) to

$$x_{n+1}x_n + Ax_{n+1} - Bx_n - C = 0 \qquad (8.6\text{-}10)$$

Equation (8.6-10) is called the Riccati equation. From the methods of Chapter 7, we can get the solution[11]

$$\frac{1}{x_n - \delta} = K\left[\frac{A + \delta}{B - \delta}\right]^n - \frac{1}{A - B + 2\delta} \qquad (8.6\text{-}11)$$

where δ is the solution of

$$\delta^2 + (A - B)\delta - C = 0 \qquad (8.6\text{-}12)$$

and K is an arbitrary constant. The constant K is evaluated by noting that at $n = 0$, $x_n = 0.02$, which yields $\delta = -0.003$ from equation (8.6-12). With this value for δ in (8.6-11), we can get $K = 42.2$. For $x_n = 0.6$ (feed plate) we can calculate that $n = 6$ from equation (8.6-11). If there are 9

[10] *Ibid.*, p. 337.
[11] *Ibid.*, p. 338.

actual plates from the bottom of the column to the feed plate, then we can refer to an efficiency of $6/9 \times 100 \cong 67\%$.

If instead of a material balance around the bottom of the column [equation (8.6-7)], a material balance is made about the top of the column including plate m, the result is

$$\bar{V}y_{m-1} - \bar{L}x_m - Dx_D = 0 \tag{8.6-13}$$

where $\bar{V} = \bar{L} + D$. The y_{m-1} term can be eliminated by using equation (8.6-1), substituting the subscript m for n. The result is equation 8.6-14

$$x_m x_{m-1} + ax_m + bx_{m-1} + c = 0 \tag{8.6-14}$$

where

$$a = \frac{1}{\alpha - 1} \tag{8.6-15}$$

$$b = \frac{Dx_D(\alpha - 1) - \alpha\bar{V}}{\bar{L}(\alpha - 1)} \tag{8.6-16}$$

$$c = \frac{Dx_D}{\bar{L}(\alpha - 1)} \tag{8.6-17}$$

Equation (8.6-14) is again a Riccati equation which is solved by the methods of Chapter 7. This solution is given by[12]

$$x_m = \delta + \frac{1}{K\left(-\dfrac{a + \delta}{b + \delta}\right)^m - \dfrac{1}{(a + \delta) + (b + \delta)}} \tag{8.6-18}$$

where δ is given by

$$\delta = \frac{-(a + b) \pm \sqrt{(a + b)^2 - 4c}}{2} \tag{8.6-19}$$

and K is an arbitrary constant as before.

8.7. Gas–Liquid Plate Reactors

Material B is formed continuously from A by reaction in a plate tower as sketched in Figure 86. The reaction occurs in the liquid phase due to the presence of a nonvolatile catalyst which is circulated through the tower

---

[12] H. S. Mickley, T. K. Sherwood, and C. E. Reid, *Applied Mathematics in Chemical Engineering*, McGraw-Hill, New York (1957), p. 328.

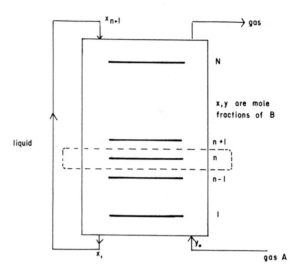

Fig. 86. Gaseous plate absorber.

in the liquid stream. The liquid-phase reaction may be written as $A \xrightarrow{k} B$. Gas A is introduced into the base of the tower essentially in the pure state. It is assumed that the gas throughout the tower is essentially pure A and that the liquid throughout the tower is saturated with respect to A. The reaction rate (production of B) is given as $r_B = kx_A$, where x_A is a constant. A steady-state material balance written for component B around plate n is obtained from equation (5.1-8). The result is

$$(Gy_{n-1} + Lx_{n+1})\big|_{\text{in}} - (Gy_n + Lx_n)\big|_{\text{out}} + kHx_A = 0 \qquad (8.7\text{-}1)$$

where H is the plate holdup volume. The relationship between x_n and y_n (mole fractions of B) is written in terms of an efficiency E, as

$$E = \frac{y_n - y_{n-1}}{y_n^* - y_{n-1}} \qquad (8.7\text{-}2)$$

where

y_n^* = gas composition of B in equilibrium with the liquid composition x_n

y_n = actual vapor composition of B from plate n

$y_n^* = mx_n$ (equilibrium relationship) $\qquad (8.7\text{-}3)$

Equations (8.7-2) and (8.7-3) may be combined to give

$$mx_n = \frac{1}{E} y_n - \frac{1 - E}{E} y_{n-1} \qquad (8.7\text{-}4)$$

Substituting equation (8.7-4) into (8.7-1), we can get, after some rearrangement,[13]

$$y_{n+1} - [2 + E(\lambda - 1)]y_n + [1 + E(\lambda - 1)]y_{n-1} = - \alpha E x_A \quad (8.7-5)$$

where

$$\alpha = \frac{mkH}{L} \quad (8.7-6)$$

$$\lambda = \frac{mG}{L} \quad (8.7-7)$$

The boundary conditions for this situation are

$$y = y_0 \qquad \text{at } n = 0 \quad (8.7-8)$$

$$x_1 = x_{N+1} \qquad \text{(recirculating liquid)} \quad (8.7-9)$$

Equation (8.7-5) can be solved by the methods of Chapter 7, giving[14]

$$y_n = C_1 + C_2 Q^n + \frac{\alpha x_A n}{(\lambda - 1)} \quad (8.7-10)$$

where $Q = 1 + E(\lambda - 1)$. Using the boundary conditions in (8.7-10), we can get the final answer

$$y_n - y_0 = \frac{\alpha x_A}{1 - \lambda} \left\{ \frac{N}{\lambda(1 - Q^N)} (1 - Q^n) - n \right\} \quad (8.7-11)$$

The corresponding equation for x_n is given by equation (8.7-4).

If the gas absorber discussed above had no chemical reaction occurring, the material balance, instead of being equation (8.7-1) would be

$$Gy_{n-1} + Lx_{n+1} = Gy_n + Lx_n \quad (8.7-12)$$

If equation (8.7-2) still holds, but if the equilibrium relationship is a bit more general than (8.7-3), such as

$$y_n^* = mx_n + b \qquad \text{(equilibrium)} \quad (8.7-13)$$

we can get the finite difference equation[15]

$$y_{n+1} - (Q + 1)y_n + Qy_{n-1} = 0 \quad (8.7-14)$$

---

[13] W. R. Marshall and R. L. Pigford, *Applications of Differential Equations to Chemical Engineering Problems*, University of Delaware, Dover (1947), p. 86.

[14] *Ibid.*, p. 88.

[15] *Ibid.*, p. 81.

where

$$Q = 1 + E(\lambda - 1) \tag{8.7-15}$$

The solution of equation (8.7-14) is

$$y_n = C_1 + C_2 Q^n \tag{8.7-16}$$

and we proceed as before to a final solution. For an efficiency of 100% ($E = 1$), the final solution for N, the number of plates, is[16]

$$N\big|_{E=1} = \frac{\ln\left[(1 - \lambda)\, \dfrac{y_0 - (mx_{N+1} + b)}{y_{N+1} - (mx_{N+1} + b)} + \lambda \right]}{\ln(1/\lambda)} \tag{8.7-17}$$

8.8. Difference Equations from Differential Equations

Some phenomena which are analyzed in terms of partial differential equations (distributed parameters) may be transferred to discrete models, characterized by finite difference equations (lumped parameter systems). For example, consider unsteady-state one-dimensional heat conduction in a slab, where the governing equation is

$$\frac{\partial T}{\partial t} = \alpha \frac{\partial^2 T}{\partial x^2} \tag{8.8-1}$$

with

$$\alpha = k/\varrho C_p$$

$$k = \text{thermal conductivity}$$

$$\varrho = \text{density}$$

$$C_p = \text{heat capacity}$$

The boundary conditions might be

$$T(x, t) = T_i \qquad \text{at } t = 0 \tag{8.8-2}$$

$$-k\,\frac{\partial T}{\partial x} = S \qquad \text{at } x = 0 \qquad 0 < t < t_1 \tag{8.8-3}$$

$$-k\,\frac{\partial T}{\partial x} = 0 \qquad \text{at } x = 0 \qquad t > t_1 \tag{8.8-4}$$

$$-k\,\frac{\partial T}{\partial x} = 0 \qquad \text{at } x = L \tag{8.8-5}$$

[16] *Ibid.*, p. 82.

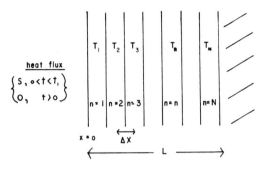

Fig. 87. Finite difference approach to physical problems.

Equations (8.83) and (8.84) indicate that the heat source at $x = 0$ ceases for times greater than t_1. Equation (8.8-5) implies that no heat flows past the surface $x = L$. An analytical solution for this mathematical situation is available[17]

$$T - T_i = \frac{St}{\varrho C_p L} + \frac{SL}{k} \left[\frac{3(L - x)^2 - L^2}{6L^2} \right.$$
$$\left. - \frac{2}{\pi^2} \Sigma \frac{(-1)^n}{n^2} e^{-\alpha n^2 \pi^2 t / L^2} \cos \frac{n\pi}{L} (L - x) \right] \quad (8.8\text{-}6)$$

In finite difference notation we can break down the derivatives by using Taylor series expansions. Figure 87 illustrates the division of the slab into representative units. A Taylor series expansion for $T_{n-1}(x)$ about point x_n gives[18]

$$T_{n-1} = T_n + \left(\frac{\partial T}{\partial x} \right)\Big|_n (x_{n-1} - x_n) + \frac{1}{2!} \left(\frac{\partial^2 T}{\partial x^2} \right)\Big|_n (x_{n-1} - x_n)^2$$
$$+ O(x_{n-1} - x_n)^3 \quad (8.8\text{-}7)$$

where $0(x_{n-1} - x_n)^3$ reads "terms of order $(x_{n-1} - x_n)^3$." Solving equation (8.8-7) for the first derivative yields

$$\left(\frac{\partial T}{\partial x} \right)\Big|_n = \frac{T_n - T_{n-1}}{\Delta x} - \frac{1}{2!} \left(\frac{\partial^2 T}{\partial x^2} \right)\Big|_n (\Delta x) + O(\Delta x)^2 \quad (8.8\text{-}8)$$

where $\Delta x = x_{n-1} - x_n = $ constant. Usually the third term of the right-side

[17] D. M. Himmelblau and K. B. Bischoff, *Process Analysis and Simulation*, John Wiley, New York (1968), p. 224.

[18] *Ibid.*, p. 221.

of equation (8.8-8) is neglected and often the second term is also neglected. Equation (8.8-8) is sometimes called the backward difference formula. Similarly, a forward difference formula may be derived by rewriting equation (8.8-7) as

$$T_{n+1} = T_n + \left(\frac{\partial T}{\partial x}\right)\Big|_n (x_{n+1} - x_n) + \frac{1}{2!}\left(\frac{\partial^2 T}{\partial x^2}\right)\Big|_n (x_{n+1} - x_n)^2$$
$$+ O(x_{n+1} - x_n)^3 \qquad (8.8\text{-}9)$$

which yields the forward difference form

$$\left(\frac{\partial T}{\partial x}\right)\Big|_n = \frac{T_{n+1} - T_n}{\Delta x} - \frac{1}{2!}\left(\frac{\partial^2 T}{\partial x^2}\right)\Big|_n (\Delta x) + O(\Delta x)^2 \qquad (8.8\text{-}10)$$

where $\Delta x = x_{n+1} - x_n = $ constant. Adding equations (8.8-8) and (8.8-10) yields the central difference formula

$$\left(\frac{\partial T}{\partial x}\right)\Big|_n = \frac{T_{n+1} - T_{n-1}}{2\Delta x} + O(\Delta x)^2 \qquad (8.8\text{-}11)$$

The finite difference representation of the second derivative can be obtained by adding equations (8.8-7) and (8.8-9) to get

$$T_{n+1} + T_{n-1} = 2T_n + \left(\frac{\partial T}{\partial x}\right)\Big|_n (-\Delta x + \Delta x) + \left(\frac{\partial^2 T}{\partial x^2}\right)\Big|_n (\Delta x)^2 + O(\Delta x)^3 \qquad (8.8\text{-}12)$$

which, upon rearrangement, becomes

$$\left(\frac{\partial^2 T}{\partial x^2}\right)\Big|_n = \frac{T_{n+1} - 2T_n + T_{n-1}}{(\Delta x)^2} + O(\Delta x) \qquad (8.8\text{-}13)$$

We usually neglect $O(\Delta x)$.

Now applying the finite difference equation (8.8-13) to (8.8-1), for $n = 1$, we get

$$\left(\frac{\partial T}{\partial t}\right)\Big|_n = \alpha\left(\frac{T_{n+1} - 2T_n + T_{n-1}}{(\Delta x)^2}\right) \qquad (8.8\text{-}14)$$

or, for $n = 1$,

$$\left(\frac{\partial T}{\partial t}\right)\Big|_1 = \alpha \frac{(T_2 - T_1)}{(\Delta x)^2} - \alpha \frac{(T_1 - T_0)}{(\Delta x)^2} \qquad (8.8\text{-}15)$$

The finite difference representation of the boundary conditions must also be used. Thus from equation (8.8-3) we get

$$-k \frac{\partial T}{\partial x} \simeq -k \frac{T_1 - T_0}{\Delta x} = S \qquad \text{at } x = 0 \qquad (8.8\text{-}16)$$

where equation (8.8-11) was used. Rearranging equation (8.8-16), we get

$$-\frac{T_1 - T_0}{\Delta x} \simeq \frac{S}{k} \qquad (8.8\text{-}17)$$

which can be substituted into (8.8-15) to get

$$\left(\frac{\partial T}{\partial t}\right)\bigg|_1 = \alpha \frac{(T_2 - T_1)}{(\Delta x)^2} + \frac{S}{\varrho C_p \Delta x} \qquad (8.8\text{-}18)$$

We can write similar expressions for $n = 2, 3, \ldots, N$ from equation (8.8.14). With $n = N$,

$$\left(\frac{\partial T}{\partial t}\right)\bigg|_N = \alpha \frac{(T_{N+1} - 2T_N + T_{N-1})}{(\Delta x)^2} \qquad (8.8\text{-}19)$$

The term T_{N+1} has no physical significance since it lies beyond the slab. However, if there is symmetry about $x = L$, then $T_{N+1} = T_{N-1}$, so that equation (8.8-19) simplifies to

$$\left(\frac{\partial T}{\partial t}\right)\bigg|_N = 2\alpha \frac{(T_{N-1} - T_N)}{(\Delta x)^2} \qquad (8.8\text{-}20)$$

From equation (8.8-2) we can infer another boundary condition for each section:

$$T_n = T_i \qquad \text{at } t = 0 \qquad (8.8\text{-}21)$$

It is now possible to numerically compute temperature profiles as a function of time and distance. For $t > t_1$ (when the heat source has been stopped and the slab insulated), the temperature gradients disappear and the slab reaches a uniform temperature. The final temperature T_∞ can be obtained from an overall energy balance,

$$SA(t_1 - t_0) = C_p \varrho AL(T_\infty - T_i) \qquad (8.8\text{-}22)$$

where

$$S = \text{heat source at } x = 0$$
$$A = \text{area of the slab}$$
$$T_\infty = \text{final temperature of the slab}$$
$$L = \text{length of slab}$$
$$C_p = \text{heat capacity of the slab}$$

Rearranging equation (8.8-22), we get

$$T_\infty = T_i + \frac{St_1}{\varrho C_p L} \tag{8.8-23}$$

which can be used to calculate T_∞. We then check this value against the finite difference result.

Assignments in Chapter 8

8.1. Verify equations (8.1-3), (8.1-4), (8.2-2), (8.2-7), (8.2-8), (8.3-1), (8.3-3), (8.3-8), (8.3-9), (8.3-15), (8.3-18), (8.3-19), (8.4-3), (8.4-5), (8.5-2), (8.5-3), (8.5-4), (8.5-8), (8.6-6), (8.6-7), (8.6-8), (8.6-13), (8.6-14), (8.7-4), (8.7-10), (8.7-16), (8.8-7), (8.8-9), and (8.8-12).

8.2.[19] Butyl acetate is to be produced in a battery of continuous stirred tank reactors operating in series by reacting 2000 lb/hr of butanol with 326 lb/hr of glacial acetic acid containing sufficient sulfuric acid to catalyze the reaction at 100°C. Under these conditions the rate of reaction can be expressed by $r = kC_A^2$, where $k = 0.28$ ft^3/lb mole·min and C_A is the concentration of acetic acid in lb moles/ft^3.

If the effective volume of each tank is 10 ft^3 and the density of the reaction mixture is assumed constant at 48 lb/ft^3, estimate the number of reaction vessels required if the concentration of acetic acid in the final discharge is not to exceed 3 lb/ft.

8.3.[20] It is often of interest to investigate the start-up characteristics of cascaded systems by analyzing their unsteady-state behavior. Suppose we have a series of stirred tanks filled with pure solvent. At time zero, a flow

[19] V. G. Jenson and G. V. Jeffreys, *Mathematical Methods in Chemical Engineering*, Academic Press, New York (1963), p. 525.

[20] H. S. Mickley, T. K. Sherwood, and C. E. Reed, *Applied Mathematics in Chemical Engineering*, McGraw-Hill, New York (1957), pp. 325, 332.

L of concentration C_0 is introduced into the first tank. There are only two solute components A and B.

(a) Show that the governing equation is

$$LB_n - L_{n+1}B_{n+1} + kC_{n+1}V = \frac{\partial(B_{n+1}V)}{\partial t}$$

where B_n is the concentration of component B, and $A \xrightarrow{k} B$.

(b) Show that the material balance of component A around tank $n + 1$ is

$$LC_n - LC_{n+1} - kC_{n+1}V = \frac{\partial(C_{n+1}V)}{\partial t}$$

(c) From (a) and (b), show how we can get

$$F_{n+1} - F_n = -\theta \frac{\partial F_{n+1}}{\partial t}$$

where

$$F_n = C_n + B_n$$
$$\theta = V/L$$

(d) With all initial concentrations equal to zero, take the Laplace transform $\mathscr{L}_{t \to p}$ to then produce

$$\bar{F}_n = G\left(\frac{1}{1 + \theta p}\right)^n$$

where G is an arbitrary constant

(e) Show how to get

$$G = \frac{C_0}{p}$$

(f) Similarly, produce

$$\bar{B}_n = \frac{H}{[1 + (k + p)\theta]^n} + \frac{C_0}{p(1 + \theta p)^n}$$

with

$$H = -\frac{C_0}{p}$$

(g) Finally, invert $\bar{B}_n$ to yield

$$B_n = C_0\left\{1 - \exp\left(\frac{-Lt}{V}\right)[\cdots]\right\}$$

For Further Reading

Mathematical Methods in Chemical Engineering, by V. G. Jenson and G. V. Jeffreys, Academic Press, New York, 1963.

Applied Mathematics in Chemical Engineering, by H. S. Mickley, T. K. Sherwood, and C. E. Reed, McGraw-Hill, New York, 1957.

Applications of Differential Equations to Chemical Engineering Problems, by W. R. Marshall and R. L. Pigford, University of Delaware, Dover, Delaware, 1947.

Process Analysis and Simulation, by D. M. Himmelblau and K. B. Bischoff, John Wiley, New York, 1968.

Subject Index